AKADEMIYA NAUK SSSR · INSTITUT OKEANOLOGII IM. P. P. SHIRSHOVA

Academy of Sciences of the USSR · P. P. Shirshov Institute of Oceanology

M. E. Vinogradov, Editor

LIFE ACTIVITY OF PELAGIC COMMUNITIES IN THE OCEAN TROPICS

(Funktsionirovanie pelagicheskikh soobshchestv tropicheskikh raionov okeana)

Based on Data of the 44th Cruise
of the R/V "Vityaz"

Izdatel'stvo "Nauka"
Moskva 1971

Translated from Russian

Israel Program for Scientific Translations
Jerusalem 1973

TT 72-50035

Published Pursuant to an Agreement with
THE NATIONAL MARINE FISHERIES SERVICE,
NATIONAL OCEANIC AND ATMOSPHERIC ADMINISTRATION,
U.S. DEPARTMENT OF COMMERCE
and
THE NATIONAL SCIENCE FOUNDATION, WASHINGTON, D.C.

IPST Cat. No. 60094 1
UDC 524.12

ISBN 0 7065 1274 X

Translated by N. Kaner
Edited by H. Mills

Printed in Jerusalem by Keter Press
Binding: Wiener Bindery Ltd., Jerusalem

Available from the
U.S. DEPARTMENT OF COMMERCE
National Technical Information Service
Springfield, Va. 22151

XII/6/3

Contents

STUDIES OF THE LIFE ACTIVITY OF OCEANIC BIOLOGICAL SYSTEMS

M. E. Vinogradov
Shirshov Institute of Oceanology of the Academy of Sciences of the USSR

NATURAL BIOLOGICAL PRODUCTIVITY OF THE OCEAN

Utilization of the biological resources of the ocean is as yet almost exclusively by partial tapping of natural pelagic or benthic communities in the form of one of the terminal links in the trophic chains, i. e. fish or whales. In 1967 the world catch of marine fish reached 45 million tons, having trebled over the last 20 years. Commercial fishing is gradually extending to more and more regions so that by now practically the entire World Ocean is being fished.

The need to increase catches called for scientific research in order to determine the most efficient fishing methods. Of prime concern was elucidation of the quantitative distribution pattern of forage items, the ocean's productivity, and the relation between the distribution of the foraging fields of fish and hydrological and other physicochemical factors.

Most marine biological research teams worked along these general lines for decades and worked out ways for better and more economical fishing. They clarified the main aspects of the ocean's overall biological structure, estimated its productivity, and gathered and systematized abundant data on the specific composition and biology of the organisms inhabiting it. In the USSR these quantitative investigations of biological processes were headed by Academician L. A. Zenkevich.

One fact that emerged was that the ocean's commercial resources are limited and the maximum permissible fishing quota of marine items was established at 70 — 100 million tons, which is slightly in excess of the present-day world catch. The approach to this limit is naturally already affecting ocean fishing.

The highest growth rate of the world catch was seen in the period 1958 — 1961, when the average annual increment was 3.4 million tons, or 9%, with respect to the preceding year. The average catch increment was 3.1 million tons, or 6.7%, in 1961 — 1964, and 2.6 million tons, or 4.7%, in 1964 — 1967. Catches in the traditional fishing regions, primarily in the shelf zones, have more or less ceased to increase in recent years, in spite of more vessels and improved gear. Moreover, the catches of certain items have begun to decline, together with a marked increase in the relative proportion of inferior species which were previously regarded as trash. The overall growth of the world catch in recent years is mainly on account of new fisheries that have opened up in the southern hemisphere.

Obviously, further, even very energetically intensified "free-lance" fishing extended to the few still unexploited areas and species cannot produce the desired increase of catches.

Catches may increase only briefly, until the ceiling of "free-lance" fishing is reached. Any further intensification is liable to undermine the ocean's resources until the situation may become irreversible, and then neither restrictions nor even total long-term prohibition will achieve restocking; this has in fact already happened with several commercial species.

Detailed assessment of the ocean's resources and studies of its biological productivity do lead to recommendations for rationalization of fisheries, but they cannot increase the overall yield of biological produce above a certain ceiling. It is imperative to find a way to increase the amount of protein products obtained from the ocean.

HOW TO INFLUENCE THE OCEAN'S PRODUCTIVITY?

The inevitable question is how to effect from transition "free-lance" fishing to cultivation of the ocean's resources, in other words, how to artificially improve the productivity of biological communities, as is done on dry land.

A fundamentally new formulation of the basic task calls for a new scientific methodology. The marine "farmer" must study the structure and life activity of biological communities and the patterns of their utilization of energy, i. e. the processes of control in the ocean's biological systems.

The importance and urgency of this is self-evident. Studies of the structure of oceanic biological communities, their thermodynamics and their control processes are now becoming one of the principal trends in marine biology. They are of great theoretical interest in that they reveal the system of causal relationships that govern the formation of the ocean's population.

The international oceanic research program recently published by the United Nations (Global Oceanic Research, 1969) emphasizes the need to study the production and functioning of oceanic ecosystems as the basis for controlled change. Decision No. 752 of the Presidium of the Academy of Sciences of the USSR of September 8, 1967 states that rational utilization of the food resources of seas and oceans must be based on in-depth studies of the biological productivity of marine communities and recommends elaboration of methods for improving the ocean's biological productivity by controlling the transformation of matter and energy in the communities of marine organisms in the World Ocean.

CONTROL PROCESSES IN ECOLOGICAL SYSTEMS

Marine communities are complex biological systems composed of populations of individual species in a state of dynamic equilibrium. The production level of the terminal links of a community depends on the amount of energy influx and the efficiency of utilization of the energy in the community's nutritive network. According to MacArthur, the more stable a system, the more energy is expended within it, and the smaller is the fraction of energy reaching the terminal trophic levels which are utilized by man as useful products. Multispecific oceanic communities are fairly stable, reacting relatively insignificantly to external factors; as a rule, their food chains are long, the utilization of energy within them is efficient, and consequently their economic productivity is low. Their ultimate yield (useful to man) can be increased by introducing an artificial disturbance in the regulatory relationships in the community, by modifying and shortening the food chains, and thus by lowering the community's stability. Maintenance of the activity of such a community necessitates the expenditure of a definite amount of work (energy).

An essential feature of marine communities is that the produce of the upper trophic levels (usually the third — fifth) is utilized by man, in contrast to terrestrial communities, in which the produce of the first and the second level is utilized. This in itself makes for a lower productivity of marine communities, and therefore development of techniques for direct exploitation of the lower trophic levels may be of considerable economic value.

The structure of communities and the regulation processes in them consist of an extremely complex gamut of interpopulational relationships against a background of evolutionary adaptation of the organisms to existence under the conditions of a particular community. Attention must be paid to the morphophysiological features of animals, their genetic characteristics, behavioral reactions, etc. However, ecological investigations clearly show food relationships to be the main factors in integrating communities as a whole and controlling their fundamental makeup and productivity. Therefore, studies of the trophic relationships within a community, of the flow of energy through the biological system and its utilization on the different trophic levels must be directed toward working out the best ways to control modifications of structure.

The investigation should run along two related lines: development of a theory and methods of controlling the terminal product yield in biological systems of coastal waters and development of a theory for controlling biological systems of the open ocean.

The first trend involves the following (enumerated in order of increasing complexity and efficiency): a) optimization of the exploitation regime for commercial populations (scientifically validated regulation of fisheries); b) acclimatization of new commercial species; c) hydrological and biological amelioration of the reproduction, growth and feeding areas of commercial populations, i. e. transition to controlled biological systems; d) artificial propagation of commercial animals; e) artificial cultivation of commercial species under controlled conditions; f) breeding of new races of commercial species.

These tasks are already being realized with respect to some commercially valuable species of fish and invertebrates. Artificially "farmed" oysters, mussels, trepangs, kelp, and other marine animals and algae are well known instances. This method of marine husbandry will evidently be more extensively developed eventually.

Realization of the possibilities of modifying biological systems in open seas and oceans (pelagic biological systems in the first place) could yield a still greater effect. However, this is a much more complex problem. which will be solved only at great effort and cost. Nevertheless, the need to work in this direction is obvious, and the search for approaches to a solution is already in evidence in several countries.

Work is being done on experimental methods for determination of the main parameters of communities, a stepwise quantitative assessment of the flow of energy through a system, methods for establishing a detailed picture of the distribution of biological items and for simulation of processes taking place in nature. As yet, the experimental research is mostly concerned with the initial links in the production process.

PURPOSE OF THE SPECIALIZED CRUISE OF THE "VITYAZ" DEVOTED TO STUDIES OF PELAGIC ECOSYSTEMS

One of the first steps in a comprehensive investigation of ocean communities in the USSR was represented by the 44th cruise of the R/V "Vityaz" from December 1968 to March 1969 which was especially devoted to quantitative studies of various aspects of the structure, energy exchange and functional regularities of oceanic pelagic ecosystems. Experts from different institutions took part in the expedition. The vessel was converted to an open-ocean experimental biological laboratory. Practically all the participants of the cruise worked on the common project or on affiliated aspects.

This was the first time that the comprehensive task of studying the structure of oceanic pelagic communities was undertaken. Of course, certain stages of the production process, e. g. production of organic matter by phytoplankton, had been the subject of fairly detailed studies previously. However, in order to characterize the process in its entirety from biogenic salts and phytoplankton to predatory invertebrates and fish and in order to attempt to elucidate the ecological mechanism of its realization, several new studies were necessary to provide quantitative estimates of energy expenditure and of the production of previously uninvestigated links. This called for new techniques and apparatus.

The expedition was involved not only in pure research but also in scientific-methodological projects, the principal one being elaboration of techniques for comprehensive studies of the process of exchange of biological energy and mass in the ocean and coordination techniques for the teams engaged in synchronous measurements of the characteristics of these processes on board. Therefore the testing of several new techniques that either had not been used previously or else had been applied to ecological investigations only in fresh or rich coastal waters and on dry land, and their application to biological oceanographic research must be regarded as an important result of the cruise.

The investigations were conducted at prolonged (up to two weeks) stations — polygons. The chief requirement regarding the location of these polygons was a state of equilibrium of the biological systems and a comparatively rich population permitting a quantitative appraisal of energy and mass exchange in the components of an ecosystem by modern methods.

In other words, the location of the polygons had to satisfy the following four requirements as a minimum:

1. Minimal seasonal long-term variations in the plankton; otherwise, a system could not be regarded as an equilibrium system over the observation period even in the first approximation. Therefore, the observations had to be carried out in the tropical zone of the ocean.

2. Minimal horizontal water transport, making it possible to assume all observations at a prolonged station to be made in the same waters. This meant that the observations could not be conducted in productive regions of high-gradient currents, for instance at the equator or in coastal regions.

3. The plankton population had to be sufficiently rich to lend itself to determination of its total production by modern methods and to provide sufficient material for experimental determination of the energy balance of individual animals, their rates of feeding and production, etc. Consequently, the work could not be performed in the most severely impoverished halistatic regions.

4. Weather conditions were to approach the ideal ones, wind force not exceeding 3 or 4 throughout the observation period.

The nearest accessible region of this kind was the tropical part of the Pacific Ocean south of the equator and west of 180° E, and the bulk of the work was carried out in this area (see figure).

The observations were conducted on three polygons: No. 2, station 6033 at 4° S, 168°30' E in the waters of the South Equatorial Current relatively poor in plankton, for 7 days and 20 hours; polygon No. 3, station 6052 at 9°48'S, 156°40'E in the divergence zone in the Coral Sea relatively rich in plankton, for 9 days and 8 hours; polygon No. 4, station 6064 at 9°48' S, 160°10' E in the waters of the South Equatorial current, for 6 days and 7 hours. Two additional polygons were executed, No. 1, station 6029 in the lagoon of the Butaritari Atoll (Gilbert Islands) and No. 5, station 6065 in Kieta Bay of Bougainville Island (the Solomon Islands).

In accordance with the general project, the following main aspects were investigated on the cruise.

1. Studies of the process of creation of primary production; evaluation of the utilization of solar energy in a pelagic community; vertical variations of primary production; analysis of the pigment system of phytoplankton and its variability.

2. Studies of the abundance and production of the microbial population and its role in zooplankton alimentation; studies of the ways of incorporating dissolved organic matter in the production process by bacteria.

3. Investigation of the specific composition of pelagic communities in tropical waters.

4. Studies of the trophic structure of plankton communities; determination of the production intensity of zooplankton; evaluation of the efficiency of utilization of energy flowing from the first trophic level; quantitative evaluation of metabolic processes in planktonic organisms.

5. Studies of the vertical distribution of zooplankton in relation to external factors; diurnal variations in the vertical distribution of plankton.

6. Evaluation of the part played by ichthyofauna (fish larvae) in pelagic food chains.

7. Analysis of the distribution and composition of organic detritus and its mineralization characteristics.

8. Studies of bioluminescence as a manifestation of the life activities of pelagic communities and a method for quantitative studies of the microdistribution of plankton.

9. Studies of patterns governing the distribution of radiant energy and of the optical characteristics of water responsible for the distribution of this radiation.

10. Studies of a) hydrological and b) hydrochemical environmental parameters (including the distribution of nitrogen, phosphorus, silicon, and iron).

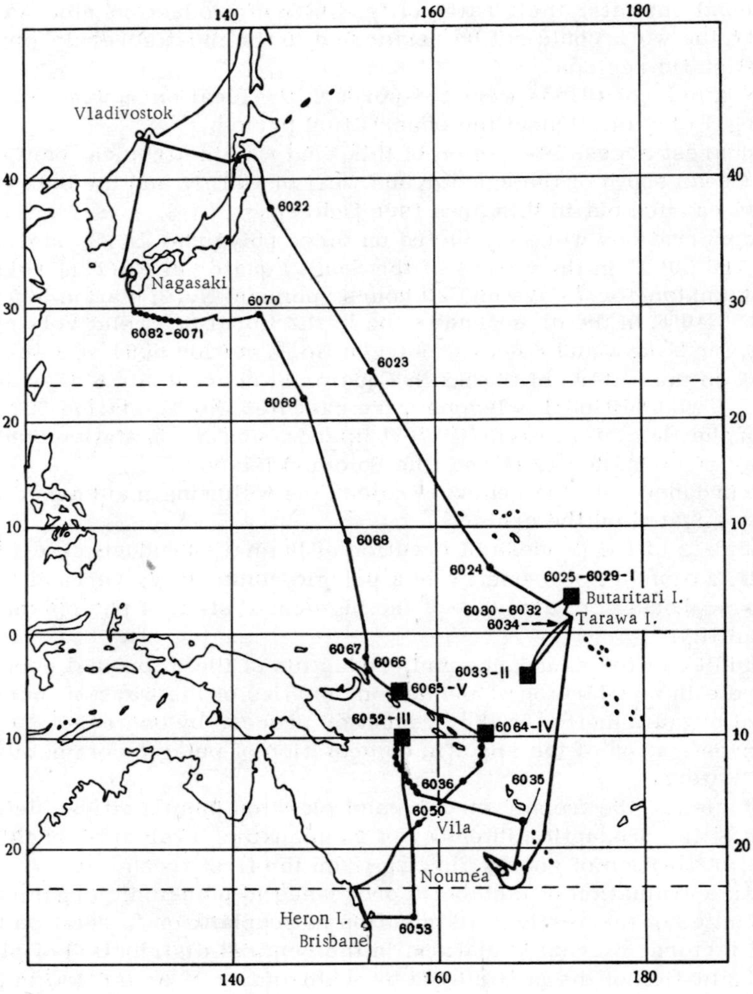

Diagram of the 44th cruise of the "Vityaz."

Circles and squares denote cruise stations and prolonged polygons respectively.

In addition, special investigations were made of the vertical and horizontal microdistribution of plankton. It was assumed that the pelagic ecosystem can only function in oligotrophic regions in the presence of a nonuniform distribution of organisms, i. e. the existence of layers and patches with higher concentrations, since the average values of the biomass of forage organisms are often lower than the threshold concentration below which the filter feeders no longer replenish their metabolic losses. Consequently, the system could not exist if the distribution of organisms were uniform or close to uniform.

In the course of preparations for the expedition, A. A. Lyapunov, associate member of the Academy of Sciences, working in collaboration with several future cruise participants, developed a mathematical model of the oceanic pelagic ecosystem. Work on the model was not the primary aim of the cruise, but it served as a pivot bringing together the different aspects of the investigation.

MAIN RESULTS OF THE CRUISE

The scientific results of the cruise are described in detail in the papers of this collection. We shall therefore list here only those that are of fundamental importance for future work.

1. The rate of consumption and assimilation of bacterioplankton by zooplankton in the surface waters of the tropical ocean regions was shown (Sorokin, Petipa, and Pavlova) to be commensurate with the rate of its consumption of algae.

Reliable determinations were made of the abundance and productivity of bacterioplankton, the degree of its aggregation, and the capacity of mass planktonic species (filter feeders) for consuming and assimilating the aggregated bacterioplankton. About one-third of the bacteria in the natural bacterioplankton was found to be in aggregates larger than 3 microns, which can be directly caught and eaten not only by filter feeders but even by fish larvae, whereas unaggregated bacterioplankton is scarcely eaten by such filter feeders as calanids. Copepod filter feeders subsisting on bacterial and plant food have a similar organic matter balance.

Hence, the pelagic communities inhabiting the surface zone in oligotrophic tropical regions are not exclusively dependent upon the phytoplankton. There is another influx of food resources into the community, namely allochthonous dissolved organic matter and bacterioplankton supported by it. Consequently, the productivity of the zooplankton and higher trophic levels in a community is not solely determined by phytoplankton production (as has generally been assumed), but also by bacterioplankton production. This finding substantially modifies previous notions concerning the entire process of biological production in the tropical pelagic zone.

2. Stable functioning of the ecosystem of surface layers, at an insignificant current velocity was assumed to be ensured, inter alia, by the influx of biogenic salts from below by turbulent diffusion compensating for the loss of these salts with organic residues (corpses, animals' metabolites). Such an equilibrium was found to be nonexistent, and the

population of the upper layer (0 — 50 m) in oligotrophic regions was found to exist largely owing to horizontal transport of the biogens from areas of upwelling to those with sinking waters. The biological systems are out of equilibrium in any area of the movement of water in the course of its "aging." An equilibrium system should include the entire cycle, more precisely that part in which the water is near the surface, i. e. from the moment of its upwelling to its sinking in the convergence zone. This conclusion is especially significant for devising functional models of a community, calculation of its energy balance, and also for the planning of further research expeditions.

3. A new method of studying the vertical microdistribution of plankton bathyphotometric sounding revealed the existence of two stationary layers of high plankton concentration, at depths of 25—40 m and (especially intensive) at 70—90 m. Sometimes, the vertical extent of the high concentration layers does not exceed 5 m. The layer of the lower maximum of bioluminescent organisms is simultaneously the layer of concentration of active phytoplankton and bacterioplankton and high gradient of critical biogenic elements (nitrogen primarily). Such narrow zones of high concentration of food items and consumers appear to be a prerequisite for the existence of tropical pelagic ecosystems. They must be taken into account in constructing any model of an ecosystem.

4. The use of radioisotopic techniques elucidated the quantitative characteristics of the energy balance of mass zooplankton species (phytophages, euryphages, predators) and permitted determination of not only the production of organic matter by algae (Koblents-Mishke), but also the production of phytophagous animals (Shushkina) and bacteria (Sorokin).

It may be claimed that the investigations for the first time furnished a quantitative evaluation of the consumption and production of the principal trophic levels of the oceanic pelagic ecosystem and offered a novel view of certain crucial aspects of its functioning.

FUTURE INVESTIGATIONS

Only the first step has been made in studies of the dynamics of energy and mass exchange in marine biological systems. The unsolved problems are much more numerous than the solved ones. Above all, this applies to studies of the terminal links of communities that are the most important economically, i. e. macroplankton and fishes.

Apparatus must be devised for continuous automatic gathering of information on the populations of planktonic organisms and the parameters of the aqueous environment. The apparatus should be of a fundamentally new type, making use of telemetric techniques and autonomous and towed devices for collecting the macroplankton and nekton. Improvements must be made in the experimental techniques for investigations of the alimentation, production and energy balance of different components of communities. Obviously, several problems related to the influence of different environmental parameters on communities can be best investigated by multifactorial experimental methods. There is a need for detailed investigation of the microdistribution of populations making up the natural system. Methods of studying plankton microdistribution are still little developed.

The difficulties involved in studying natural communities render simulation of processes taking place in them especially important. Working mathematical models must be created that reflect the principal aspects in the functioning of biological communities. As pointed out by Vinberg, a model of an ecosystem is a powerful tool for testing the mutual consistency of different experimental facts and observations. Another important trend toward attaining the goal could be biological simulation, i. e. setting up small, isolated ecosystems under controlled conditions in the ocean.

In the course of further development of research in control processes in biological systems, besides the key problem of energy flow through a system, a broad cycle of related aspects will inevitably come under examination, including the behavior of individual components of a community, the factors governing communities, adaptation, genetic and morphological characteristics, and finally evolution of the communities themselves.

Work on the problem of control in oceanic biological systems may be expected to furnish a theoretical basis for rational, controlled modification of the communities and increased yields of commercially valuable products; it will lead to efficient farming of the ocean and to a greater harvest of animal protein.

CONSTRUCTION OF A MATHEMATICAL MODEL OF BALANCE RELATIONSHIPS IN THE ECOSYSTEM OF TROPICAL OCEANIC WATERS

A. A. Lyapunov
Institute of Hydrodynamics of the Siberian Branch of the Academy of Sciences of the USSR

The impetus for this work was provided by a discussion of mathematical simulation of life activity in the ocean, in which V. G. Bogorov, M. E. Vinogradov, O. I. Koblents-Mishke, Yu. A. Rudyakov, E. A. Shushkina (Institute of Oceanology of the Academy of Sciences of the USSR), I. I. Gitel'zon (Institute of Physics of the Siberian Branch of the Academy of Sciences of the USSR), O. S. Kulagina (Institute of Applied Mathematics of the Academy of Sciences of the USSR), and T. I. Bulgakova (All-Union Research Institute of Marine Fisheries and Oceanography) participated.

The aim was to concretize the sketches of mathematical models of an oceanic biogeocenosis for the ecosystem of the upper pelagic layers of the tropical oceanic regions, where a special investigation of the structure of pelagic communities was planned on board the "Vityaz." Extensive help in preparation of this paper was received from M. E. Vinogradov.

INTRODUCTION

Studies of the ocean's biological productivity involve research into the life activities of a multitude of living organisms and their interaction both among themselves and with their environment. These studies necessitate gathering a very large amount of empirical evidence and its systematization from some unifying standpoint. Elucidation of the whole complex of metabolic processes taking place in an ecosystem appears important to us. A list should be compiled of the main components of the system which play a significant part in biomass production and may be critical for the process. On the other hand, there is the need for a catalog of metabolic processes in the ecosystem bringing their empirical investigation up to the point making a quantitative evaluation of their intensity possible.

Constructing a model of balance relationships in a community is laborious and complicated and can only be done by several successive approximations. However, it may be useful in the early stages of the work by playing an organizing role as regards the gathering and primary processing of data.

Studying a system from a macroscale approach (Lyapunov and Yablonskii, 1963), including investigations of the interrelationship of the system in its entirety with the environment, its inputs (incoming items, their form and their entrance points) and outputs (items liberated into the environment, their form and emergence points). This is followed by the microscale approach. A given system is represented as a set of relatively autonomously functional elements connected by certain communication channels. Signals arriving at the system's inputs are transmitted to some of these elements, where they are processed. Their processing results in new signals proceeding along communication channels to other elements, and so on. Finally, some ultimate signals appear at the system's output. Some portions of matter or information may serve as the signals. In studies of balance relationships in an oceanic community, the system in its entirety consists of the community and its habitat, the points of entry of matter and energy into the community serve as its inputs, while the ways by which matter and energy leave the community are the output. Hence, in the first approximation (in the absence of systematic exploitation of the community by man), the community is regarded as a relatively autonomous system entered only by energy (sunlight) which is subsequently dissipated.

The following two aspects are important. A mathematical model of balance relationships in an ecosystem cannot be constructed unless studies of the factual data have attained a certain degree of completeness. One must have an idea of the distribution of matter and energy entering the community among the pertinent components, the exchange processes among these components, and the community's output. If some actually occurring processes are ignored or their quantitative evaluation is faulty, attempts at calculating the community's kinetics will immediately indicate either excessive accumulation of matter in some of its units, or, on the contrary, lack of raw material for some of the processes taking place in it. Furthermore, failure to include a process of transition of matter or energy from one component to another will make it impossible to compile a complete scheme of balance relationships.

A mathematical model plays an important organizing role in the gathering of data. Since a mathematical model describes the interrelationships of different components of a system, while the factual data are gathered by different specialists, each of them being professionally oriented toward a certain single component, the availability of a mathematical model makes for a significant rapprochement between the different groups of researchers. It stimulates endeavors toward adjustment of data obtained by the different groups at a possibly earlier working stage and toward avoiding lacunas in the gathering of information that are inevitable when specialists in different fields work independently.

FIRST VARIANT OF THE MODEL

One must determine which components of the ecosystem must be taken into account and which of their interactions are significant. Furthermore, and this is especially important, one must be able to evaluate the possible effect of human interference on the rhythm of vital processes in the community and to establish the rational forms and norms for such an interference. The only way of answering these requirements is to construct mathematical models of processes in the ecosystem that can yield a quantitative evaluation of the intensity of exchange processes among its components.

In addition to quantitative evaluation of the basic components (living organisms and inert elements), studies of the structure and productivity of oceanic biological communities call for knowledge of their interrelationships and the intensities of the processes of formation and destruction of components.

The construction of such a mathematical model in the broad sense of the word involves the following principal stages.

1. Pinpointing of the ecosystem's essential components.
2. Elucidation of the complex of exchange processes in which they participate.
3. Formulation of the laws governing these exchange processes.
4. Mathematical description of the complex of processes occurring in the system, i. e. construction of the mathematical model in the strict sense of the word.
5. Qualitative comparison of the model constructed with reality and (in the case of discrepancies) improvement of the model.
6. Establishment of techniques for determination of the model's numerical parameters.
7. Performance of the necessary experiments and calculation of numerical parameters.
8. Computations of the model's kinetic properties for different initial and boundary data.
9. Comparison of the results with reality and (when necessary) refinement of the original model followed by repetition of the above stages until satisfactory agreement is reached.

In preparation toward the 44th cruise of the "Vityaz," a sketch variant of the mathematical model of the pelagic ecosystem was constructed for tropical regions of the ocean. The data obtained by the expedition revealed the sketch model to be unsatisfactory. Nevertheless, it was useful to a certain extent, by focusing the different specialists' opinions, making it possible to discover some phenomena that had not attracted attention previously but that are undoubtedly essential for construction of mathematical models of oceanic communities.

We shall start by describing the initial mathematical model of the ecosystem of the surface layers of pelagic waters in the tropical regions and by formulating the assumptions on which it was based. We shall then provide a purely qualitative comparison of these assumptions with the facts established by the expedition.

In constructing the mathematical model we were guided by the following general hypotheses.

All processes occur without time lag (this assumption is not always justified). They are described by differential equations with linear or bilinear terms in their right-hand sides, and only the term describing the photosynthesis includes a logical condition (see below). This is because photosynthesis may be limited both by sunlight intensity and by the availability of chemical raw materials (biogenic elements).

The coefficients of the equations are assumed to be functions of depth and time. In this way it is assumed that the coefficients of the equations (or the relationships determining these coefficients) are invariable in horizontal planes.

The initial distribution of all the components of the system depends exclusively upon depth and is assumed to be constant in the horizontal planes.

This hypothesis will have to be discarded in subsequent models.

The model took into account the following ecosystem components: L, volume density of luminous energy; C_N, nitrogen concentration (meaning available nitrogen, NH_4^+); C_P, phosphorus concentration; φ_1, phytoplankton concentration; φ_2, zooplankton concentration; ψ, detritus concentration.

The following assumptions are made concerning each of these components.

L. Sunlight is absorbed by water (including neutral turbidity) — coefficient a. It is absorbed by phytoplankton, zooplankton, and detritus — coefficients a_1, a_2, a_3. It participates in photosynthesis — coefficient A describing the intensity of photosynthesis (its determination will be discussed below). The fraction of luminous energy expended in photosynthesis is negligibly small, as demonstrated by experiments.

C_N. Nitrogen is consumed in the biosynthesis of the phytoplankton. (The total intensity of biosynthesis is assumed to be always proportional to that of photosynthesis.) Notation: h_N, nitrogen consumption coefficient; v_N, coefficient of nitrogen liberation by lysis of detritus; Δ, coefficient of diffusion due entirely to turbulence.

C_P. Phosphorus participates in the same processes as nitrogen, and the coefficients are h_p, v_p, and Δ.

φ_1. Phytoplankton is synthetized in the process of photosynthesis — coefficient α. The phytoplankton does not die but is entirely consumed by zooplankton on the "bimolecular" scheme — coefficient β. It settles under the influence of the force of gravity — convection coefficient ω_1. Turbulent diffusion coefficient Δ.

φ_2. Proliferation of zooplankton follows Volterra's scheme. The zooplankton is not differentiated into phytophages and predators belonging to different orders. Zooplanktonic animals are assumed to feed on phytoplankton, one another (cannibalism), and detritus — coefficients $\gamma_1, \gamma_2, \gamma_3$. The zooplankton dies to the extent of fraction ε, undergoes diffusion with the coefficient $\Delta^* > \Delta$ because it has movements of its own that can be classified with diffusion and is furthermore subject to turbulent diffusion.

Finally, the zooplankton performs oscillatory turbulent migrations, upward at night and downward in the daytime. We introduce the simplifying assumption that the velocity of zooplankton convection flow is $\omega_2(X, t)$.

ψ. Detritus is produced by dead zooplankton — coefficient ν. It is dissolved — coefficient μ. In addition, it is formed from the zooplankton's excreta, the amount of these being proportional to the food consumed by the zooplankton in the form of phytoplankton, zooplankton, and detritus — coefficients $\theta_1, \theta_2, \theta_3$. The detritus settles under the action of the force of gravity — coefficient ω_3 and is subject to turbulent diffusion — coefficient Δ.

For simplicity it is assumed that the total intensity of biosynthesis (or photosynthesis) is determined in the following manner. Participants in the biosynthesis include solar energy as well as several source nutrients, including N, P, CO_2, Si, Ca, and trace elements. It may be assumed that only nitrogen, phosphorus, and solar energy may act as a critical factor in biosynthesis, while all the other components are available to excess. It is necessary to introduce coefficients characterizing the utilization of N, P, and solar energy per unit synthetized biomass. The actual rate of synthesis is determined by the critical factor, i. e. the component that is capable of ensuring the least intensity of synthesis. Hence, according to Poletaev (1966), the critical coefficient characterizing the rate of photosynthesis has the form

$$A = \min\{lL, g_N C_N, g_P C_P\},$$

where l, g_N, g_P are various coefficients.

Let us now construct the system of equations.

Model I

$$\frac{\partial L}{\partial x} = -L(a + a_1\varphi_1 + a_2\varphi_2 + a_3\psi)$$

$$\frac{\partial C_N}{\partial t} = -h_N A\varphi_1 + v_N\psi + \frac{\partial}{\partial x}\left(\Delta\frac{\partial C_N}{\partial x}\right)$$

$$\frac{\partial C_P}{\partial t} = -h_P A\varphi_1 + v_P\psi + \frac{\partial}{\partial x}\left(\Delta\frac{\partial C_P}{\partial x}\right)$$

$$\frac{\partial\varphi_1}{\partial t} = \alpha A\varphi_1 - \beta\varphi_1\varphi_2 + \frac{\partial}{\partial x}\left(\Delta\frac{\partial\varphi_1}{\partial x}\right) - \frac{\partial}{\partial x}(\omega_1\varphi_1)$$

$$\frac{\partial\varphi_2}{\partial t} = \gamma_1\varphi_1\varphi_2 - \gamma_2\varphi_2^2 + \gamma_3\varphi_3\psi - \varepsilon\varphi_2 + \frac{\partial}{\partial x}\left(\Delta\frac{\partial\varphi_2}{\partial x}\right) - \frac{\partial}{\partial x}(\omega\varphi)$$

$$\frac{\partial\psi}{\partial t} = \nu\varphi_2 - \mu\psi + \theta_1\varphi_1\varphi_2 + \theta_2\varphi_2^2 - \theta_3\varphi_2\psi + \frac{\partial}{\partial x}\left(\Delta\frac{\partial\psi}{\partial x}\right) - \frac{\partial}{\partial x}(\omega_3\psi).$$

The minus sign preceding the term $\gamma_2\varphi_2^2$ implies that the cannibalism of the zooplankton modifies its mass by forming detritus from zooplankton.

Let us formulate the boundary conditions for this system of equations. The solution is sought in the band $t>0$, $X_2>x>0$, where X_2 is the lower boundary of the layer under investigation. At depths $x=X_1\leqslant X_2$ (upper boundary of the discontinuity layer) many equation coefficients suffer a discontinuity, including the diffusion coefficient Δ. Here, X_1 is the depth of the mixing layer. If $x=X_1$, the second derivatives of unknown functions may not exist. The only presumption is continuity of the unknown functions themselves and their derivatives with respect to x.

If $x=0$, the quantities $C_N(0, t), \frac{\partial C_N}{\partial x}$ must be given; moreover it must be assumed that

$$\frac{\partial C_P}{\partial x} = \frac{\partial\varphi_1}{\partial x} = \frac{\partial\varphi_2}{\partial x} = \frac{\partial\psi}{\partial x} = 0.$$

If $x=X_2$, the following quantities should be given:

$$C_N(X_2, t),\ C_P(X_2 t),\ \varphi_1(X_2, t),\ \varphi_2(X_2, t),\ \psi(X_2, t).$$

If $t=0$, the following quantitites should be given: $C_N(x, 0)$, $C_P(x, 0)$, $\varphi_1(x, 0)$, $\varphi_2(x, 0)$, $\psi(x, 0)$.

Model II

This model differs from model I in the assumptions $a=$ const and $a_1=a_2=a_3=0$; then $L=L(0, t)e^{-ax}$. All the other relationships of model I remain valid.

Let us transform the equations of model I. If n signifies the direction making angle α with the t-axis, then

$$\frac{\partial\varphi}{\partial t}+\omega\frac{\partial\varphi}{\partial x}=\frac{1}{\cos\alpha}\frac{\partial\varphi}{\partial n}.$$

If it is assumed that

$$\omega_1=\operatorname{tg}\alpha_1;\quad \omega_2=\operatorname{tg}\alpha_2;\quad \omega_3=\operatorname{tg}\alpha_3;\quad \frac{1}{\cos\alpha_i}=\eta_i,$$

then the equations for L, C_N, C_P remain unchanged. The equations for φ_1, φ_2 and ψ become

$$\eta_1\frac{\partial\varphi_1}{\partial n_1}=\alpha A\varphi_1-p\varphi_1\varphi_2+\frac{\partial}{\partial x}\left(\Delta\frac{\partial\varphi_1}{\partial x}\right)-\omega_1\varphi_1$$

$$\eta_2\frac{\partial\varphi_2}{\partial n_2}=\gamma_1\varphi_1\varphi_2-\gamma_2\varphi_2^2+\gamma_3\varphi_2\psi-\varepsilon\varphi_2+\frac{\partial}{\partial x}\left(\Delta^{\bullet}\frac{\partial\varphi_2}{\partial x}\right)-\omega_2\varphi_2$$

$$\eta_3\frac{\partial\psi}{\partial n_3}=\nu\varphi_2-\mu\psi+\theta_1\varphi_1\varphi_2+\theta_2\varphi_2^2-\theta_3\varphi_2\psi+\frac{\partial}{\partial x}\left(\Delta\frac{\partial\psi}{\partial x}\right)-\omega_3\psi.$$

It is seen that the following lines characterize the six unknown functions: straight lines t = const for function L; straight lines x = const for functions C_N, C_P; lines that are orthogonal with respect to the lines n_1 = const for function φ_1; lines that are orthogonal with respect to the lines n_2 = const for function φ_2; lines that are orthogonal with respect to the lines n_3 = const for function ψ.

Proceeding from this it is possible to construct the simplest finite-difference schemes for solving these equations, as follows (in the case of constants α_1, α_2, α_3). Let points with abscissas $x_0=0$, $x_1=\delta$, ..., $x_i=i\delta$, ..., $x_q=q\delta=X_2$ be selected on some straight line $t=t_m$. For simplicity we shall assume that X_1 is among these points and that the values of functions C_N, C_P, φ_1, φ_2, ψ are known at all points with coordinates $(x_i, t_m, i=0, 1, \ldots, q)$, and that the value of function L is known at the point (x_0, t_{m+1}). Moreover, the normal derivatives of all the unknown functions (except L) are known at the points (x_0, t_{m+1}). Then the calculation can proceed in line with the following scheme.

F i r s t s t e p. Starting from equations for L, the values of function are calculated at all the points (x_i, t_m) by finite-difference calculation with respect to x.

S e c o n d s t e p. Finite-difference calculation with respect to t is used to calculate the values of functions C_N and C_P at the points (x_i, t_{m+1}).

T h i r d s t e p. Finite-difference calculation with respect to n_1, n_2 and n_3 is used to calculate the values of functions φ_1, φ_2 and ψ at poinst with coordinates

$$[x_i + \omega_1(t_{m+1} - t_m),\ t_{m+1}],\ [x_i + \omega_2(t_{m+1} - t_m),\ t_{m+1}]$$

and

$$[x_i + \omega_3(t_{m+1} - t_m),\ t_{m+1}].$$

Fourth step. Interpolation of functions φ_1, φ_2 and ψ to points with the coordinates (x_i, t_{m+1}).

Remark. The equations for $C_N, C_P, \varphi_1, \varphi_2$ and ψ contain diffusion terms which, generally speaking, cannot be ignored, and therefore the finite-difference scheme for these functions must be implicit. In this case the boundary conditions yield special terms for the extreme points corresponding to $i=0$ and $i=q$. Equations relating the values of functions at three neighboring points are derived in order to determine the other $q-1$ values. Such equations are always correct provided the diffusion coefficient is constant.

The above considerations make it possible to hope that the problem in question is mathematically quite reasonable. At the same time, it requires painstaking elaboration.

The following questions arise: a) final shaping of the finite-difference schemes; b) elucidation of the form assumed by the finite-difference schemes at $x=X_1$; c) establishment of relationships between the coefficients satisfying the natural conservation laws; d) study of the system of characteristics and construction of finite-difference schemes for a more complete system of equations; e) the problem of the experimental determination of the coefficients, possibly by using methods for solving inverse problems in mathematical physics.

Remarks

Remarks concerning solutions of equations for φ_1, φ_2, and ψ. Let us examine equations of the form

$$\frac{\partial\varphi}{\partial t} + \omega\frac{\partial\varphi}{\partial x} = F(x_1, t_1, \varphi_1, \ldots).$$

All the above equations are of this form, except for the diffusion equation. The characteristics are the solutions of equations

$$\frac{dt}{\varphi_1} = \frac{dx}{\omega_2}.$$

Therefore, the finite-difference calculation should be carried out along these lines.

Example. In the case of φ_2 it can be assumed that

$$\omega_2 = \omega x^{a} \sin\frac{2\pi}{T}.$$

The equation of the characteristics is

$$\omega \sin \frac{2\pi t}{T}\, dt = \frac{dx}{x^{\alpha}}.$$

If $\alpha \neq +1$, then $\frac{T\omega}{2\pi} \cos \frac{2\pi}{T} t = \frac{1}{x^{\alpha-1}(\alpha-1)} - C.$

If $\alpha = +1$, then $x = Ce^{\frac{\omega T}{2\pi} \cos \frac{2\pi t}{T}}$.

Remark concerning the balance relationship for φ_1, φ_2, and ψ. Let all these quantities be measured in identical units. Then the conservation law demands fulfillment of the balance of participating substances in each elementary act. This applies to the following acts:

$\varphi_2 \rightarrow \psi$: death of zooplankton and formation of detritus; the relationship is $\varepsilon = \nu$.

$\varphi_1 \rightarrow [\varphi_2, \psi]$: consumption of phytoplankton by zooplankton with detritus formation; the relationship is $\beta = \gamma_1 + \theta_1$.

$\varphi_2 \rightarrow [\varphi_2, \psi]$: cannibalism of zooplankton with detritus formation; the relationship is $\gamma_2 = \theta_2$.

$\varphi_1 \rightarrow [\varphi_2, \psi]$: consumption of detritus by zooplankton with detritus formation; the relationship is $\gamma_3 = \theta_3$.

These relationships make it possible to reduce the number of coefficients to be determined experimentally. Furthermore, $h_N + h_P = a$ and $v_N + v_P = \nu$.

Assuming a constant P/N ratio in all the organic components (φ_1, φ_2, ψ), the following relationships can be established between the coefficients:

$$h_N : h_P = v_N : v_P = q_P : q_N = C$$

The value of constant C can be determined by chemical analyses. Finally, fixing all the units of measurement (the biomass should be measured by its constituent nitrogen or phosphorus), it is possible to determine, for instance, quantities h_N and l.

Hence it follows that quantities h_N, h_P, v_N, v_P, q_P, q_N, and l can be determined purely analytically.

Remarks concerning the diffusion and convection coefficients. Coefficient Δ is determined from studies of ocean water turbulence. Coefficient Δ^* is the sum of Δ and diffusion of the Brownian component of the zooplankton's own movement. Coefficients ω_1 and ω_3 are determined from the viscosity of the water and specific gravity and dimensions of the particles that are present, or else experimentally. The coefficient ω_2 requires an independent measurement in the course of studies of the vertical migrations of the zooplankton.

DETERMINATION OF THE COEFFICIENTS OF THE EQUATIONS

We shall now demonstrate with the aid of the above model how all the coefficients of the equations can be obtained from measurements performed by oceanographic vessels.

The diffusion coefficient Δ is determined in the course of studies of oceanic water turbulence. Determination of the remaining coefficients falls into two parts. The first part consists in establishing the form of a coefficient, based on general considerations, accurate to within a numerical factor. This is possible because the physicochemical and biological processes, the rates of which are expressed by these coefficients, depend upon temperature, pressure, water salinity, and possibly illuminance. The form of these relationships can be determined. Then these coefficients will become known to within numerical factors, which generally speaking may be dependent upon the specific composition of flora or fauna or some trace elements that are ignored by the model in question. These unknown numerical factors can be determined from survey data by integrating the model's initial equations over the rectangle $0 \leqslant x \leqslant X,\ 0 \leqslant t \leqslant T$.

Let the coefficients v_N, v_P, h_N, h_P, etc. be designated as $v_N f_{V_N}(x, t), v_P f_{V_P}(x, t)$, and so on. It is noted that all the functions $f_{V_N}, f_{V_P}, \ldots$ may be regarded as known.

Let it now be assumed that

$$\int_0^T \int_0^X f_{V_N}(x, t)\,\psi(x, t)\,dx\,dt = H_{V_N}.$$

Under our conditions, the ratios $l: g_N : g_P$ are known, and it is therefore possible to determine the value of A. With the survey data available, it is possible to calculate the following integrals:

$$\int_0^T \int_0^x f_{h_N}(x, t)\,A\varphi_1(x, t)\,dx\,dt = H_{h_N};$$

$$\int_0^T \int_0^x f_{h_P}(x, t)\,A\varphi_1(x, t)\,dx\,dt = H_{h_P};$$

$$\int_0^T \int_0^x f_{V_P}(x, t)\,\psi(x, t)\,dx\,dt = H_{V_P};$$

$$\int_0^T \int_0^x f_{\alpha}(x, t)\,A\varphi_1(x, t)\,dx\,dt = H_{\alpha};$$

$$\int_0^T \int_0^x f_{\beta}(x, t)\,\varphi_1(x, t)\,\varphi_2(x, t)\,dx\,dt = H_{\beta};$$

$$\int_0^T \int_0^x f_{\gamma_1}(x, t)\,\varphi_1(x, t)\,\varphi_2(x, t)\,dx\,dt = H_{\gamma_1};$$

$$\int_0^T \int_0^x f_{\gamma_2}(x, t)\,\varphi_2^2(x, t)\,dx\,dt = H_{\gamma_2};$$

$$\int_0^T\int_0^x f_{\gamma_3}(x,t)\,\varphi_2(x,t)\,\psi(x,t)\,dx\,dt = H_{\gamma_3};$$

$$\int_0^T\int_0^x f_{\varepsilon}\varphi_2(x,t)\,dx\,dt = H_{\varepsilon};$$

$$\int_0^T\int_0^x f_{\nu}\varphi_2(x,t)\,dx\,dt = H_{\nu};$$

$$\int_0^T\int_0^x f_{\mu}\psi(x,t)\,dx\,dt = H_{\mu};$$

$$\int_0^T\int_0^x f_{\theta_1}\varphi_1(x,t)\,\varphi_2(x,t)\,dx\,dt = H_{\theta_1};$$

$$\int_0^T\int_0^x f_{\theta_2}\varphi_2^2(x,t)\,dx\,dt = H_{\theta_2};$$

$$\int_0^T\int_0^x f_{\theta_3}\varphi_2(x,t)\,\psi(x,t)\,dx\,dt = H_{\theta_3}.$$

In terms of this notation and proceeding from the initial equations, we obtain the following relationships:

$$\int_0^X [C_N(T,x) - C_N(0,x)]\,dx = -h_N H_{h_N} + v_N H_{v_N} +$$

$$+ \int_0^T \left[\Delta(t,X)\frac{\partial C_N}{\partial x}\bigg|_{x_t=X} - \Delta(t,0)\frac{\partial C_N}{\partial x}\bigg|_{x_t=0}\right] dt;$$

$$\int_0^X [C_P(T,x) - C_P(0,x)]\,dx = -h_P H_{h_P} + v_P H v_P + \int_0^T \Delta(t,X)\frac{\partial C_P}{\partial x}\bigg|_{\substack{x=X\\t}} dt;$$

$$\int_0^X [\varphi_1(T,x) - \varphi_1(0,x)]\,dx = \alpha H_\alpha - \beta H_\beta + \int_0^T \Delta(t,X)\frac{\partial \varphi_1}{\partial x}\bigg|_{\substack{x=X\\t}} dt -$$

$$- \int_0^T [\omega_1(t,X)\,\varphi_1(t,X) - \omega_1(t,0)\,\varphi_1(t,0)]\,dt;$$

$$\int_0^X [\varphi_2(T,x) - \varphi_2(0,x)]\,dx = \gamma_1 H_{\gamma_1} - \gamma_2 H_{\gamma_2} + \gamma_3 H_{\gamma_3} - \varepsilon H_\varepsilon +$$

$$+ \int_0^T \Delta(t,X)\frac{\partial \varphi_2}{\partial x}\bigg|_{\substack{x=X\\t}} dt - \int_0^T [\omega(t,X)\,\varphi_2(t,X) - \omega(t,0)\,\varphi_2(t,0)]\,dt;$$

$$\int_0^X [\psi(T,x) - \psi(0,x)]\,dx = \nu H_\nu - \mu H_\mu + \theta_1 H_{\theta_1} + \theta_2 H_{\theta_2} - \theta_3 H_{\theta_3} +$$

$$+ \int_0^T \Delta(t,X)\frac{\partial \psi}{\partial x}\bigg|_{\substack{x=X\\t}} dt - \int_0^X [\omega_3(T,x)\,\psi(T,x) - \psi_3(0,x)\,\psi(0,x)]\,dx.$$

Besides these five equations, our coefficients are related to each other by the following eight equations:

$$\varepsilon = \nu; \quad \beta = \gamma_1 + \theta_1; \quad \gamma_2 = \theta_2; \quad \gamma_3 = \theta_3;$$
$$h_N + h_P = \alpha; \qquad V_N + V_P = \nu,$$
$$h_N : h_P = v_N : v_P = C.$$

The relationship

$$g_P : g_N = C.$$

must be additionally used to determine A.

On the whole, there are fewer relationships than coefficients. However, we may derive integral relationships for different time intervals $(0, T_1)$, (T_1, T_2), etc.; consequently, we can derive more equations than coefficients. The latter must be calculated by the method of least squares.

We now come to the accuracy of determination of these coefficients, since we may maneuver the time invervals with a view to deriving the equations. We have to find a scheme of calculation that will yield the most accurate values of the coefficients, starting from a given set of measurements.

Finally, discarding the assumption that the process proceeds identically in all the horizontal planes but retaining the hypothesis that the equation coefficients depend solely upon depth and time, it is possible to use the same scheme to calculate the coefficients, the only modification being that the volume integrals be calculated with respect to three-dimensional volumes, not depth alone.

CAUSES OF THE INADEQUACY OF THE FIRST VARIANT OF THE MODEL

It is useful to examine certain new data obtained during the 44th cruise of the "Vityaz" that were disregarded in the first variant of the model. Most important is the significant role played by microflora in trophic chains of communities inhabiting the upper pelagic layers of the tropical ocean; the nonuniform distribution of life and the formation of thin layers with a high concentration of life not always related to thermoclines or pycnoclines; the substantial part played by horizontal transport of biogens from upwelling zones of deep waters to convergent zones in the life of the upper tens of meters of waters in oligotrophic tropical regions.

It was demonstrated that at least one-third of the mass of bacterial plankton forms natural aggregates larger than $3-5\mu$ that are consumed and assimilated by filter feeders (see Petipa, Pavlova, and Sorokin, 1971). Production of bacteria in the surface pelagic layers of tropical oligotrophic regions is comparable to, and possibly even exceeds, the phytoplankton production. In this case, development of zooplankton is no longer entirely (or even chiefly) limited by development of phytoplankton, as was assumed in the first variant of the model. The bacterial count depends not so much upon the biogenic elements (although the possibility of their being critical cannot be excluded a priori), as upon dissolved organic matter.* How this limiting takes place, to what extent it depends on the composition of organic matter, and which of the latter components are critical, is obscure and calls for special investigation. Nevertheless it is obvious that an ecosystem model cannot be complete if it ignores the activity of the microflora and its role in the food chains.

The concentrations of available inorganic forms of nitrogen phosphorus are very low (close to analytical zero) in the upper mixed layer at a sufficient distance from the upwelling zones of deep waters. The depth at which the amount of light is still sufficient for photosynthesis but the slowly ascending biogens have not yet been consumed in the underlying layer proves to be optimal for development of phytoplankton and bacteria. In regions with weak wind mixing and a comparatively constant temperature regime, the layer with a high concentration of bacterioplankton and phytoplankton and zooplankton subsisting on them need not be located at the upper pycnocline boundary, its depth being controlled by the critical amount of light (Vinogradov, Gitel'zon, and Sorokin, 1971). In the region explored by the "Vityaz" on its 44th cruise, this layer was constantly located at depths of 70—90 m and although its vertical extent was very small, 5—15 m, the numbers of bacteria in this layer were so high that they could serve as the staple forage resource for the zooplankton accumulating there.

According to Vinogradov, Gitel'zon, and Sorokin (see this collection), the upper water layers (about 50 m) in oligotrophic halistatic regions of the ocean are separated by the accumulation of bacteria and phytoplankton from the biogens arriving from below. The life activity of the plant and animal population in the surface layer is largely supported by biogenic elements and dissolved organic matter brought by water flowing from the divergence zone.

Intensive upwelling of deep, biogen-rich waters giving rise to large quantities of phytoplankton and zooplankton occurs in several regions in the ocean tropics. Owing to the intensity of life processes in the warmed surface water, organic residues and products of life activity do not completely settle out of the surface layer and are mineralized, and the liberated mineral salts re-enter the life cycle of the surface layer. Consequently, the biogenic elements have sufficient time to pass through several cycles while the water travels from the divergence to the convergence zones. Naturally, some of these elements leave the surface layer with corpses or migrating animals and the water is gradually depleted — it "ages." Nevertheless, the existence of phytoplankton and zooplankton in the uppermost layers of oligotrophic regions, where the upwelling of biogenic elements from the deep layers is so weak as to be completely consumed by concentrations of bacterioplankton and phytoplankton at the lower boundary of the photic layer, is apparently supported by horizontal influx of biogens with the flow of "aging" water.

Thus, the true picture of the phemoneon is seen to be much more complex than was provided for in the model examined above. The question arises of the necessity of such preparation of information on the simulated ecosystem that would make possible the construction of an improved

* This circumstance, however, is partially taken into account by the assumption that the zooplankton consumes detritus.

variant of the mathematical model of balance relationships. Instead of constructing a single model for the entire ecosystem, it is probably simpler and more useful in the initial stage to construct separate models for certain sufficiently autonomous subdivisions of the ecosystem.

Bibliography

Lyapunov, A. A. O matematicheskikh modelyakh zhiznedeyatel'nosti (Mathematical Models of Life Activity). — In Sbornik: "O matematicheskom modelirovanii zhiznedeyatel'nosti." Moskva, Izdatel'stvo "Mysl'." 1968a.

Lyapunov, A. A. Ob izuchenii balansovykh sootnoshenii v biogeotsenoze (popytka matematicheskogo analiza) (Studies of Balance Relationships in a Biogeocenosis (an Attempt at Mathematical Analysis)). — Zhurnal Obshchei Biologii, Vol. 29, No. 6. 1968b.

Lyapunov, A. A. and S. V. Yablonskii. Teoreticheskie problemy kibernetiki (Theoretical Problems of Cybernetics). — Problemy kibernetiki, No. 9. 1963.

Petipa, T. S., E. V. Pavlova, and Yu. I. Sorokin. Izuchenie pitaniya massovykh form planktona tropicheskoi oblasti Tikhogo okeana radiouglerodnym metodom (Radiocarbon Studies of the Feeding of Mass Forms of Plankton in the Tropical Pacific). (In this collection.)

Poletaev, I. A. O matematicheskikh modelyakh elementarnykh protsessov v biogeotsenozakh (Mathematical Models of Elementary Processes in Biogeocenoses). — Problemy kibernetiki, No. 18. 1966.

Vinogradov, M. E., I. I. Gitel'zon, and Yu. I. Sorokin. O prostranstvennoi strukture soobshchestv evfoticheskoi zony tropicheskikh vod okeana (The Spatial Structure of Communities in the Euphotic Zone of the Ocean Tropics). (In this collection.)

OPTICAL STUDIES IN THE WEST PACIFIC

V. I. Voitov and O. V. Kopelevich
Shirshov Institute of Oceanology of the Academy of Sciences of the USSR

Studies of the transfer and conversion of radiant energy are important for investigations of energy exchange in the ocean from the standpoint of energy utilization in a pelagic community. Comprehensive measurements of the optical properties of seawater (attenuation and scattering coefficients, characteristic scattering curves) and of the horizontal and spatial irradiances in different parts of the visible region of the spectrum were conducted in December 1968 – February 1969 from the "Vityaz" in five polygons* in the southwestern Pacific.

Polygons Nos. 2 – 4 were in the open ocean (Figure 1), and No. 5 was located in Kieta Bay (Solomon Islands).** The thermohaline structure of the waters and hydrodynamic conditions were approximately the same in polygons Nos. 2 – 4, and the optical properties of these waters were also found to be closely similar. Polygon No. 3, situated in the southernmost part of the Solomon Sea, had water of a somewhat lower transparency, owing to the abundance of plankton (mainly blue-green algae). Another noteworthy feature of this sea is the abrupt drop in transparency in the 3,000-m horizon.†

Investigation of the vertical optical structure with an FPR immersion transparency meter †† revealed a transparency layer 15 – 25 m thick in the depth range 75 – 130 m in all polygons. This layer shifts up and down with an approximately six-hour period (Figure 2). Since the pycnoclines underwent analogous shifts, it may be assumed that the diurnal oscillations of the low transparency layer are caused by internal waves of a tidal nature. In many cases the low transparency layer was coextensive with the high bioluminescence layer and sonic scattering layer. Here it is probably a question of concentrations of zooplankton (mainly crustaceans with the property of luminescence) in a layer that is rich in phytoplankton, detritus, and bacteria. The layer of maximum luminescence layer and that of low transparency separated in the daytime; for instance, at 1800 hr on

* [Test areas.]

** For the coordinates of the polygons see the paper by Vinogradov in this collection.

† It is not clear why this is so. An appreciable decrease in transparency in the 2,500 – 3,000-m horizons had also been observed during other cruises of the "Vityaz," in the Pacific. Analysis of vertical density distribution at stations where this was observed revealed a certain increase in the density gradients at these depths.

†† Modernized variant of the FPM photoelectric immersion meter (Kozlyaninov, 1961).

December 12, 1968, the low transparency layer was situated at 80 — 85 m, while the layer of maximum bioluminescence and sonic scattering (according to the echogram) were in the 100 — 110-m horizon.

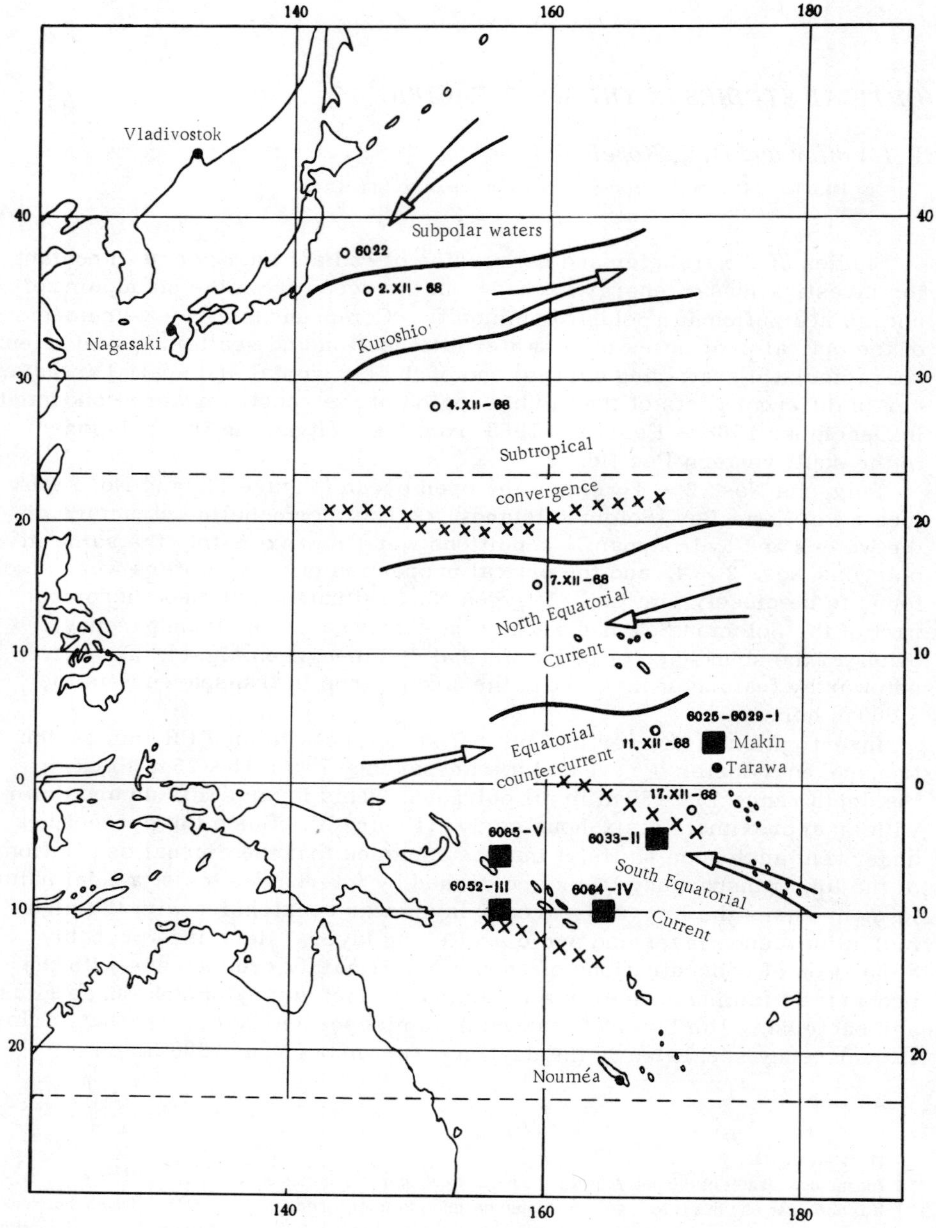

FIGURE 1. Map of hydrooptical stations with boundaries of the main currents and convergences.

Heavy lines and arrows mark the boundaries and directions of currents; crosses — convergence lines; circles — sampling sites.

Extensive studies of the propagation of solar energy in the ocean were conducted in polygons Nos. 2—4. A LYuPO underwater irradiance meter* was used to measure the downward and upward irradiances in five regions of the visible spectrum. Subaerial irradiance variations were monitored by the device's subaerial transducer. Total energy measured by the subaerial transducer in the range 390—650 nm was 55—60% of the total amount of energy recorded by the pyranometer in the range 300—3,000 nm (Figure 3).

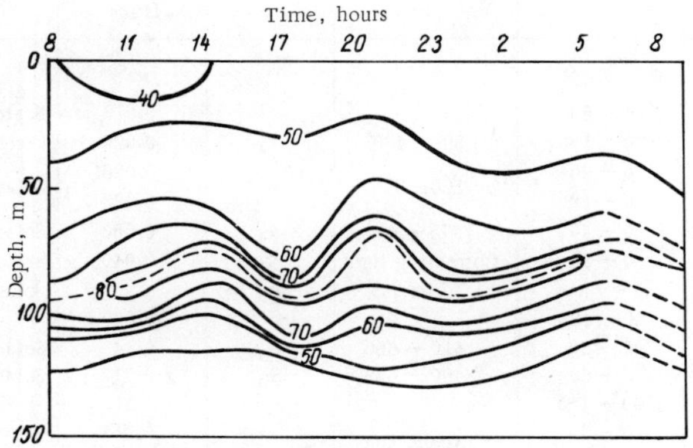

FIGURE 2. Diurnal oscillations of the low transparency layer on February 8—9, 1969 (polygon No. 4).

Dashes mark the position of the layer axis. Numbers represent readings of the instruments which are proportional to attenuation.

A special spherical attachment permitted measurements of the spectral values of spatial irradiance. A correcting filter made it possible to obtain the underwater illuminance values in luxes.

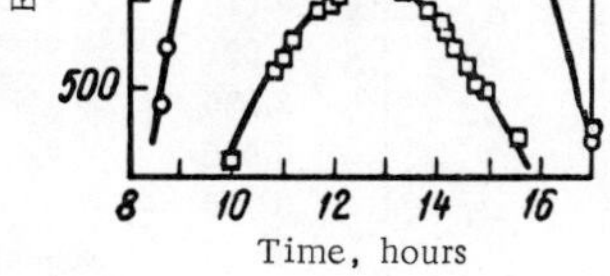

FIGURE 3. Diurnal variation of the total energy incident on the sea surface (January 20, 1969)

Circles represent pyranometer measurements (after Romanov); squares — measurements performed with a selenium photocall (390—680 nm)

* Modernized variant of the photometric underwater irradiance meter FMPO-64 (Karelin and Pelevin, 1970).

The results of the measurements (values of decimal vertical attenuation coefficients in m^{-1} for downward, upward and spatial irradiances, diffuse reflection coefficients of the sea) are given in Tables 1 — 3. The values of these characteristics are approximately the same for all three polygons.

TABLE 1. Vertical attenuation coefficients for downward (α_1), upward (α_2), and spatial (α_3) irradiances and diffuse reflection coefficient of the sea (ζ) measured in polygon No. 2 (station 6033) in December 1968

Observation hours	Depth, cm	Spectral region, nm	Optical character-istics	Values of the character-istics	Observation conditions
December 20					
7.30 — 8.00	0 — 90			0.013	Scattered cumuli, calm,
	90 — 150	390—460	α_1	0.022	solar elevation 12—18°30
8.05 — 8.35	0 — 40	520— 575	α_1	0.028	Solar elevation 21 — 27°
	40 — 120			0.023	
8.35 — 8.45	0 — 40	575 — 610	α_1	0.052	Solar elevation 27 — 28°30'
8.55 — 9.35	0 — 20	Correcting light filter	α_1	0.040	Solar elevation 29 — 40°
	20 — 130			0.020	
9.35 — 9.55	0 — 20	610 — 680	α_1	0.14	Solar elevation 40 — 45°
11.00 — 13.35	0 — 60	460 — 520	α_1	0.012	Solar elevation 59 — 67°
	60 — 185			0.017	
	10 — 75	460 — 520	α_2	0.013	
	75 — 110			0.022	
	5 — 100	460 — 520	ζ	0.04	
	0 — 40	460 — 520		0.012	
	40 — 110		α_3	0.017	
December 23					
8.15 — 8.40	0 — 50	460 — 520	α_1	0.014	Nonuniform cloud cover, sea state 1, solar elevation 20 — 28°
	50 — 150			0.017	
8.40 — 9.05	0 — 50	520 — 575	α_1	0.028	Solar elevation 28 — 34°
	50 — 150			0.020	
9.10 — 10.00	0 — 60	460 — 520	α_2	0.012	Solar elevation 34 — 47°
	0 — 60	460 — 520	α_3	0.015	
10.45 — 12.00	0 — 30	Correcting light filter	α_1	0.024	
	30 — 170			0.018	Solar elevation 55°30'
	0 — 90		α_2	0.016	
	0 — 10		ζ	0.01	
	10 — 40			0.02	
	40 — 80			0.03	
	0 — 10		α_3	0.043	
	10 — 100			0.019	
13.30 — 13.45	0 — 40	460 — 520	α_1	0.012	Overcast, sea state 1
16.00 — 16.20	0 — 50	390 — 460	α_1	0.010	
	50 — 130			0.021	
16.35 — 16.50	0 — 20	520 — 575	α_1	0.030	
	20 — 100			0.022	
16.50 — 17.00	0 — 70	520 — 575	α_1	0.028	
17.05 — 17.10	0 — 20	Correcting light filter	α_1	0.033	Nonuniform cloud cover, sea state 1
	20 — 70			0.023	Solar elevation 25 — 24°

TABLE 2. Vertical attenuation coefficient for downward (α_1) upward (α_2), and spatial (α_3) irradiances and diffuse reflection coefficient of the sea (ζ) measured in polygon No. 3 (station 6052) in January 1969

Observation hours	Depth, m	Spectral region, nm	Optical characteristics	Values of the characteristics	Observation conditions
January 16					
9.25 — 9.55	0 — 25	Correcting light filter	α_1	0.034	Nonuniform cloudcover, sea state 1 — 2. Solar elevation 41 — 48°
	25 — 130			0.022	
	130 — 160			0.016	
10.00 — 11.15	0 — 80	460 — 520	α_1	0.012	Solar elevation 48° 30' — 66°
	80 — 130			0.022	
	130 — 160			0.014	
12.30 — 14.00	0 — 20	610 — 680	α_1	0.123	Solar elevation 79 — 68°
13.25 — 13.50	0 — 25	575 — 610	α_1	0.056	
	25 — 60			0.042	Solar elevation 74°30' — 70°30'
14.05 — 15.35	0 — 40	390 — 460	α_1	0.012	Solar elevation 67 — 44°
14.30 — 15.20	0 — 70	520 — 575	α_2	0.022	Solar elevation 60°30' — 48°30'
	0 — 70	520 — 575	α_3	0.026	
January 18					
13.30 — 14.00	0 — 20	610 — 680	α_1	0.132	Nonuniform cloudcover, sea state 1 — 2; solar elevation 73°30' — 68°
January 19					
9.20 — 9.45	0 — 50	390—460	α_1	0.014	Nonuniform cloudcover, sea state 1 — 2, solar elevation 40 — 46°
	50 — 120			0.025	
9.45 — 10.05	0 — 60	460 — 520	α_1	0.016	
	60 — 120			0.022	Solar elevation 46 — 51°
10.20 — 11.00	0 — 30	520 — 575	α_1	0.030	
	30 — 110			0.023	Solar elevation 54 — 62°
11.00 — 11.40	0 — 10	575 — 610	α_1	0.078	
	10 — 60			0.045	Solar elevation 66 — 71°
11.50 — 13.10	0 — 10	Correcting light filter		0.032	
	10 — 150		α_1	0.021	
	0 — 70		α_2	0.017	Solar elevation 79°30' — 78°
	0 — 40			0.029	
	40 — 70		α_3	0.021	
13.20 — 15.05	0 — 70			0.017	
	70 — 104	460 — 520	α_1	0.023	Mist, sea state 1 — 2, solar elevation 70 — 54°
	104 — 150			0.018	
13.20 — 15.05	0 — 30			0.009	
	30 — 60	460 — 520	α_2	0.019	
	60 — 80			0.031	
	0 — 10	460 — 520	ζ	0.05	Mist, sea state 1 — 2, solar elevation 70 — 54°
	10 — 70			0.06	
	0 — 80	460 — 520	α_3	0.017	

TABLE 2 (continued)

Observation hours	Depth, m	Spectral region, nm	Optical characteristics	Values of the characteristics	Observation conditions
January 20					
9.25 — 9.55	0 — 30	520 — 575	α_1	0.031	Mist, sea state, solar elevation 40 — 49°
	30 — 110			0.023	
10.00 — 10.20	0 — 70	390 — 460	α_1	0.014	Solar elevation 49 — 53°30'
	70 — 110			0.024	
10.50 — 15.25	0 — 60	390 — 460	α_1	0.015	Solar elevation 60 — 80°
	60 — 140			0.025	
10.50 — 15.25	0 — 80	460 — 520	α_1	0.014	Solar elevation 60 — 80°
	80 — 150			0.022	
10.50 — 13.05	0 — 20	575 — 610	α_1	0.058	Solar elevation 60 — 80°
	20 — 50			0.042	
10. 50 — 12.00	0 — 20	610 — 680	α_1	0.136	Solar elevation 60 — 78°
11.00 — 15.25	0 — 160	Correcting light filter	α_1	0.024	Solar elevation 60 — 80°
15.30 — 15.55	0 — 50	460 — 520	α_1	0.014	Solar elevation 50 — 43°
	50 — 100			0.021	
	100 — 160			0.017	
15.55 — 16.00	0 — 15	610 — 680	α_1	0.149	Solar elevation 45 — 44°
16.05 — 16.10	0 — 20	575 — 610	α_1	0.067	Solar elevation 44°30' — 40°

Analysis of the results leads to the following conclusions:

1. The maximum spectral transmission of natural daylight for oceanic waters lies in the blue-green region of the spectrum (light filters Nos. 1 and 2; see also Table 4).

2. The curves for vertical variation of underwater irradiance for these light filters (see Figure 4) display a characteristic discontinuity at depths of 70 — 80 m, where the vertical attenuation coefficients change from 0.012 —0.017 m^{-1} to 0.022— 0.025 m^{-1}. The changes should evidently be ascribed to the presence of the low transparency layer, the upper boundary of which lies at approximately these depths.

3. The red spectral region (610 — 680 nm) is subject to the strongest attenuation in oceanic waters. Already at depths of about 20 m the sensitivity of the instrument is inadequate for recording radiation in this spectral band (light filter No. 5).

4. The curves for the vertical variation of underwater irradiance for light filters Nos. 3 and 4 similarly display characteristic discontinuities. The vertical attenuation coefficient changes from 0.030 to 0.023 m^{-1} at a depth of about 30 m for light filter No. 3; it changes from 0.056 — 0.078 to 0.042 — 0.047 m^{-1} at about the 20-m depth for light filter No. 4. These changes are associated with the Forbes effect; because of the wide transmission band of the light filters used in the investigations, the vertical attenuation coefficient diminishes at great depths which are reached by only the wavelengths that are less strongly attenuated by seawater. The vertical attenuation coefficient for total energy (390 — 680 nm) decreases gradually with increasing depth (as the water filters out, first the red and then the yellow portion of the emission spectrum). However, the vertical

attenuation coefficient again increases beginning with the depth of 60 — 70 m (owing to the presence of the low transparency layer).

TABLE 3. Vertical attenuation coefficients for downward (α_1), upward (α_2), and spatial (α_3) irradiances and diffuse reflection coefficient of the sea (ζ) measured in polygon No. 4 (station 6064) in February 1969

Time	Depth, m	Spectral region, nm	Optical characteristics	Values of the characteristics	Observation conditions
February 9					
15.10 — 15.45	0 — 100	460 — 520	α_1	0.013	Overcast, sea state 1 — 2
15.45 — 16.05	0 — 20			0.015	Same
	20 — 90	460 — 520	α_1	0.011	
	90 — 100			0.028	
February 10					
7.50 — 8.50	0 — 15	Correcting light filter		0.040	Variable cloudcover, sea state 2, solar elevation 38 — 53°
	15 — 140		α_1	0.023	
8.50 — 9.25	0 — 60			0.013	Solar elevation 53°30' — 62°30'
	60 — 110	460 — 520	α_1	0.023	
	110 — 140			0.018	
9.35 — 10.30	0 — 40			0.016	Solar elevation 62°30' — 77°30'
	40 — 60	Correcting light filter		0.021	
	60 — 80		α_2	0.027	
	0 — 100		α_3	0.024	
10.30 — 12.50	0 — 56		α_2	0.012	
	56 — 90	460 — 520		0.026	
	0 — 70		α_3	0.014	Solar elevation 70°30' — 79°
	70 — 90			0.025	
12.50 — 13.50	0 — 20	610 — 680	α_1	0.130	Solar elevation 79 — 78°
	0 — 10		α_3	0.094	
	0 — 10		ζ	0.003	
	0 — 10		α_3	0.130	
13.00 — 13.20	0 — 40	575 — 610	α_1	0.055	Solar elevation 79 — 78°
February 12					
8.40 — 9.20	0 — 40			0.012	Mist, sea state 1 — 2, solar elevation 52 — 62°
	40 — 70	390 — 460	α_2	0.020	
	70 — 90			0.035	
	0 — 60			0.012	
	60 — 90		α_3	0.020	

5. The investigators did not succeed in establishing visible effects of variation of the vertical attenuation coefficients due to changes in radiation structure and their dependence upon the solar elevation and state of the sky. Investigation of these effects probably calls for an instrument that would take direct measurements of the vertical attenuation coefficient (for instance, two transducers making simultaneous illuminance measurements at the two different depths), with the obligatory use of monochromatic filters.

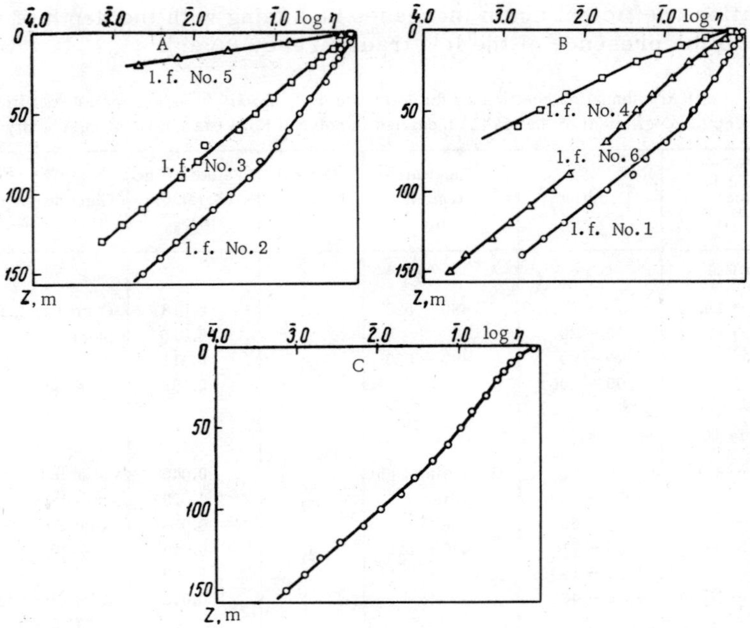

FIGURE 4. Vertical variation of the underwater irradiance coefficient (polygon No. 3, January 20, 1969):

A — for light filters (l. f.) No. 2 (460 — 520 nm), No. 3 (520 — 575 nm), and No. 5 (610 — 680 nm); B — for light filters No. 1 (390 — 460 nm), No. 4 (575 — 610 nm), and No. 6 (correcting); C — for total energy (390 — 680 nm). Vertical variation of the vertical attenuation coefficient α_1: 0–1 m — 0.080; 1–5 m — 0.038; 5–10 m — 0.024; 10–20 m — 0.016; 20–30m — 0.014; 30–60 m — 0.015; 60–70 m — 0.019; 70–190 m — 0.029.

TABLE 4. Vertical variation of the relative (%) spectral composition of sunlight (polygon No. 3, solar elevation 80°)

Depth, m	Spectral range, nm					E/W, lumens/watts
	390 — 460	460 — 520	520 — 575	575—610	610—680	
0	23%	22%	20.6%	11.4%	23%	278
1	25.8	27	20.2	9.3	17.7	260
5	33.5	32.9	21.4	6.3	5.9	278
10	38.7	32.0	23.4	4.5	1.4	274
15	39.2	36.9	20.6	2.9	0.4	231
20	40.9	38.4	18.8	1.7	0.2	182
30	42.4	42.1	14.5	0.9	0.1	154
40	47.9	38.9	12.7	0.5	—	136
50	48.3	40.9	10.5	0.3	—	116
60	48.8	41.8	9.2	0.2	—	103
70	47.0	47.6	5.4	—	—	106
80	45.2	47.4	7.4	—	—	102
90	42.7	50.4	6.9	—	—	85
100	38.5	54.1	7.4	—	—	98
110	38.3	55.0	6.7	—	—	82
120	34.4	58.2	7.4	—	—	93
130	33.8	59.3	6.9	—	—	97
140	30.2	63.2	6.6	—	—	107

Optical measurements in Kieta Bay were performed down to 30 m. Water transparency in the bay was significantly lower than in the oceanic polygons Nos. 2 — 4, but appreciably higher than in the lagoon of Tarawa Atoll. The very elongated characteristic curve of scattering denotes the presence of coarse particles. The vertical attenuation coefficients also proved to be higher than in the oceanic polygons. The transmission maximum shifted from the blue-green to the green region of the spectrum (light filters Nos. 2 and 3) (Table 5). The extreme right columns in Tables 4 and 6 show the ratio of illuminance to total irradiance in the spectral region 390 — 680 nm, varying by a factor > 5 depending on optical properties and depth. Therefore, the choice of a universal conversion factor between energy and light units does not seem possible.

TABLE 5. Vertical attenuation coefficients for downward (α_1), upward (α_2), and spatial (α_3) irradiances and diffuse reflection coefficient of the sea (ζ) measured in Kieta Bay in the Solomon Islands on February 14, 1969 (observation conditions: overcast, calm)

Time	Depth, m	Spectral region, nm	Optical characteristics	Values of the characteristics
12.30 — 12.40	0 — 30	Correcting light filter	α_1	0.059
12.40 — 12.50	0 — 30	520 — 575	α_1	0.056
12.55 — 13.00	0 — 30	460 — 520	α_1	0.056
13.00 — 13.10	0 — 30	390 — 460	α_1	0.071
13.10 — 13.15	0 — 30	575 — 610	α_1	0.082
13.15 — 13.25	0 — 15	610 — 680	α_1	0.164
13.25 — 13.45	0 — 30	Correcting light filter	α_2	0.046
	0 — 30		α_3	0.059

Investigations of the optical properties of seawater in the vicinity of coral atolls and directly in their lagoons were of considerable interest for studying the conditions of formation of the optical structure of Pacific tropical waters. No previous optical observations had been carried out in lagoons of coral atolls, except for Jerlov's (1951) UV radiation measurements in the Kapingamaranga Atoll. During the 44th cruise of the "Vityaz" measurements were achieved of the optical characteristics of lagoon waters of the Butaritari and Tarawa atolls, revealing a significant difference in their optical properties due to structural difference between the atolls. The Butaritari Atoll is an insular half-ring closed on the south, but there is a fairly extensive water exchange with the ocean on the north. Consequently, the optical properties (including spectral transparency) of Butaritari lagoon water are very similar to those of the surrounding oceanic water. The optical properties of the water are different in the lagoon of the Tarawa Atoll, which is practically encircled by the insular ring. The Tarawa lagoon water is much more turbid than the Butaritari water.

Studies performed on the approach to the Butaritari Atoll (at distances of 10.6, 5, 2.3, and 1 mile from it) in order to determine the boundaries of distribution of the lagoon waters showed that atolls with open water exchange do not have any appreciable influence on the surrounding waters.

TABLE 6. Vertical variation of the relative (%) spectral composition of sunlight in Kieta Bay under a cloudy sky

Depth, m	Spectral region, nm					E/W, lumens/watts
	390 — 460	460 — 520	520 — 575	575 — 610	610 — 680	
0	22.3	22.7	21	11.6	22.4	297
1	23	24.2	22.4	11.7	18.7	314
5	24.1	29.3	27.2	11.1	8.3	368
10	23.3	33.4	31.1	9.4	2.8	410
15	21.4	36.4	33.7	7.6	0.9	468
20	19.1	38.8	35.9	5.9		439
30	14.8	42.3	39.4	3.5		453

On her way to the main working region, the "Vityaz" traversed all the major currents of the temperate and tropical latitudes in the West Pacific (see Figure 1). No simultaneous measurements of the optical properties of these currents had been made before (Kozlyaninov, Voitov, and Khalemskii, 1968), and we were therefore curious to carry out a comparison of the optical characteristics determined by the method of quasi-synoptic survey (Table 7) over a large area. As was to be expected, the subpolar waters formed in the Bering Sea and Sea of Okhotsk proved to be the most turbid.

TABLE 7. Optical characteristics in the western Pacific

Region	Transparency θ, %						Scattering coefficient for λ = 546 nm	Parameter of the characteristic scattering curve $\log \beta^{(15)}/\beta^{(90)}$
	λ_{eff} = 473 nm	520	546	568	601	649		
Subpolar waters	67	66	64.5	63.5	48	45	0.19	1.68
Kuroshio	72	71	67.5	66.5	58	48	0.12	2.10
Subtropical convergence	80	77.5	72.5	69	62	51	0.14	1.76
North Equatorial Current	82	82.5	81	73.5	63	54	0.07	1.81
Equatorial Countercurrent	75.5	75.5	74	70	59.5	56	0.12	1.83
South Equatorial Current	73	77.5	69	66	56	49.5	0.12	1.96

The characteristic scattering curve of these waters was less elongated than the curves of the other waters. The waters of the subtropical divergence, North Equatorial Current, and Equatorial Countercurrent differed among themselves insignificantly. Somewhat higher transparency values were obtained for a water sample taken from the North Equatorial Current. This result contradicts the conventional notion that the most transparent waters in the World Ocean are those in the regions of subtropical convergences (Sargasso Sea, the Central anticyclonic area in the Indian Ocean, etc.). It is fully possible that the turbidity in the upper water layer in the region of tropical convergence was temporary, resulting from storms. The low value of transparency in the waters of the South Equatorial Current was probably due to biological factors (a vigorous bloom of blue-green algae was noted here). The more elongate characteristic scattering curve of this current denoted a relative predominance of coarse particles evidently of biological origin. However, the most elongate scattering curve was that of the Kuroshio, possibly owing to the presence of coarse mineral particles of terrigenous origin.

Bibliography

Publications in Russian

Burkov, V. A. Tsirkulyatsiya vod (Circulation of Waters). — In: Tikhii okean, Vol. 2. Gidrologiya Tikhogo okeana. Moskva, Izdatel'stvo "Nauka." 1968.

Karelin, A. K. and V. N. Pelevin. Morskoi izmeritel' podvodnoi obluchennosti FMPO-64 i ego primenenie v gidroopticheskikh issledovaniyakh (The FMPO-64 Underwater Irradiance Meter and its Application to Hydrooptical Research). — Okeanologiya, Vol. 10, No. 1. 1970.

Kozlyaninov, M. V. Rukovodstvo po gidroopticheskim izmereniyam v more (Manual of Marine Hydrooptical Measurements). — Trudy Instituta Okeanologii Akademii Nauk SSSR. No. 47. 1961.

Kozlyaninov, M. V., V. I. Voitov, and E. N. Khalemskii. Opticheskie kharakteristiki vod (Optical Characteristics of Waters). — In: Tikhii okean, Vol. 2. Gidrologiya Tikhogo okeana. Moskva, Izdatel'stvo "Nauka." 1968.

Publications in Other Languages

Jerlov, V. G. Optical Studies of Ocean Waters. — Reports Swed. Deep-sea Exped. 1947—1948, Vol. 3, Fasc 1. 1951.

ABSORPTION OF SOLAR ENERGY IN THE SEA

O. V. Kopelevich
Shirshov Institute of Oceanology of the Academy of Sciences of the USSR

The purpose of this paper is calculation of the amount of solar energy absorbed by seawater in different spectral intervals in relation to depth. Calculations of this kind are absent in the literature. The calculation is based on the well-known Gershun formula

$$\omega = \operatorname{div} \vec{\underline{E}} = \varkappa \underline{E},$$

where ω is the spatial absorption density; $\vec{\underline{E}}$, light vector; $\underline{E}$, spatial irradiance; $\varkappa$, absorption density of the medium (Gershun, 1936, 1945). If the sea is regarded as a plane-parallel medium with a uniformly illuminated surface, the amount of radiant energy absorbed per unit volume can be expressed by the formula

$$\frac{d\underline{W}_{abs}}{dz} = \varkappa \underline{E}(\bar{z}) = \underbrace{\varkappa_w \cdot \underline{E}(\bar{z})}_{\text{absorption by pure water}} + \underbrace{\varkappa_m \cdot \underline{E}(\bar{z})}_{\text{absorption by solutes and suspended matter}} \quad (1)$$

where $\varkappa_w$ is the absorption index of pure water and $\varkappa_m$ is the combined index of absorption by solutes and suspended matter in seawater. Assuming that scalar irradiance attenuation with increasing depth in an optically isotropic seawater layer obeys the Bouguer law $\underline{E}(\bar{z}_2) = \underline{E}(\bar{z}_1) \cdot e^{-\alpha_3(z_2 - z_1)}$ (where α_3 is the vertical attenuation coefficient for spatial irradiance), * it is possible to determine the energy absorbed in the layer $\Delta z = z_2 - z_1$, thus:

$$\Delta \underline{W}_{abs} = \frac{\varkappa \underline{E}(\bar{z}_1)}{\alpha_3} \cdot (1 - e^{-\alpha_3 \Delta z}) = \underbrace{\varkappa_w \cdot \frac{\underline{E}(\bar{z}_1)}{\alpha_3}(1 - e^{-\alpha_3 \Delta z})}_{\text{energy absorbed by pure water}} + \underbrace{\varkappa_m \cdot \frac{\underline{E}(z_1)}{\alpha_3} \cdot (1 - e^{-\alpha_3 \Delta z})}_{\text{energy absorbed by solutes and suspended matter}}. \quad (2)$$

* Strictly speaking, this assumption is valid only for deep layers. However, measurements show that it is well justified (in the absence of stratification) also for the upper layers of the sea. For instance, calculation of α_3 based on the data reported by Tyler (1960), who measured brightness distribution at different depths, yields the values 0.072 ($\sigma = 0.005$) for the layer 10.4 – 66.1 m (clear, sunny sky) and 0.084 ($\sigma = 0.007$) for the layer 6.1 – 55.0 m (dull sky). Although the tridimensional brightness distributions in these depth intervals are subject to considerable variation, the coefficient of variation was only 7% in the first case and 8% in the second.

Thus, if the index of light absorption by seawater is known, determination of radiant energy absorbed in the sea requires only the measurement of spatial irradiance.

However, accurate determination of the absorption index of seawater is a very difficult experimental task. The determination is usually made from measurements of spatial, downward, and upward irradiances by the formula

$$\varkappa = \frac{E_1(\bar{z})}{\underline{E}(z)} \cdot (\alpha_1 - \zeta \cdot \alpha_2), \qquad (3)$$

where α_1 and α_2 are vertical attenuation coefficients for downward and upward irradiances; $\zeta = \frac{E_2(\bar{z})}{E_1(\bar{z})}$, diffuse reflection coefficient of the sea; $\underline{E}_1(\bar{z})$ and $\underline{E}_2(\bar{z})$ are downward and upward irradiances (Pelevin, 1965).

In the Institute of Oceanology of the Academy of Sciences of the USSR these characteristics are determined using the FMPO-60 underwater irradiance photometer with a special attachment for scalar irradiance measurements (Karelin and Pelevin, 1970), and recently by means of its modernized variant the instrument LYuPO. In the absence of waves, the standard deviation may amount to approximately 3% of the measured quantity. The accuracy of determination of the absorption index and absorbed energy depend on the vertical interval Δz. If the vertical structure of the medium permits the choice of a 10-m interval, the standard deviation for absorption index $\varkappa$ determined by formula (3) and the amount of radiant energy absorbed in the sea determined by formula (2) may amount to 10 and 25 – 30%, respectively, of the measured quantities.

The values of energy absorbed in the sea calculated from measurements of downward, upward and scalar irradiances performed in the Solomon Sea during the 44th cruise of the "Vityaz" (Voitov and Kopelevich, in this collection) are listed below. The values are stated in percentages of the total energy in the interval 390 – 680 nm incident on the sea surface.*

Δz, m	Spectral interval, nm					
	390 – 460	460 – 520	520 – 575	575 – 610	610 – 680	390 – 680
0 – 10	5.6	5.3	8.8	6.7	20.3	46.7
10 – 20	4.2	3.7	4.8	1.3	0.6	14.6
20 – 30	3.0	2.6	2.6	0.3	–	8.5
30 – 40	2.2	2.1	1.1	–	–	5.4
40 – 50	1.8	1.3	0.6	–	–	3.7
50 – 60	2.1	1.0	0.4	–	–	3.5
60 – 70	1.6	0.7	0.2	–	–	2.5
70 – 80	1.0	0.5	–	–	–	1.5

* On the sea surface the energy was distributed by the spectral intervals in the following manner: 390–460 nm, 23‰; 460 – 520 nm, 22%; 520–575 nm, 20%; 575 – 610 nm, 12%; 610 – 680 nm, 23%; its spectral composition was scarcely affected by solar elevations The total energy in the interval 390–680 nm was 55–60% of the total amount of energy recorded by the pyranometer (in the range 300 – 3, 000 nm).

By resorting to absorption indexes of pure distilled water (Joseph, 1952), it is possible to estimate the fraction of radiant energy absorbed by the water itself. Absorption by the latter accounts for 15 — 17% of the total amount of energy absorbed in the spectral interval 390 — 460 nm and 460 — 520 nm. This percentage decreases to 10 — 12% in the low transparency layer (at depths 50 — 80 m). Absorption by the pure water accounts for 40 — 50% of entire absorption in the interval 520 — 575 nm (25 — 30% in the low transparency layer). In the intervals 575 — 610 and 610 — 680 nm, practically all the absorbed energy is absorbed by the water itself. The energy in these spectral intervals decreases rapidly with increasing depth, and therefore the fraction of total energy absorbed by pure water (in the range 390 — 680 nm) also decreases appreciably with increasing depth, thus:

ΔZ, m	0—10	10—20	20—30	30—40	40—50	50—60	60—70	70—80
%	80	44	30	26	20	14	11	10

In marine biological investigations, for which the absorption of solar energy in the sea is of considerable interest, the incidental optical measurements are limited, as a rule, to downward irradiance. Substituting (3) in (2) and assuming $\alpha_3 = \alpha_1$, *

$$\Delta \underline{W}_{abs} = k[\underline{E}_1(\bar{z}_1) - \underline{E}_1(\bar{z}_2)]. \qquad (4)$$

The coefficient $k = 1 - \zeta \frac{\alpha_2}{\alpha_1}$ is close to 1 (since $\alpha_1 \approx \alpha_2$, while ζ does not usually exceed 0.05 — 0.07). Analysis of the results of optical measurements performed in the Pacific Ocean, the Mediterranean, and the Black Sea during expeditions of the Institute of Oceanology of the Academy of Sciences of the USSR shows that the coefficients of variation of the values of k do not exceed 2%.

Spectral interval, nm	390 – 460	460 – 520	520 – 575	575 – 610	610 – 680
Mean value of coefficient k	0.96	0.97	0.98	0.99	1.00
Number of measurements	108	93	265	37	32
Standard deviation	0.022	0.016	0.007	0.012	0.002

The approximate formula (4) permits calculation of absorbed energy values in different spectral intervals from measurements of only one quantity, i. e. the downward irradiance.

* This assumption is borne out by practice. For instance, for the above data reported by Tyler (1960) the ratio of the vertical attenuation coefficient for downward irradiance α_1 to vertical attenuation coefficient for spatial irradiance α_3 was 1.01 (σ = 0.061) for the layer 10.4 – 66.1 m and 1.00 (σ = 0.008) for the layer 6.1 – 55.0 m.

Bibliography

Publications in Russian

Gershun, A. A. Svetovoe pole (Light Field). Moskva — Leningrad. ONTI. 1936.

Gershun, A. A. K teorii svetovogo polya v rasseivayushchei srede (Contribution to the Theory of a Light Field in a Dispersive Medium). — Doklady Akademii Nauk SSSR, Vol. 49, No. 8. 1945.

Karelin, A. K. and V. N. Pelevin. Morskoi izmeritel' podvodnoi obluchennosti FMPO-64 i ego primenenie v gidroopticheskikh issledovaniyakh (The FMPO-64 Underwater Irradiance Meter and its Application to Hydrooptical Research). — Okeanologiya, Vol. 10, No. 1. 1970.

Pelevin, V. N. 1965. Ob izmerenii pokazatelya istinnogo pogloshcheniya sveta v more (Measurements of the True Light Absorption Index in the Sea). — Izvestiya Akademii Nauk SSSR, Fizika atmosfery i okeana, Vol. 1, No. 5. 1965.

Voitov, V. I. and O. V. Kopelevich. Opticheskie issledovaniya v zapadnoi chasti Tikhogo okeana (Optical Studies in the West Pacific). (In this collection.)

Publications in Other Languages

Joseph, J. Meeresoptik. Berlin-Göttingen-Heidelberg. 1952.

Tyler, J. Radiance Distribution as a Function of Depth in an Underwater Environment. Bull. Scripps Inst. Oceanogr. Vol. 7, No. 5. 1960.

MAJOR HYDROLOGICAL AND HYDROCHEMICAL FEATURES OF THE REGION INVESTIGATED DURING THE 44th CRUISE OF THE "VITYAZ"

V. S. Arsen'ev, L. I. Galerkin, E. A. Plakhin, and V. V. Sapozhnikov
Shirshov Institute of Oceanology of the Academy of Sciences of the USSR

The general trend of the hydrological and hydrochemical investigations performed during the 44th cruise of the "Vityaz" was determined by the specific biological specialization of the cruise (Vinogradov, this collection); attention here was focused mainly on the upper layers, which are the most interesting for studies of biological production processes.

All the main hydrological work was carried out in three polygons* (Figure 1) in the region of the Pacific approximately demarcated by 4 and 10° S and 156 and 169° E.

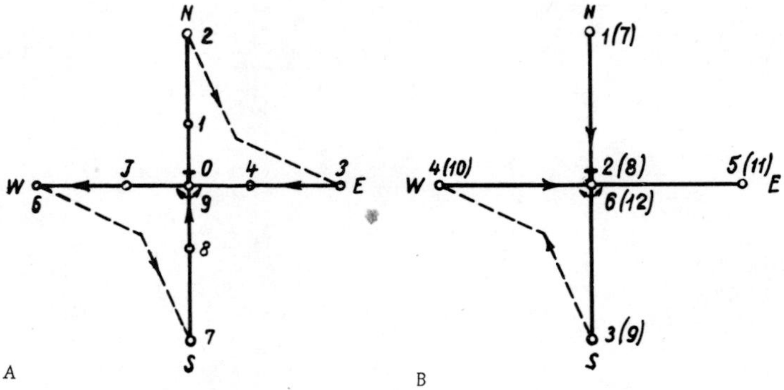

A B

FIGURE 1. Diagram of polygons Nos. 2 and 3 (A) and 4 (B).

An anchored buoy is in the center; arrows indicate the direction of movement of the vessel around the buoy with the stations; circles — position of the hydrological stations; numbers — the successive order in which they were executed.

The water masses in the region investigated were analyzed by conventional methods and techniques (Dobrovol'skii et al., 1960; Radzikhovskaya and Leont'eva, 1968; Leont'eva and Radzikhovskaya, 1962).

* [Test areas.]

The analytical data reveal the presence of three water layers in all polygons (stations 6033, 6052, 6064): the surface layer (relatively homogeneous), an intermediate layer of high salinity, and an intermediate layer of low salinity. The layers have the following distinctive characteristics.

The upper mixed layer (or surface water mass), is distributed vertically down to 40 — 50 m or 80 — 100 m (Figure 2). Differences in the thermohaline characteristics of the different polygons are slight (Tables 1 and 2).

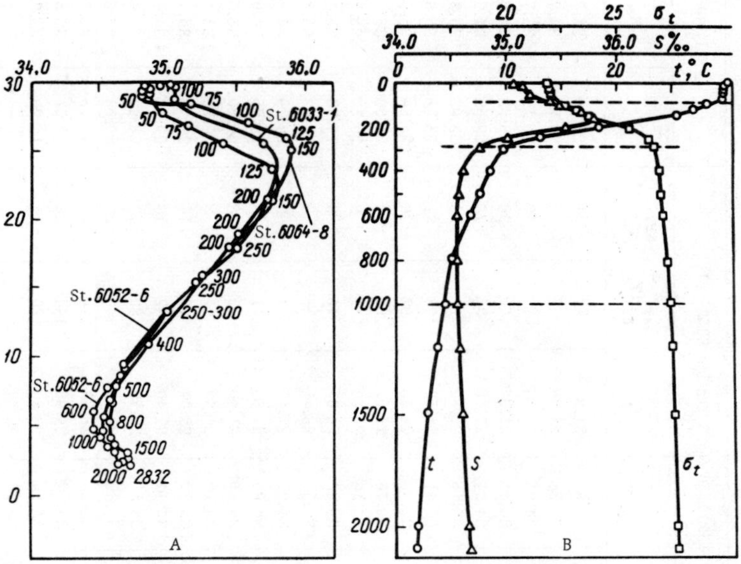

FIGURE 2. A — TS curves for polygons Nos. 2 (station 6033), 3 (station 6052), and 4 (station 6064); B — vertical distribution of temperature, salinity, density and boundaries of water masses at station 6033—0.

The underlying intermediate layer of high salinity spreads to depths of 270 — 300 m. The core of this layer with maximum salinity 35.75 — 35.95‰ occupies a stable position at 125 — 150 m throughout the region. There are some differences in the thermohaline characteristics of this layer according to observations in the different polygons. The differences between the oceanic polygons are small, but the high salinity layer in the Coral Sea (polygon No. 3) is formed against a background of appreciably lower temperature than in the oceanic region (polygons Nos. 2 and 4). The high salinity layer is the most heavily stratified one (see Figure 2). The temperature and density gradients in this layer are 1 — 2 orders of magnitude higher than in the underlying and overlying water layers (see Table 1).

The next underlying intermediate layer of low salinity (water mass C) possessing fairly homogeneous characteristics reaches to 900 — 1,100 m in the region studied. Its core with minimum salinity about 35.50‰ (see Table 2) lies at 600 — 800 m.

TABLE 1. Temperature, salinity, density, and temperature and density gradients at the stations of multidiurnal polygons

Station No. / Horizon	Polygon No. 2 (station 6033–0)					Polygon No. 3 (station 6052–6)					Polygon No. 4 (station 6064–10)				
	$T°$	$S‰$	σ_t	$\frac{\delta T}{\delta Z}\cdot 10^3$	$\frac{\delta \sigma_t}{\delta Z}\cdot 10^3$	$T°$	$S‰$	σ_t	$\frac{\delta T}{\delta Z}\cdot 10^3$	$\frac{\delta \sigma_t}{\delta Z}\cdot 10^3$	$T°$	$S‰$	σ_t	$\frac{\delta T}{\delta Z}\cdot 10^3$	$\frac{\delta \sigma_t}{\delta Z}\cdot 10^3$
0	30.16	35.08	21.82	41	12	29.42	34.81	21.85	1	1	29.30	34.76	21.80	4	0
10	29.75	35.15	21.94	2	0	29.41	34.86	21.84	3	0	29.26	34.72	21.79	14	9
20	29.73	35.16	21.94	4	0	29.38	34.85	21.84	28	10	29.12	34.78	21.88	4	6
30	29.77	35.17	21.94	4	4	29.10	34.85	21.94	70	36	29.08	34.84	21.94	3	6
50	29.69	35.21	22.01	2	7	27.70	34.96	22.48	35	16	29.02	34.96	22.05	38	25
75	29.64	35.42	22.18	60	22	26.83	35.15	22.89	53	26	28.06	35.39	22.68	51	26
100	28.14	35.50	22.74	46	21	25.50	35.41	23.53	76	31	26.78	35.68	28.32	47	23
125	27.00	35.69	23.26	55	20	23.60	35.76	24.35	92	26	25.60	35.91	28.87	64	20
150	25.63	35.80	23.77	142	37	21.30	35.76	25.01	68	13	24.00	36.93	24.36	70	18
200	18.52	35.52	25.58	109	17	17.90	35.46	25.67	51	8	20.50	35.79	25.25	56	9
250	13.08	35.03	26.41	64	6	15.30	35.22	26.09	45	6	17.70	35.47	25.72	55	9
300	9.86	34.76	26.70	—	—	13.04	35.01	26.41	—	—	14.93	35.27	26.21		
0	29.87	34.73	21.53	3	7	29.57	34.91	21.83	1	2	29.79	34.91	21.76	4	7
10	29.84	35.10	21.90	1	4	29.58	34.95	21.85	1	5	29.75	34.98	21.88	5	6
20	29.83	35.17	21.94	3	1	29.69	35.05	21.90	2	8	29.80	35.07	21.89	0	3
30	29.80	35.18	21.95	9	5	29.71	35.18	21.98	5	3	29.80	35.12	21.92	1	1
50	29.62	35.21	22.04	6	8	29.62	35.22	22.04	23	14	29.78	35.15	21.94	1	4
75	29.48	35.42	22.24	57	21	29.05	35.44	22.40	42	16	29.76	34.42	22.04	52	24
100	28.05	35.48	22.76	38	23	28.01	35.51	22.79	30	13	28.47	35.49	22.63	38	14
125	26.71	35.67	23.33	33	12	27.27	35.62	28.12	47	20	27.53	35.58	22.97	55	24
150	25.88	35.70	23.62	119	33	26.10	35.81	28.63	114	27	26.15	35.76	28.58	139	36
2000	19.95	35.63	25.27	133	21	21.42	35.77	24.98	144	26	19.21	35.55	25.40	104	15
250	13.32	34.98	26.33	73	10	14.20	35.18	26.30	80	9	14.20	35.02	26.17	—	—
300	—	—	—	—	—	10.18	34.78	26.77	—	—	—	—	—	—	—

Note. The temperature and density gradients are stated in degrees and arbitrary units of density/meter. The station numbers consist of the cluster number (four digits) followed by the number of stations within the cluster (the single digit following the dash).

TABLE 2. Principal thermohaline characteristics of the surface and intermediate water masses in the polygons

Polygon No. 12 (station 6033–0)		Polygon No. 3 (station 6052–6)		Polygon No. 4 (station 6064–10)	
thermohaline characteristics	lower boundary, m	thermohaline characteristics	lower boundary, m	thermohaline characteristics	lower boundary, m
Surface water mass (A)					
Temperature, salinity, density at the surface: 30.16°, 35.08‰, 23.82; at the lower boundary: 28.90°, 35.44‰, 22.46	85	Temperature, salinity, density of the surface: 29.42°, 34.87‰, 21.85; at the lower boundary: 27.70°, 34.96‰, 22.48	50	Temperature, salinity, density at the surface: 29.30°, 34.76‰, 21.80; at the lower boundary: 28.06°, 35.38‰, 22.68	75
Intermediate water mass of high salinity (B)					
Maximum salinity 35.8‰; temperature and density, 25.63°, 23.77	300	Maximum salinity 35.76‰; temperature and density, 21.30°, 25.01	300	Maximum salinity 35.93‰; temperature and density 24.00°, 23.36	300
Intermediate water mass of low salinity (C)					
Minimum salinity, 34.54‰; temperature and density 5.23°, 27.32		Minimum salinity 34.4‰; temperature and density 5.29°, 27.24		Minimum salinity 34.56‰; temperature and density 4.58°, 27.40	

Although it is common practice to distinguish also between deep bottom waters (Leont'eva and Radzikhovskaya, 1968), this differentiation seems too arbitrary in the absence of special observations because of the smooth variation of the thermohaline and chemical characteristics at considerable ocean depths, without any natural boundary layers or substantial vertical gradients.

This structure of the waters, common to the entire region, is also reflected by the vertical distribution of hydrochemical characteristics (Table 3).

The surface waters contain practically no mineral forms of biogenic elements: phosphorus, nitrogen, and silicon. The intermediate waters of high salinity are distinguished by their low oxygen concentrations. Typical of the low salinity layer (especially at its lower boundary) is a maximum concentration of phosphates and nitrates; the main oxygen minimum and its maximum deficit also occur at the same depths.*

The lower boundary of the surface layer is determined by the depth at which a dynamic equilibrium is established between the processes of assimilation of biogenic elements and the liberation of oxygen in the course of photosynthesis and the process of decomposition and mineralization of dead organic matter. It is not accidental that the lower boundary of the surface layer coincides with zero values of oxygen deficit** and is underlain by waters with high concentrations of biogenic elements.

* Here, oxygen deficit is defined as the difference between its maximum content (100% saturation) and the actual concentration.

** The lower boundary of the surface layer seems to coincide with the depth at which oxygen-liberating processes are compensated by processes of oxygen consumption for oxidation of the decaying organic matter and respiration of marine organisms.

TABLE 3. Principal hydrochemical characteristics of the surface and intermediate water masses in the polygons

Polygon No. 2. station 6033–0-9)	Lower boundary of the layer, m	Polygon No. 3 (station 6052–0-9)	Lower boundary of the layer, m	Polygon No. 4 (station 6064–7-12)	Lower boundary of the layer, m
Surface water mass (A)					
Subsurface oxygen at depths 50–75 m. The concentration of all biogenic elements is practically zero. At the lower boundary of the surface layer there is a marked maximum of nitrates up to 1.0 μg-at/l and the largest vertical gradient of phosphates and nitrates	70 – 90	A very distinct oxygen maximum at 50 m. Oxygen minimum at 10 m. The homogeneous surface layer is without biogenic elements. The concentration of nitrites in the layer of the maximum at the lower boundary of the surface layer reaches 1.3 μg-at/l	50 – 100	Subsurface oxygen maximum at 25 m, 10, and 50 m. Concentration of phosphates in the surface layer about 0.2 μ g-at/l, that of nitrates up to 1.0 μg-at/l. Maximum of nitrites 0.6 μg-at/l at depth 100 m	50 – 100
Intermediate water mass (B)					
Upper oxygen maximum (up to 32‰). The concentrations of phosphates and nitrates increase steadily, reaching 2.2 and 28 μ g-at/l. There is a small intermediate maximum of organic phosphorus	250 – 300	Upper oxygen minimum at 250 m, often extended (150–300 m). Occasionally there is an upper intermediate maximum of phosphates and nitrates coinciding with the upper oxygen minimum. One or two intermediate organic phosphorus maxima	250–300	Upper oxygen minimum at 300 m, occasionally two and more minima at 150–400 m. Organic phosphorus concentration in the upper minimum lower than in the main minimum	50 – 300
Intermediate water mass (C)					
Main lower oxygen minimum (< 2.0 mg/l). The concentrations of nitrates and phosphates reach their maximum	800–1200	Main oxygen maximum at 1,500 m, with a concentration of 2.8–3.0 ml/l, which is higher than in the upper minimum. The concentrations of phosphates and nitrates rise to a maximum at the lower boundary of the layer	1500	Oxygen minimum at the lower boundary, its concentration ≤3.0 ml/l. The concentrations of phosphates and nitrates attain their maximum at 800 – 1,200 m, remaining constant farther down	1200

The large vertical gradients of phosphates and nitrates at the lower boundary of the surface layer are apparently due entirely to biochemical processes and are unrelated to the pycnocline, which is situated farther down and is as a rule indistinct; density gradients of the same order of magnitude occur throughout a wide depth interval from 75–100 to 200–250 m (see Table 1).

The oxygen concentration in the surface layer is 4.2 — 4.4 ml/l. Oxygen saturation of the water is 100± 2%. The concentration of organic phosphorus in the surface layer is fairly high, 0.6 — 1.0 mg-atom P per liter, testifying to active processes of assimilation and destruction of organic matter.

The absence of mineral forms of phosphorus in the presence of a primary production of 100 — 200 C mg/m²·day suggests that the phytoplankton partially utilizes also phosphorus from organic compounds.

Dissolved carbon dioxide in the surface waters is practically in equilibrium with the atmospheric carbon dioxide. The distribution of biogenic elements and oxygen in intermediate waters is complex and unclear. There are several alternating extremes of these characteristics through the depth interval 100 — 1,200 m.

Regarding the overall structure and characteristics of its water masses, the region investigated is the closest to the tropical structure of waters in the southern hemisphere (Radzikhovskaya and Leont'eva, 1968; Rotschi and Lemasson, 1967), as is most clearly shown by the characteristic shape of the TS curve (see Figure 2). Another characteristic feature is the presence of two oxygen minima, two maxima of nitrates and phosphates, and an abrupt change of the vertical gradients of these characteristics at the depth of the upper oxygen minimum. However, the relatively low salinity and high temperature at the surface as well as certain chemical characteristics suggest that the region is under the influence of equatorial waters and is genetically closely related to these. This applies primarily to the oceanic polygon No. 2 (station 6033), which is located near the boundary of the Equatorial Countercurrent (Burkov, 1968).

The appearance of the main oxygen minimum, the maximum of the preforms of phosphates (1.4 μg-at P/l), and minimum salinity at depths of 1,000 — 1,200 m is apparently due to the influx of antarctic intermediate waters into the region. The upper oxygen minimum at 250 m and the high concentrations of silica, nitrates, and phosphates are due to advection of eastern equatorial waters. This is clearly demonstrated by the vertical distribution of organic phosphorus (Figure 3), since the complete mineralization and liberation of phosphorus take place in the upper 0 — 300-m layer. Organic phosphorus is practically absent below 300 m.

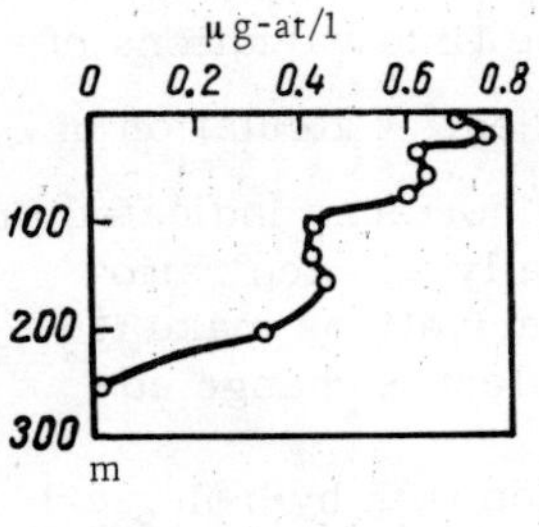

FIGURE 3. Vertical distribution of organic phosphorus at station 6064 — 10

One of the tasks of the hydrological observations during the 44th cruise of the "Vityaz" was collection of data for subsequent calculation of the vertical velocity components and vertical movements of waters in the polygons. In accordance with the specific conditions of this cruise, from all the methods available for calculating vertical velocities, that by equation of advectional water transport and turbulent diffusion was chosen, thus:

$$\frac{\delta k}{\delta t} + u\frac{\delta k}{\delta x} + v\frac{\delta k}{\delta y} + w\frac{\delta k}{\delta z} = \frac{\delta}{\delta x}\left(A_x\frac{\delta k}{\delta x}\right) + \frac{\delta}{\delta y}\left(A_y\frac{\delta k}{\delta y}\right) + \frac{\delta}{\delta z}\left(A_z\frac{\delta k}{\delta z}\right),$$

where k is the concentration of an element or a quantitative characteristic of a property transported by the seawater; u, v and w are projections of the current velocity vector on the corresponding axes x, y, and z; t is time; A_x, A_y, and A_z are the turbulent exchange coefficients along the respective axes.

Since the equation contains the first and second derivatives of the concentrations of elements (properties) along the three coordinate axes, a "cross" pattern of hydrological stations was chosen. The center of the cross was marked with an anchored buoy with BPV-2 current meters. The arms of the cross ran from north to south and from west to east, and hydrological sampling stations with a complete set of hydrological-hydrochemical determinations were located along each arm at distances of 10 and 25 miles from the center.

Assuming all the velocity components and all the turbulent exchange coefficients to be unknown, it is necessary to solve a system of six equations with six different elements (k). Usually such a system is not solved, not only because its solution is difficult and time-consuming but also because it is hardly ever possible to carry out simultaneous determinations of the concentrations of all the elements. Moreover, these elements must satisfy the conservativity conditions, i. e. their concentrations at different points must be controlled solely by the movement of water and turbulent mixing. The hydrochemical and biological determinations showed that biogenic elements (nitrogen and phosphorus) and oxygen in the upper 500-m layer can by no means be regarded as conservative (Sapozhnikov and Galerkin, this collection).

Solving the system of equations based on the initial equation involves various assumptions and simplifications related to taking into account the peculiarities of the movement of waters in the region. Thus, diurnal sampling stations at the center of the polygons (crosses) indicated insignificant time variations of temperature and salinity. Consequently the process could be regarded as quasi-stationary and the time variations of temperature and salinity $\frac{\delta T}{\delta t}$ and $\frac{\delta S}{\delta t}$ could be disregarded. Calculation of the second derivatives of temperature and salinity by horizons indicated their order of magnitude to be $10^{-15} - 10^{-16}$. With the usually adopted ratio $A_z : A_l = 1:10^6$, the orders of magnitude of the second derivatives make it possible to ignore the terms with the horizontal turbulent exchange coefficients.

Instrumental observations of currents in conjunction with hydrological data at the cross stations make it possible to substitute the observed values of horizontal velocity components u and v in the initial equation and to solve the system with respect to the vertical velocity component w. The variable properties, temperature and salinity, were taken from water sampling observations. After all the assumptions, the calculation system of equations takes the form

$$u\frac{\delta T}{\delta x} + v\frac{\delta T}{\delta y} + w\frac{\delta T}{\delta z} = A_z\frac{\delta^2 T}{\delta z^2}$$
$$u\frac{\delta S}{\delta x} + v\frac{\delta S}{\delta y} + w\frac{\delta S}{\delta z} = A_z\frac{\delta^2 S}{\delta z^2}.$$

It is assumed here that the x-axis is directed eastward, the y-axis northward, and the z-axis from the sea surface downward.

When this system is solved by means of determinants, the A_z values do not directly participate in the calculation of w, and there is therefore no need to assume a vertical turbulent exchange coefficient or to make any assumptions concerning its vertical variation.

All the calculations were performed on a "Minsk-22" electronic digital computer for both scales of distances between the stations, $\Delta x = \Delta y = 10$ miles and $\Delta x = \Delta y = 25$ miles. The observed components of currents (u and v) were averaged by diurnal periods in order to exclude the fluctuations of tidal periods. The calculation included averages for individual diurnal periods and also daily averages for the entire period of observations (6 — 10 days).

The best results during calculation of the vertical velocities were obtained for distances of 25 miles between the stations of the cross and by averaging u and v for two days, i. e. for the period during which all the hydrological stations on the cross were executed. The results of the calculation are given in Table 4.

TABLE 4. Vertical velocity components (cm/sec) in the polygons*

Horizon, m	Station 6033	Station 6052	Station 6064
25	$+1.5 \cdot 10^{-3}$	$-2.8 \cdot 10^{-2}$	$-5.6 \cdot 10^{-3}$
50	$+1.7 \cdot 10^{-2}$	$+6.2 \cdot 10^{-4}$	$+1.7 \cdot 10^{-3}$
75	$-1.4 \cdot 10^{-3}$	$-3.6 \cdot 10^{-3}$	$+1.7 \cdot 10^{-3}$
100	$-2.1 \cdot 10^{-4}$	$-2.4 \cdot 10^{-3}$	$+2.9 \cdot 10^{-3}$
125	$+1.4 \cdot 10^{-5}$	$-2.9 \cdot 10^{-3}$	$-1.2 \cdot 10^{-3}$
150	$+3.8 \cdot 10^{-4}$	$-4.6 \cdot 10^{-3}$	$-4.8 \cdot 10^{-4}$
200	$+3.5 \cdot 10^{-3}$	$-2.4 \cdot 10^{-3}$	$-1.7 \cdot 10^{-4}$
250	$+3.6 \cdot 10^{-3}$	$+9.6 \cdot 10^{-3}$	$+4.1 \cdot 10^{-3}$
300	$-8.5 \cdot 10^{-3}$	$+5.1 \cdot 10^{-3}$	$+1.2 \cdot 10^{-4}$
400	$-4.3 \cdot 10^{-3}$	$-2.0 \cdot 10^{-3}$	$+1.0 \cdot 10^{-3}$
500	$-5.5 \cdot 10^{-3}$	$-3.4 \cdot 10^{-3}$	$+4.0 \cdot 10^{-3}$

* Plus and minus signs indicate sinking and upwelling of waters respectively.

The observations in all three polygons were carried out under very mild wind or in complete calm. It was not possible to check the calculation results for the vertical velocity by other methods (reckoning with the wind), and therefore the only possible test of the trueness of the calculated values was evaluation of the regularity in the distribution of their signs and orders of magnitude.

Of the 33 calculated and accepted values (see Table 4), 3% correspond to velocities of the order of 10^{-5} cm/sec, 18% to 10^{-4}, 73% to 10^{-3}, and 6% to 10^{-2} cm/sec.

The predominant value, 10^{-3} cm/sec, may seem unusually large. Values of 10^{-4} cm/sec were reported by Chekotillo (1961) for the northwestern Pacific. However, velocity values of the order of 10^{-3} cm/sec appear on his maps between 10 and 20° N, at all depths, but especially distinctly in maps for the 500-m depth constructed for the gradient-convection component of the current. Our results were obtained directly from water sampling observations without preliminary fitting or averaging, since such operations are known to result inevitably in diminished horizontal and vertical calculated velocities.

The entire 500-m-thick column is divided vertically into 4 or 5 layers; the vertical velocities in the upper layer from the surface to 25 – 50 m may be assumed to be due to pure drift currents, those in the intermediate layers down to 200 – 250 m to combined drift and gradient currents, and those in the lower layers to pure gradient currents.

The convergence and divergence zones of vertical velocities are situated between these layers. The boundary between the lower and the intermediate layer is the most stable and distinct one. It passes between the horizons 250 and 300 m in the form of convergence at station 6033, and between horizons 200 and 250 m in the form of divergence at stations 6052 and 6064. This boundary may be assumed to separate the intermediate water masses of low and high salinities (see Table 2).

Intensification of velocities in the direction toward the convergence of vertical movements may be regarded as the common feature of all three stations. In the divergence region the velocities begin, as it were, from zero and attain their maximum values in the direction toward the convergence. At station 6033 the velocities increase by a factor of 7 – 10 at the first convergence (50 – 75 m), by factors of 200 and 2, respectively, on top (from $1.4 \cdot 10^{-5}$ at 125 m to $3.6 \cdot 10^{-3}$ at 250 m) and below at the second convergence (250 – 300 m). The acceleration is not very large at the convergence at station 6064, but it can be traced distinctly from 50 to 200 m in the upper and lower layers.

The field of vertical velocities contains a peculiar discontinuity at the convergence. Therefore, the transition from the top down from minus to plus is shown by a continuous line in the graph of the vertical variation of w (Figure 4), but the reverse transition (from plus to minus) is shown as discontinuous. The layer between the two convergences (75 – 250 m) at station 6033 is very well approximated by a straight line. The analogous layer above and below the divergence (125 – 500 m) at station 6064 can also be represented approximately by a straight line, but in this case the point in the horizon 250 m ($+4.1 \cdot 10^{-3}$ cm/sec) must be regarded as erroneous, and the approximation to the straight line is in itself much coarser. The graph in Figure 4 is constructed on a semilogarithmic scale, i. e. the relationship represented by the straight line in this graph is described by an exponential function of the form $w = ae^{bz}$, where a and b are coefficients, z is the depth, and e is the base of natural logarithms. For instance, the straight line for station 6033 (upper) is satisfactorily described by the formula $w = 7.1 \cdot 10^{-3} e^{0.0065x}$.

As yet, we are in possession of only two instances (out of the total three) revealing this relationship, so that any attempt at generalizing and explaining the nature of the phenomenon would be premature.

The transport of biogenic elements to the surface photosynthetic layer occurs only in the intermediate subsurface layers, which are separated from the deep waters by zones of descending movements. On the whole, the intensity of transport increases upward, the velocities (at stations 6033 and 6064) increasing sevenfold, whereas the concentrations of biogenic elements and their gradients increase only 1.5 – 2 times. The situation and thickness of these layers differ at the different stations, 75 – 100 m at station 6033, 75 – 200 m at station 6052, and 125 – 200 m at station 6064.

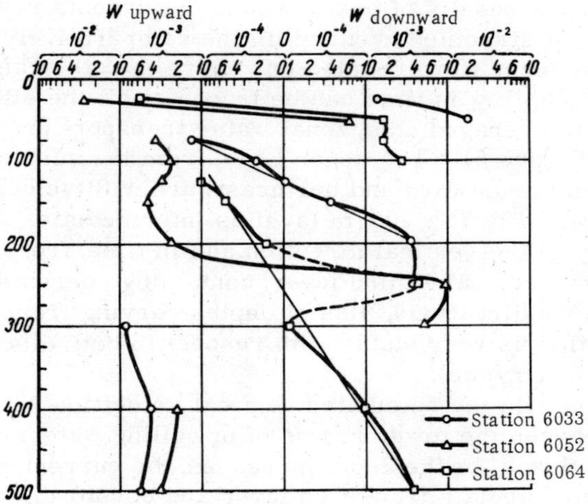

FIGURE 4. Vertical distribution of velocities.
The depth scale (y-axis) is uniform, the velocity scale logarithmic. The velocities increase downward to the right of the zero line and upward to its left.

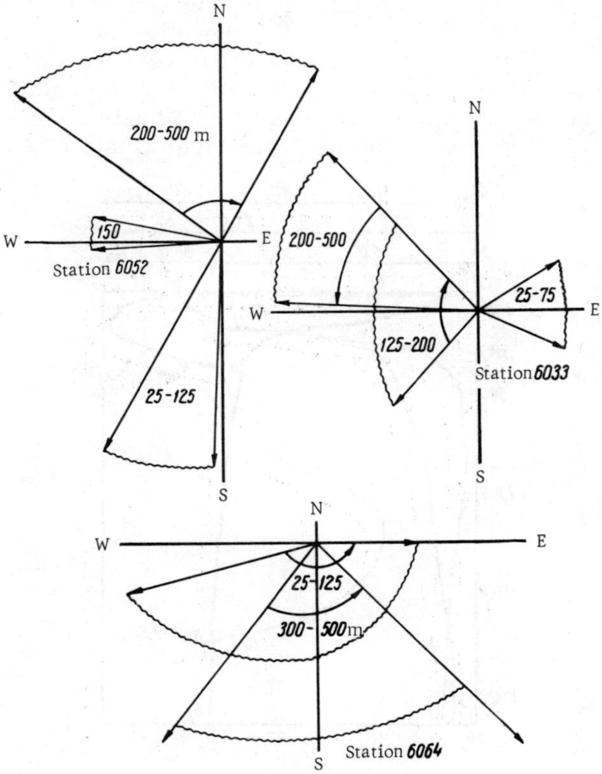

FIGURE 5. Schematic hodographs of the velocity vector of currents in polygons No. 3 (station 6052), No. 2 (station 6033), and No. 4 station 6064).

Analysis of the results of observations of currents revealed their very complex vertical structure even within the comparatively limited 500-m layer. Two or three flows located one above the other and velocity, direction, and stability in time can be traced at all the stations.

According to averaged data, zonal water transport predominates at station 6033 (Figure 5). The upper layer from the surface down to 50 m is carried mainly eastward and northeastward with velocities of 10 — 20 cm/sec. The 75 — 200-m layer is intermediate. The currents in it vary markedly in the vertical direction and in time, from SE to NW and from 5 to 25 m/sec. The third layer, containing currents of western and northwestern directions, lies at depths varying from 200 — 250 to 500 m. This flow is very stable with respect to both direction and velocity (25 — 30 cm/sec).

Comparison with the calculated vertical velocities shows satisfactory agreement between the lower region of upwelling waters and the stable deep northwestern flow. As the depth increases, the current rotates mainly clockwise in the intermediate transition layer and counterclockwise in the deep layer. The change of the sign of rotation of the current occurs near the location of the change of the sign of the vertical velocity (200 — 250 m).

At station 6052 the upper layer extending from the surface down to 125 m is under stable southward movement. The maximum average diurnal velocities (35 — 40 cm/sec) are consistently observed at the horizons 50 — 75 m, the velocities dropping to 10 — 12 cm/sec above and below. The transition layer, with its core at a depth about 150 m, has velocities of 8 — 11 cm/sec and a stable eastward direction. The lower layer extending from 200 to 500 m moves mainly northward (from NW to NE) with a velocity of 15 — 25 cm/sec. The vectors in the lower layer rotate clockwise.

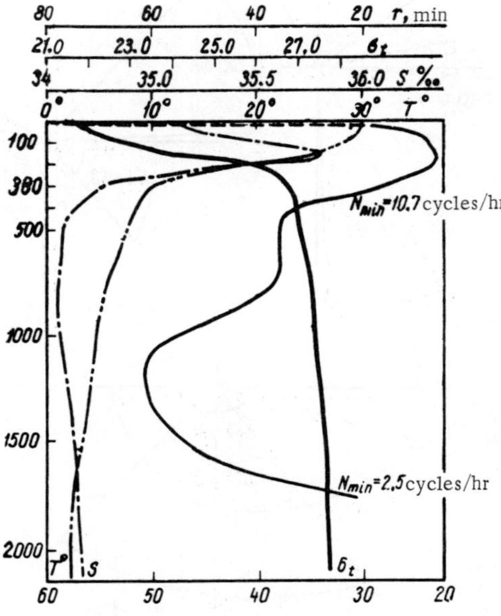

FIGURE 6. Vertical distribution of temperature, salinity, density, and Vjaisjal-Brunt frequency at station 6033-8

At station 6064 the current in the 25 — 125-m layer shifts smoothly from near-western through southern to eastern (i. e. counterclockwise) and has velocities of 10 — 20 cm/sec. The currents of the intermediate layer (150 — 250 m) have fairly high velocities (up to 20 cm/sec at 150 m) but their direction is unstable. The current in the lower layer (300 — 500 m) changes direction from SW to SE, the velocities increasing from 30 to 35 cm/sec.

Stations 6052 and 6064 are situated near the Solomon Islands and the effect of the coast evidently manifested itself in the predominantly meridional transport at these stations in comparison with station 6033 which shows a more pronounced flow zonality.

In conclusion it should be noted that the existence of a stratification of density in the upper (bilogically active) layer calls for more detailed hydrological observations for evaluation of the internal wave parameters. Figure 6 shows, besides the curves of the vertical variation of hydrological characteristics, the Vjaisjal-Brunt frequency distribution curve. As is known, this frequency N is related to density gradients by the formula $N^2 = \frac{g}{\rho}\frac{d\rho}{dz}$ and determines the upper limit of oscillation frequencies during the generation and development of internal waves in a stratified layer (Sabinin, 1966). The frequency N maximum coincides satisfactorily with the depth of the maximum salinity layer and is of the order of 10.7 cycles per hour (period ~ 6 min).

The graph can provide only an estimate of the possible range of oscillation frequencies in the absence of special observations for determination of the periods and amplitudes of internal waves. It should be noted that density gradients of this order of magnitude may be accompanied by internal waves with relatively large oscillation amplitudes; this must be reckoned with in the analysis, for instance, of diurnal migrations of the plankton.

For this purpose, while working in a region with conditions of this kind it is necessary to perform more careful measurements in the stratified layer with continuous recorders or instruments with low-inertia transducers (of the type STG or ChBTZ) and special observations for estimation of the internal wave parameters.

Bibliography

Publications in Russian

Bruevich, S. V. et al. — In: Tikhii Okean, Vol. 3. Khimiya Tikhogo okeana. Moskva, Izdatel'stvo "Nauka." 1966.

Burkov, V. A. Tsirkulyatsiya vod (Circulation of Waters). — In: Tikhii Okean, Vol. 2. Gidrologiya Tikhogo okeana. Moskva, Izdatel'stvo "Nauka." 1962.

Chekotillo, K. A. Raschet vertikal'nykh dvizhenii vod v severo-zapadnoi chasti Tikhogo okeana (Calculation of Vertical Water Movements in the Northwestern Pacific). — Okeanologiya, Vol. 1, No. 6.

Dobrovol'skii, A. D., V. V. Leont'eva, and V. I. Kuksa. K kharakteristike struktur i vodnykh mass zapadnoi i tsentral'noi chasti Tikhogo okeana (Description of the Structures and Water Masses in the Western and Central Pacific). — Trudy Instituta Okeanologii Akademii Nauk SSSR. No. 40. 1960.

Leont'eva, V. V. and M. A. Radzikhovskaya. O vydelenii gidrologicheskikh struktur i vodnykh mass v okeane (Identification of Hydrological Structures and Water Masses in the Ocean). — Trudy Instituta Okeanologii Akademii Nauk SSSR. No. 66. 1962.

Radzikhovskaya, V. V. and M. A. Leont'eva. In: Tikhii Okean, Vol. 2. Gidrologiya Tikhogo okeana. Moskva. Izdatel'stvo "Nauka." 1968.

Sabinin, K. D. O svyazi korotkoperiodnykh vnutrennikh voln s vertikal'nym gradientom plotnosti v more (Relationship between Short-Term Internal Waves and Vertical Density Gradient in the Sea). — Izvestiya Akademii Nauk SSSR. Fizika atmosfery i okeana, Vol. 2, No. 8. 1966.

Sapozhnikov, V. V. and L. I. Galerkin. K raschetu pervichnoi produktsii tropicheskikh vod po vertikal'nomu vynosu biogennykh elementov (Calculation of Primary Production in Tropical Waters from the Upward Influx of Biogenic Elements). (In this collection.)

Vinogradov, M. E. Issledovaniya funktsionirovaniya okeanicheskikh biologicheskikh sistem (Studies of the Life Activity of Oceanic Biological Systems). (In this collection.)

Publications in Other Languages

Rotschi, H. and L. Lemasson. Oceanography of the Coral and Tasman Seas. — Oceanogr. Mar. Biol. Ann. Rev. 5. 1967.

PELAGIC BATHYMETRIC SOUNDING AND ITS POSSIBLE APPLICATION TO STUDIES OF THE SPATIAL STRUCTURE OF BIOCENOSES

I. I. Gitel'zon, L. A. Levin, A. P. Shevyrnogov, R. N. Utyushev, and A. S. Artemkin

Kirenskii Institute of Physics of the Siberian Branch of the Academy of Sciences of the USSR, Krasnoyarsk

A feature of pelagic marine communities that is possibly decisive for their very existence is the spatially nonuniform distribution of their population, in particular, marked vertical stratification. The methods formerly used for studying this phenomenon, i. e. layer by layer investigation of the water column with plankton nets, or series of point samples taken with bathometers, have several fundamental shortcomings (Vinogradov, Gitel'zon, and Sorokin, in this collection).

Studies of the vertical microdistribution of plankton require continuous recording of a biological parameter which permits valid judgment of the distribution of the entire mass of plankton. The luminescence of organisms inhabiting the water column may serve as this parameter.

Measurement of bioluminescence intensity as a rapid method for studying the distribution of life in the ocean is made possible by the fact that the organisms capable of responding to external stimuli with luminescent flashing constitute a numerically important part of the community and obey the same distribution patterns as those governing its general structure. Along with this, their photic responses provide an opportunity of studying the laws governing their distribution and their relationships to the environment.

The method of bathymetric sounding, in which luminescence is recorded with a moving bathyphotometer, has several advantages over the conventional method of recording bioluminescence flashes with a fixed bathyphotometer. For instance it takes about one hour to obtain the vertical distribution pattern for bioluminescence in the 0 – 150-m layer by the method of fixed horizons with intervals of 10 – 15 m between observation points. During this time the ship will shift appreciably even with a small drift, and the resultant distribution pattern of bioluminescence cannot be referred to any definite point. Sounding to the same depth requires only 5 or 6 min, i. e. the resolving power of this method is much higher.

The method of fixed horizons clearly cannot elucidate the detailed stratification of bioluminescence, being especially unsuitable for clear-cut determination of the position of luminescence maxima. With this method it is impossible to distinguish between the variations of intensity of the bioluminescence field that are related to the latter's own nonuniformity and the variations that are related to the migration of plankters and to the diurnal rhythm of bioluminescence.

Finally, an instrument immersed at a certain depth stimulates luminescence on the part of surrounding organisms as a result of the rolling of the ship. The level of excitation may differ at different depths, making the results of different measurements incomparable. In contrast, the level of excitation is more constant if the instrument responsible moves at a constant and fairly high speed.

The sounding method permits the vertical variations of the characteristics of the bioluminescence field to be compared with data on plankton abundance obtained by layer by layer vertical catches with a plankton net. The method facilitates isolation of the astronomical component constituting the background against which the signal is recorded (Rudyakov, 1967). A distinctive feature of the method is the need for a continuous flow of information on the depth of location of the instrument for accurate tying of the values observed to the horizon.

APPARATUS AND OBSERVATION PROCEDURES

The use of the bathyphotometer as a sounding instrument has been described by Clarke and Hubbard (1959) and Rudyakov (1967). Clarke used the same instrument for sounding and for measurements in fixed horizons. Rudyakov did his soundings with the bathyphotometer developed in our laboratory (Gitel'zon et al., 1965; Gitel'zon, 1969) for measurements in fixed horizons.

However, the sounding method makes several specific demands on the instrument. We developed the instrument on the basis of the following considerations.

1. The intensity of bioluminescence stimulated by the bathyphotometer itself is known to depend upon its geometry, the speed of its movement related to the amplitude of the ship's rolling, and the depth of immersion of the instrument, etc. Obviously, once a certain level of excitation has been reached (in general this is different for each individual organism) the photic flux per unit volume of water will not increase with a further increase of the intensity of the excitation. In this case, the parameters characterizing the photic field generated in the excitation zone (irradiance at a given distance, etc.) will assume certain definite values, and the field itself with respect to the point and time of measurement will be steady-state, on account of the constant level of excitation and continuous replenishment of plankton in the ambient volume. Therefore, the amount of radiant energy reaching the phototransducer is capable of characterizing the maximum luminescence of all the luminescent organisms in the given volume. In other words, the instrument must ensure excitation of all the organisms that are capable of luminescence and that respond to mechanical stimuli in the volume scanned by the photomultiplier.

2. For isolation of the bioluminescent signal from the astronomical background it is important that the tridimensional brightness distribution of the astronomical component in the seawater in the lower hemisphere have a minimum of about 0° and that it increase as the angle increases to $\frac{\pi}{2}$, whereas the bioluminescence attendant on the movement of the instrument

is concentrated along the axis of its movement. Therefore, it is possible to improve the recording efficiency for bioluminescence to the detriment of the astronomical component by decreasing the instrumental field of vision.

3. At a comparatively high sounding speed, when the duration of the bioluminescent flash (τ) is comparable with the duration of the instrument's stay at the point of measurement, it is necessary to take into account the possible flash-recording error due to the relative movement of instrument and organism.

Let us examine the magnitude of this error for a triangular pulse with a vertical front and a slanting trailing edge determined by the law $1-\frac{t}{\tau}$ (this pulse shape is quite close to the actual) moving away from the surface of the bathyphotometer's window.

If the maximum flash brightness is B_0 and it occurs at the distance r_0 from the phototransducer, then when the instrument and source are fixed with respect to each other the phototransducer will record the following energy per unit area:

$$u_0=\frac{1}{2}\frac{B_0\tau}{r_0^2}.$$

Taking into account the relative movement of instrument and organism and also the possibility that the flash may not be on the normal to the surface of the bathyphotometer's window, the energy reaching the phototransducer cathode is

$$u-\frac{B_0\cos\varphi}{r_0^2}\int_0^{t_p}\frac{1-\frac{t_p}{\tau}}{(1+at)^2}dt=\frac{B_0\cos\varphi}{r_0^2}\left[\frac{(a\tau+1)\ln(1+at_p)}{a^2\tau}-\frac{t_p}{a\tau}\right],$$

where $a=\frac{V}{r_0}$; V is the sounding speed; r_0 is the minimum distance between the phototransducer cathode and the organism (the minimum distance at which a flash can be recorded); φ is the angle between the normal to the surface of the window and the direction toward the organism; t_p is time of reliable recording of an individual flash.

Hence, the ratio of the energy of an isolated flash of duration τ recorded by a moving bathyphotometer to the energy recorded by a fixed transducer (distortion of recorded energy) is

$$\eta_{(a,t_p,\varphi,\tau)}=\left[\frac{(a\tau+1)\ln(1+at_p)}{a^2\tau^2}-\frac{t_p}{a\tau^2}\right]\cos\varphi.$$

The formula shows that the distortions introduced are controlled by the influence of the following parameters that are mutually independent in general: a, evaluation of sounding speed; t_p, time of reliable recording of an individual flash (for a given speed of its movement); φ, characteristic of the direction of movement of the organism; τ, duration of individual flash.

When the function is tested for its minimum with respect to each of the parameters, it is seen that the influence of these may be taken into account in the design of the instrument and the signal recording channel in the following manner: influence of parameter a, by modifying the gain factor of input amplifier depending on the sounding speed; influence of parameter t_p, by limiting the zone scanned by the photomultiplier; influence of parameter φ, by localizing the zone of stimulated bioluminescence within limits deviating only slightly from the normal to the phototransducer surface; the influence of parameter τ can be taken into account in the signal recording channel as P_{τ_i}, i. e. the statistical weight of bioluminescent energy concentrated in the flashes $\tau_i \pm \frac{h}{2}$ (h is the summation interval). This quantity can be determined from a histogram of the distribution of the duration of flashes over the measurement time.

4. Two signals must be fed simultaneously to the shipboard recording apparatus, namely that of bioluminescence intensity and that of the depth at which the instrument is located.

A set of instruments meeting the above requirements was designed and produced for the 44th cruise of the "Vityaz." It consists of a sounding bathyphotometer and shipboard recording apparatus.

The **sounding bathyphotometer** (Figure 1) is a particular variant of an integrated set of standard bathyphotometric transducers making it possible to assemble single-, two- and multichannel designs in accordance with the purpose of the experiment.

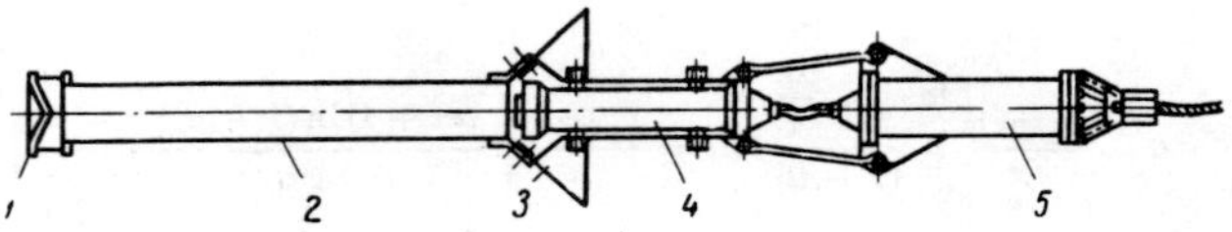

FIGURE 1. Block diagram of the sounding bathyphotometer:

1 — end cap; 2 — shielding tube; 3 — cone; 4 — bathyphotometer; 5 — distributing box.

The chief component of the instrument is the standard unit with photomultiplier, which was tested on several expeditions. The light-sensitive unit is placed in container (4). The second container (5) is the distributing box with depth sensor. The containers are connected mechanically and by cable.

The specific component of the sounding bathyphotometer is the chamber attached at its front end. Chamber cone (3) is situated so as to leave a conical slot between container (4) and chamber walls (3, 2), which is penetrated by water under pressure when the instrument is hoisted, evoking intensive bioluminescence in the zone scanned by the photomultiplier due to the pressure and acceleration gradient. Having passed through shielding tube (2), the water issues through the lateral ring-shaped opening in end-cap (1). When the instrument is lowered, the situation is reversed, the stimulated zone remaining outside the photomultiplier's field of vision, and the phototransducer can record the astronomical background, which is considerably

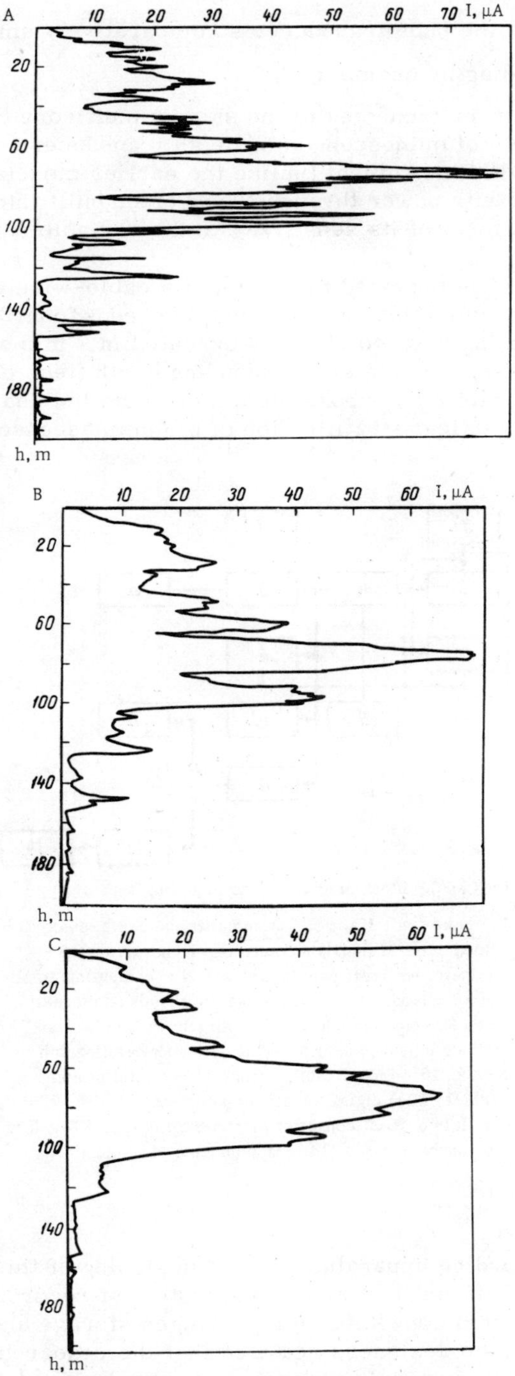

FIGURE 2. Distribution graphs for luminescence intensity:

A — time constant of the smoothing operator 4.5 sec; B — time constant of the smoothing operator 12 sec; C – time constant of the smoothing operator 18 sec.

reduced owing to the phototransducer's comparatively small angle of vision $\left(\frac{\pi}{2}\right)$ and its shielding by chamber (3, 2).

The instrument is connected to the ship by multicore cable transmitting the signals from bioluminescence and depth transducers and feeding them with power. In this instrument (unlike the earlier models) the photomultiplier is fed with power through a converter built into the instrument, permitting adjustment of its sensitivity to the variable pattern of luminescence.

The instrument is operated by means of a cable-wire winch with a 7-channel current collector. It has a range of speeds from 0.3 to 1.25 m/sec.

A single sounding down to 250 m is executed in 7 min at a sounding speed of 1.25 m/sec. For a small sounding depth (tens of meters) the expanded depth scale makes possible a more detailed analysis of the fine structure of the vertical stratification of bioluminescence.

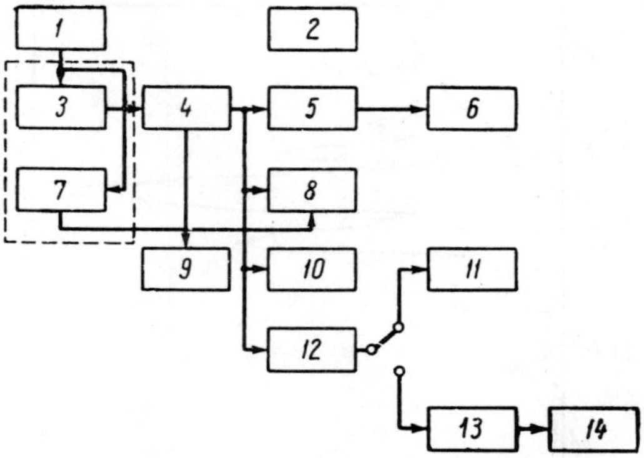

FIGURE 3. Block diagram of the recording apparatus:

1 — power unit of the phototransducer and depth sensor; 2 — demodulator of signals recorded on magnetic tape; 3 — luminescence level phototransducer; 4 — computing amplifier; 5 — modulator for recording the signals on magnetic tape; 6 — tape recorder; 7 — depth sensor; 8 — two-coordinate potentiometer PDS-021; 9 — loop oscillograph N-008; 10 — ENO-1 oscillograph; 11 — digital printer TsPM-1; 12 — digital electronic voltmeter ETsV-3; 13 — parallel-to-serial converter for transferring from ETsV-3 to tape puncher PL-80; 14 — tape puncher PL-80.

Shipboard recording apparatus. In the designing of this apparatus, attention was paid in the first place to a system of recording the incoming information in a form convenient for prolonged storage and computer input. At the same time, the designers realized that the researcher must have the possibility of continuously monitoring the experiment by already known criteria (signal level, duration of individual flashes, etc.) and of determining the basic parameters of the bioluminescence field. For this purpose, recording of luminescence intensity is best done in the coordinates depth-luminescence intensity.

Smoothed realizations of information supplied by the phototransducer are convenient for several investigations. We chose the following smoothing operator as the most easily realizable:

$$S_{x(t)} = \frac{1}{T} \int_{t-T/2}^{t+T/2} X_{(\tau)} d\tau,$$

where T is the smoothing interval (memory "duration" of operator S).

The smoothing interval must be chosen for each specific task in accordance with the desired resolution of signal structure. Figure 2 shows the distribution graphs for luminescence intensity obtained by recording the signals supplied by the phototransducer with the two-coordinate potentiometer PDS-021 in the coordinates depth-luminescence with different smoothing times (intervals).

The recording apparatus was designed in accordance with the above considerations (Figure 3). The smoothing operator (similar to the sounding speed correction) was incorporated in the computing amplifier channel (4).

RESULTS OF THE OBSERVATIONS AND THEIR STATISTICAL ANALYSIS

The general pattern of distribution of bioluminescence intensity in the region investigated is a sharp stratification (see Figures 2, 4). The main luminescence is concentrated in the upper 100 – 120 m, below which there is a sharp transition from the confluent luminescence zone to the zone with scattered flashes and small clusters that are not confined to definite horizons.

A characteristic feature of the upper 100 m is the presence of two luminescence maxima. The lower, usually more distinct maximum lies at 60 – 100 m in the region. The upper maximum lies between 20 and 50 m and is as a rule distinctly separated from the lower maximum. The upper maximum is more variable in magnitude and position than the lower maximum and did not manifest itself at all in some soundings. The intensity of luminescence in the "gap" zone between the maxima is liable to drop to 20 – 50% of the magnitude of the lower maximum.

The data examined below were obtained in polygon No. 3 (station 6052) in the divergence zone at the boundary between the Coral and Solomon seas and in polygon No. 4 (station 6064) in the waters of the South Equatorial Current east of the Solomon Islands. For a detailed hydrological description of the region the reader is referred to the paper by Arsen'ev et al. in this collection.

The measurements in both polygons were conducted over 4 – 5 days every night continuously from 1900 hr to 0430 hr local time. The interval between soundings was 6 – 7 min, but was occasionally reduced to 2 – 3 min. The total number of soundings in the 0 – 200-m layer was around 320 in polygon No. 3 and around 360 in polygon No. 4. The information supplied by the phototransducer was recorded by the two-coordinate potentiometer PDS-021 (time constant of smoothing operator 12 sec) and, for preparation

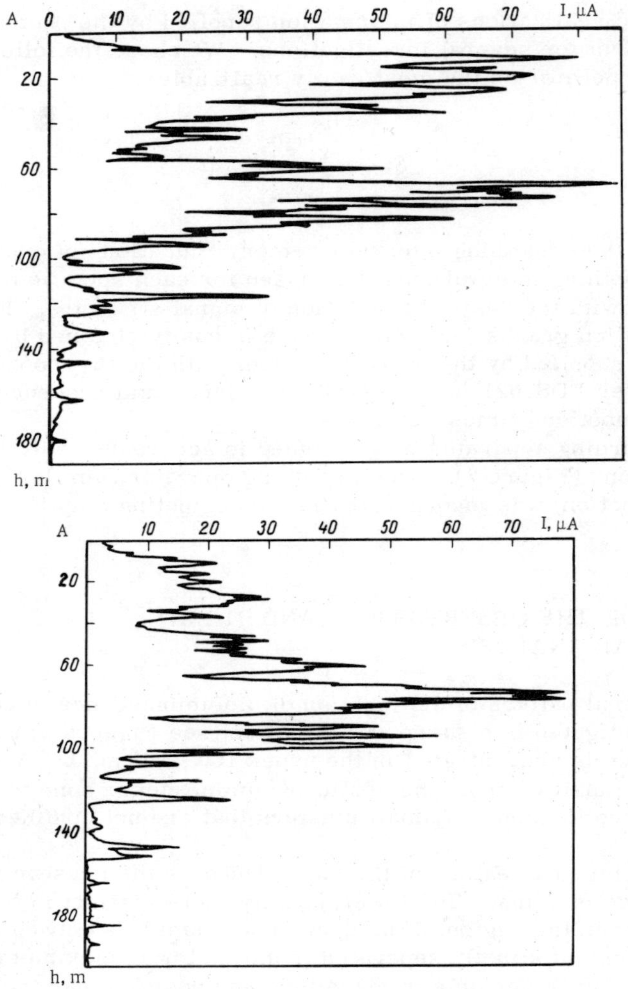

FIGURE 4. Distribution graphs for luminescence intensity
Explanation in the text.

of the information for computer input, on an MAG-59 tape recorder (frequency modulation of the input signal was used). Moreover, some signal realizations were recorded on the tape of the PL-80 tape puncher for computer input.

In considering the experimental data on luminescence stratification, the first point to be examined is that of their reliability.

In order to evaluate the variance of sampling distribution of bioluminescence, the latter's stratification was measured down to 120 m with intervals of about 2 min between measurements. With such a small time interval the measurements may be considered to have been performed practically at a single point since, with a drift of 0.4 mile/hr, the distance covered by the ship in that time was 15 – 20 m.

Depth interval, m	0 – 10	10 – 20	20 – 30	30 – 40	40 – 50	50 – 60	60 – 70
Coefficient of variation, %	56	43	56	32	25	27	20
95% confidence interval	41.3	44.5	41.3	56.8	60.5	61.5	69.5
	245	205	245	176	165	163	143

Depth interval, m	70 – 80	80 – 90	90–100	100–110	110–120	0 – 120
Coefficient of variation, %	12	14	26	35	32	34
95% confidence interval	80.5	76.5	62.5	54.6	57.6	54
	124	131	160	183	173	180

Logarithmic transformation of input data (Barnes, 1952; Van der Waerden, 1960; Rudyakov, 1967) produced distributions that were close to normal, permitting application of variance analysis techniques to evaluation of the sampling distribution reliability. A certain measure of deviation of the distribution from the normal model does not prevent the application of these techniques, because the error involved is small (Van der Waerden, 1960; Lakin, 1960).

The results of variance analysis led to the conclusion that the two distribution series do not differ in their central trend. The standard deviation is 1.326, which is 4% of the total mean (32.8). Standard deviation of one observation is 0.1273, yielding the antilogarithm of 0.1273 = 1.34 for untransformed data, and consequently the coefficient of variation is 34%.

The coefficient of variation does not remain constant throughout the depth interval 0–120 m, but decreases gradually down to 70–80 m and then is again followed by an increase.

The observed values of the coefficients of bioluminescence intensity are due more to true variation of the plankton concentration than to methodological and technical measurement errors (Cassie, 1959; Winsor and Clarke, 1940; Motoda and Anraku, 1955). The same is suggested by analysis of the results of plankton net hauls obtained with two nets operated in parallel (station 6033). In the evaluation of the results of these hauls, Vinogradov and Rudyakov write elsewhere in this collection that the volume of parallel samples taken with two nets was as a rule different despite identical networks, catching time, and conditions. The differences were especially pronounced in the 0 – 25-m layer. The following three chief concentration zones of luminescent plankton can be distinguished within the 0 – 250 m layer:

1. The 0 – 50-m layer, with extremely variable time and space characteristics of the luminescent plankton concentration.

2. The 50 – 50-m layer is characterized by a stable concentration of luminescent plankton and corresponds to the relatively stable luminescence intensity maximum. Depth fluctuations of the maximum of bioluminescence intensity evidently correspond to vertical movements of the water masses. The variation of its amplitude is primarily determined by the nonuniformity of the vertical and horizontal distribution of the plankton.

3. At depths below 90 – 100 m a clear transition is noted from the overlying confluent luminescence zone with a high concentration of luminescent organisms to the zone with a random vertical distribution of luminescent flashes.

The comparatively low representativeness of signals at depths below 120 m prevents the drawing of any specific conclusions concerning the distribution pattern of luminescent plankters here and its relation to the distribution of the plankton biomass and the principal biogenic elements below 120 m. We shall limit our examination to the zone 0 — 120 m, where the representativity of experimental data is adequate for a statistical analysis of the distribution of luminescence intensity.

Let us examine the relationship between luminescence intensity and the distribution of the plankton biomass and the principal hydrological and hydrochemical environmental parameters. It is assumed that, under the condition of transformation of initial data, all the premises on which regression analysis is based are fulfilled (Van der Waerden, 1960; Snedecor, 1961; Lakin, 1968). It is also assumed that the chosen minimal discretization intervals ensure the independence of measurements of all the variants.

The linear correlation coefficient between luminescence intensity, plankton biomass, and the concentration of the main biogenic elements must be calculated separately because the magnitude of plankton biomass characterizes the water layer filtered by the net, while the distribution of biogenic elements (in a separate measurement) characterizes only the sampling horizon; moreover, the sizes of samples of biomass and of the main biogenic elements are not comparable.

In our determination of the partial coefficient of correlation between bioluminescence intensity and plankton biomass, we were able to use only the data on plankton hauls and luminescence intensity measurements at station 6033 (Vinogradov and Rudyakov, in this collection) and to compare the plankton biomass in the investigated layer with the mean luminescence intensity in the same layer.

Layer investigated, m	0 — 25	25 — 50	50 — 75	75 — 100
Coefficient of pair correlation	0.865	0.773	0.883	0.921

The values of the coefficient of pair correlation for the 0 — 25 and 25 — 50-m layers exceed the values 0.754 ($P > 0.05$) and 0.874 ($P > 0.01$) for the 50 — 75 and 75 — 100-m layers, i. e. the correlation between the variables may be regarded as significant.

The values obtained of the coefficient of pair correlation proved the possibility of estimating the plankton biomass from bioluminescence intensity in the region investigated. Such a correlation had previously been demonstrated for the Red Sea by Rudyakov (1967).

A close correlation between bioluminescence intensity and the distribution of the total plankton biomass may be assumed to be typical for all waters inhabited by plankton that is capable of luminescence. However, the precise limits of this correlation and its quantitative expression call for special investigations in the high latitudes, where the number of plankton species is limited and subject to profound seasonal cyclicity.

In calculating the coefficients of multiple correlation the data of luminescence intensity were compared against the ambient hydrophysical and hydrochemical parameters at fixed horizons, values averaged for a layer ± 1.5 m at the depth of hydrological measurements being taken as the intensity magnitude of luminescence.

The following hydrophysical and hydrochemical parameters were analyzed: temperature (°C), salinity (‰), arbitrary density (stated in arbitrary units), pH, dissolved oxygen, phosphate, phosphorus, silicate silicon, total nitrite and nitrate nitrogen (the concentrations of all the chemical elements are stated in μg-at/l). All the information on the distribution of these elements at stations 6033 and 6052 was kindly supplied by the hydrological and hydrochemical divisions.

TABLE 1. Multiple correlation coefficients of luminescence intensity and the main hydrological parameters at 75 m (from hydrological observations at stations 6033 and 6052)

Parameter	t, °C	S, ‰	Arbitrary density	pH	O_2	P	Si	N	Luminescence
t, °C	1	0.421	−0.424	0.511	0.391	−0.861	−0.612	0.274	0.252
S, ‰		1	0.414	−0.275	−0.517	−0.406	−0.186	−0.420	−0.377
Arbitrary density			1	−0.130	−0.504	0.357	0.312	0.378	−0.385
pH				1	+0.521	−0.496	−0.631	0.528	0.412
O_2					1	−0.596	−0.584	0.731	0.526
P						1	0.561	−0.387	−0.387
Si							1	−0.390	−0.253
N								1	0.534
Luminescence									1

Table 1 lists the values of the multiple correlation coefficients calculated for the 75-m horizon; the results for other horizons are analogous.

Preliminary analysis of the coefficients of correlation between luminescence intensity and the main hydrological parameters showed no significant direct correlation between them, although the coefficient of pair correlation exceeds the significance level at P = 0.01 in some horizons. For instance, such a correlation was discovered for oxygen and nitrogen at the 75-m horizon. It is also of interest to note that the coefficient of pair correlation reverses its sign in several cases depending on the depth of measurement. Thus, the coefficient of pair correlation between luminescence intensity and nitrogen was −0.385 (above the significance level at P = 0.05) at 25 m, but it was +0.534 (above the significance level at P = 0.01) at 75 m. This reversal of the sign of the coefficient of correlation can hardly be attributed to insufficient data, but was probably caused by differences in the composition of luminescent organisms at these depths or inhibition of their photic reaction in the upper horizons at a higher nitrogen concentration in conjunction with certain environmental parameters. Further processing of data and continued investigations are required in order to solve this question.

The correlations between environmental parameters and plankton may be investigated in greater detail by vertically aimed sampling, making it possible to assess the activity and the concentration of bacterial plankton and phytoplankton and the hydrophysical and hydrochemical parameters at precisely defined depths characterized by definite bathyphotogram features (luminescence maxima and minima, etc.). Maxima of the concentration and activity of phytoplankton and bacteria correspond to the depth of maximum luminescence. The concentrations of the main biogenic elements, nitrogen and phosphorus, also rise markedly in the same zone or directly beneath. Their concentrations are close to zero above that zone (see Figure 3 in the paper by Vinogradov, Gitel'zon, and Sorokin in this collection).

No less important for understanding the structure of oceanic communities are studies of small-scale spatial nonuniformity of plankton distribution, this chiefly referring to nonuniform horizontal distribution.

Our data permit identification of nonuniformities on the scale of tens, hundreds, and thousands of meters. For evaluation of small-scale nonuniformity of the bioluminescence field (and consequently also of the total plankton concentration), samplings of the following two types were made for each series of measurements of luminescence intensity: a) samplings for the horizons 50, 60, 70, 80, 90, 100, and 110 m; b) samplings for determination of the influence of vertical movements of water masses on luminescence intensity. Samplings of type (b) were made in the following manner. The mode of each luminescence distribution was determined and the integral reading of luminescence intensity at that point was recorded. This was followed by determination of the modal values of luminescence intensity for each half of the distribution divided by the first modal reading, and so on. In this way, eight readings distributed along the depth scale in accordance with the distribution law for aggregation of luminescent organisms were taken for each distribution of luminescence intensity.

The readings thus obtained were grouped according to the time of measurements, which was equivalent to grouping on the distance scale, once the ship's drift speed was known.

The series was smoothed by sliding averages for elimination of random fluctuations in the series and determination of their most stable points (Romanenko and Sergeev, 1968; Pugachev, 1960).

The series obtained, similar to the initial data, provide information not only on the intensity and nonuniformity of the bioluminescence field but also on the migrations of organisms and the diurnal rhythm of luminescence. For differentiation of the combined interaction of these two factors, the series were smoothed with respect to the set of realizations of one polygon (Pugachev, 1960). In so doing we ignored all random deviations (including those due to nonuniform horizontal distribution of plankton), averaging as it were the entire bundle of realizations of random function.

The simplest approach to characterizing the degree of dependence of values corresponding to two different values of the independent variable is to resort to the correlation (or autocorrelation) function of the process, thus:

$$\rho_{x(\tau)} = \frac{\sum_{i=1}^{N} (x_i - \bar{x})(x_{i+\tau} - \bar{x})}{NS|x|},$$

where $\bar{x}$ is the mathematical expectation estimate; $S|x|$, evaluation of standard deviation; N, number of shifts.

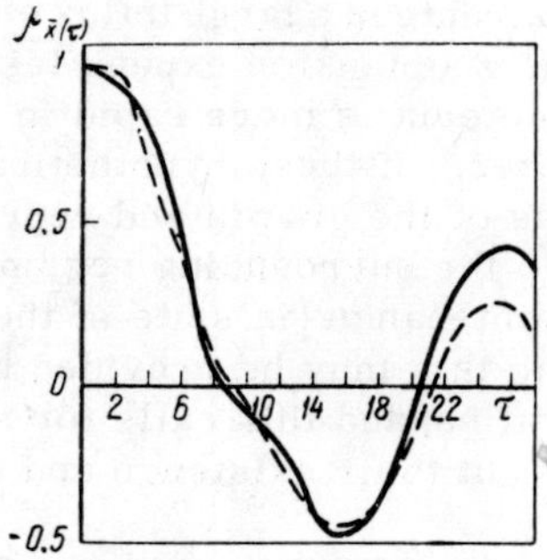

FIGURE 5. Graph of the function $\rho_{x(\tau)}$ for the 80-m horizon (solid line) and graph of the approximating function $e^{-0.1\tau}[\sin(\omega t+\frac{\pi}{4})+$ $+0.25\sin(2\omega t+\frac{\pi}{2})]$ (dashed line).

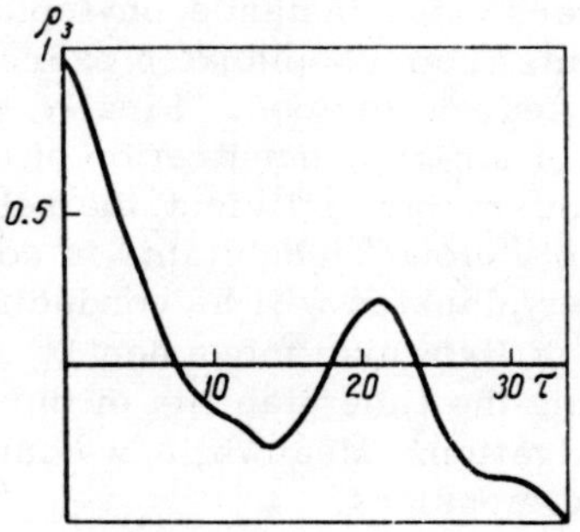

FIGURE 6. Graph of the function $\rho_{x(\tau)}$ for the 50-m horizon

The graph of functions $\rho_{x(\tau)}$ for the 80-m horizon is shown in Figure 5. A noteworthy feature is the existence of certain negative values of the correlation function, denoting some element of periodicity in the structure of this function. Such a correlation function can be approximated (Pugachev, 1960; G. A. Korn and T. M. Korn, 1968) with a fair degree of accuracy by expression of the form

$$\rho_{x(\tau)} = e^{-0.1\tau}\left[\sin\left(\omega t+\frac{\pi}{4}\right)+0.25\sin\left(2\omega t+\frac{\pi}{2}\right)\right],$$

where $\omega=\frac{2\pi}{T}$. The approximating function suggests (Pugachev, 1960) that the random function under investigation contains two sinusoidal components. At this stage it is difficult to decide which component is due to the diurnal rhythm of luminescence and which to migration of the plankters.

Figure 6 shows a graph of the autocorrelation function of one of the realizations of the 50-m horizon in the same polygon.

The correlation function of this realization contains a considerably larger number of periodic components. Its approximating function can be demonstrated to contain the harmonics $\cos\omega t$, $\cos 3\omega t$, $\cos 6\omega t$ and $\sin\omega t$.

We will not provide the values of correlation functions for all the realizations. It may be noted that their spectral composition varies from one horizon to another. In other words, the law for the horizontal distribution of plankton is significantly dependent upon depth. The highest degree of homogeneity is found in the layer of the main luminescence maximum. Its correlation function constructed from sampling data of the second group is distinguished by three main harmonics $\cos\omega t$, $\cos 2\omega t$, and $\sin\omega t$. The characteristic horizontal nonuniformities of bioluminescence intensity in the region possess dimensions ranging from 100 – 150 to 1,000 – 1,200 m.

It is not certain why these nonuniformities exist. They may be turbulent eddies causing a "cloudy" distribution of the plankton. It may also be assumed that remaining in concentrations offers certain advantages to the plankters — for instance, phytoplankton is guaranteed a large influx of biogenic elements from zooplankton excreta, while the zooplankton expend less energy in the search for food. Finally, growth of bacteria is more rapid in the presence of a high concentration of organic matter. If these hypothetical mechanisms are efficient, the relative excess of the growth and reproduction rates of "cloud" inhabitants in comparison to the surrounding regions of sparser plankton will be conducive to its maintenance (in spite of the eroding effect of Brownian movement). An answer to this may be provided by studying the time stability of the nonuniformities, and this calls for special investigation. Meanwhile, we can only point out their existence and estimate their dimensions.

Comparison between the correlation functions of the samplings of the first and second groups does not permit a conclusion to be reached concerning the influence of vertical movements of water masses on variation of the bioluminescence intensity irrespective of the depth of the layer. Such an influence can only be noted for the principal maximum layer, but for the other horizons the form of correlation function does not vary significantly with the type of sampling. At the same time, the coefficients of correlation between luminescence intensity at different horizons depend upon the type of sampling, although no clear-cut regularity can be perceived in this case either (Tables 2 and 3).

TABLE 2. Multiple correlation coefficients for luminescence intensities in different horizons for first-type sampling

Depth, m	50	60	70	80	90
50	1	0.7865	0.0453	0.2055	0.2868
60		1	0.3218	0.1752	0.4384
70			1	0.5370	0.6001
80				1	0.8252
90					1

In concluding our description of the method and first results of the sounding of the eutrophic layer with the bathyphotometer, it must be pointed out that probably the most interesting innovation introduced by the bathyphotometric sounding in hydrobiological investigations in the ocean is the possibility of detecting small-scale vertical and horizontal nonuniformities of plankton distribution, i. e. the possibility of "visualizing" the fine spatial structure of a community.

TABLE 3. Multiple correlation coefficients for luminescence intensities in different horizons for second-type sampling

	X_1	X_2	X_3	X_4	X_5	X_6	X_7	X_8
X_1	1	0.4507	0.0012	0.3648	0.5711	0.3758	0.4781	0.2008
X_2		1	−0.0198	−0.2239	0.5288	0.6063	0.2362	0.1259
X_3			1	0.3647	0.2085	−0.0327	−0.1604	0.5632
X_4				1	0.1249	−0.416	−0.1223	0.2204
X_5					1	0.5678	0.5713	0.6503
X_6						1	0.6177	0.3261
X_7							1	0.3929
X_8								1

Obviously, the sampling should be done in accordance with the structure of the system in order to obtain sufficiently reliable evaluations of balance correlations among the ecosystem parameters. The method of bathyphotometric sounding was successfully applied to elucidation of the spatial structure of an ecosystem. The rapidity of the method allows for the investigation of rapidly varying processes in oceanic ecosystems.

Bibliography

Publications in Russian

Arsen'ev, V. S., L. I. Galerkin, E. A. Plakhin, and V. V. Sapozhnikov. Osnovnye cherty gidrologii i gidrokhimii raiona issledovanii 44-go reisa "Vityazya" (Major Hydrological and Hydrochemical Features of the Region Investigated during the 44th Cruise of the "Vityaz"). (In this collection.)

Gitel'zon, I. I., O. G. Baklanov, V. S. Filimonov, A. S. Artemkin, and V. F. Shatokhin. Biolyuminestsentsiya kak gidroopticheskii i biologicheskii faktor v more (Bioluminescence as a Hydrooptical and Biological Factor in the Sea). — In Sbornik: "Biolyuminestsentsiya," Trudy Moskovskogo obshchestva ispytatelei prirody, Vol. 21. 1965.

Lakin, G. F. Biometriya (Biometry). Moskva, Izdatel'stvo "Vysshaya shkola." 1968.

Pugachev, V. S. Teoriya sluchainykh funktsii (Random Functions Theory). Moskva. Fizmatgiz. 1960.

Romanenko, A. F. and G. A. Sergeev. Voprosy prikladnogo analiza sluchainykh protsessov (Problems of Applied Analysis of Random Processes). Moskva. 1968.

Rudyakov, Yu. A. K metodike izucheniya biolyuminestsentsii morya (On Methods of Investigation of Marine Bioluminescence) — Okeanologiya, No. 4. 1967.

Sorokin, Yu. I. Kolichestvennaya otsenka roli bakterioplanktona v biologicheskoi produktivnosti tropicheskikh vod Tikhogo okeana (Quantitative Evaluation of the Role of Bacterioplankton in the Biological Productivity of Tropical Pacific Waters). (In this collection.)

Vinogradov, M. E. Issledovaniya funktsionirovaniya okeanicheskikh biologicheskikh sistem (Studies of the Life Activity of Oceanic Biological Systems). (In this collection.)

Vinogradov, M. E., I. I. Gitel'zon, and Yu. I. Sorokin. O prostranstvennoi strukture soobshchestv evfoticheskoi zony tropicheskikh vod okeana (The Spatial Structure of Communities in the Euphotic Zone of the Ocean Tropics). (In this collection.)

Vinogradov, M. E. and Yu. A. Rudyakov. Sutochnye izmeneniya vertikal'nogo raspredeleniya biomassy planktona v zapadnoi ekvatorial'noi Patsifike (Diurnal Variations of the Vertical Distribution of the Plankton Biomass in the Equatorial West Pacific). (In this collection.)

Publications in Other Languages

Barnes, H. The Use of Transformation in Marine Biological Statistics. — J. Cons. Internat. Explor. Mer., Vol. 18, No. 1. 1952.

Cassie, R. M. Microdistribution of Plankton. — New Zeal. Y. Sci., Vol. 2, No. 3. 1959.

Clarke, G. L. and C. J. Hubbard. Quantitative Records of the Luminescent Flashing of Oceanic Animals of Great Depths. — Limnol. and Oceanogr., Vol. 4, No. 2. 1959.

Korn, G. A. and T. M. Korn. Mathematical Handbook for Scientists and Engineers. New York, McGraw-Hill. 1968.

Motoda, S. and M. Anraku. The Variability of Catches in Vertical Plankton Hauls. — Bull. Fac. Fish. Hokkaido Univ., Vol. 6. No. 2. 1955.

Snedecor, G. W. Statistical Methods Applied to Experiments in Agriculture and Biology. Ames, Iowa State College Press. 1956.

Waerden, B. R. Van der. Matematische Statistik. Berlin, Springer. 1957.

Winsor, C. P. and G. L. Clarke. A Statistical Study of Variation in the Catches of Plankton Nets. — J. Mar. Res., Vol. 3, No. 1. 1940.

CALCULATION OF PRIMARY PRODUCTION IN TROPICAL WATERS FROM THE UPWARD INFLUX OF BIOGENIC ELEMENTS

V. V. Sapozhnikov and L. I. Galerkin
Shirshov Institute of Oceanology of the Academy of Sciences of the USSR

The 44th cruise of the "Vityaz" included detailed investigations of the distribution of hydrochemical characteristics at several multidiurnal stations in the western part of the tropical zone of the Pacific. A hydrological and hydrochemical description of the region and of the observation techniques is given by Arsen'ev et al. in this collection.

Simultaneously with the sounding of bioluminescence (using a bathyphotometer) and turbidity (with a hydronephelometer) the water was sampled with casts of densely spaced water samplers (at 5-m intervals), yielding vertically aimed water samples from layers of high turbidity and bioluminescence maxima. The complex of these observations revealed the existence of a thin layer of high concentration of living organisms at depths of 70 — 90 m. The same layer shows an abrupt change in the concentrations of biogenic elements, which are directly consumed in this layer by phytoplankton and bacterial cells. Their accumulation "cuts off" the upper part of the photosynthetic layer from the source of biogens.

The correlations between variations in the concentration of biogenic elements and the size of primary production have been studied by many researchers (Seiwell, 1935; Smetanin, 1956; McAllister, Parsons, and Strickland, 1961; and others). All the methods for calculating primary production from chemical characteristics have been based on fixing the initial and final concentrations of phosphorus, nitrogen, or oxygen. The time intervals of measurements could be adjusted for calculation of daily production or the "harvest" of a season.

These methods of primary production calculation cannot be used in the tropical waters of the western Pacific, because biogens are more or less absent from the photosynthetic layer throughout the year. A formalistic application of these methods to tropical waters would show zero production, which would be obviously untrue.

Observations over a 24-hour period showed that the concentrations of dissolved inorganic phosphorus and nitrates in the 0 — 50-m layer approached analytical zero. These elements were consistently absent from the upper layer at stations 6033 and 6052 (Figures 1 and 2). We therefore assumed that the entire quantity of inorganic biogenic elements flowing up into the photosynthetic layer is completely assimilated in the course of the primary synthesis of organic matter in the layer with high concentration of living organisms, after which it undergoes transformation along the food chain of the surface biocenosis. Having determined the transport of biogens effected by vertical movement of the waters, it is possible to estimate the total primary production. Assuming nitrogen and phosphorus to be the chief critical elements, we calculated their influx through the lower boundary of the surface layer.

It should be noted that the concept of a "surface layer devoid of biogenic elements" does not agree with the hydrological concept of the surface layer, the thermopycnocline lying considerably lower down, at 170 – 220 m.

Advective vertical transport of properties (k) is usually expressed by quantity $W\frac{\partial k}{\partial z}$. However, the product of concentration by velocity does not characterize the total transport of matter but the variation of its concentration at a given point per unit time under the influence of currents with dimensionality $g \cdot cm^{-3} \cdot sec^{-1}$. In order to determine the transport of matter through unit surface, we must use the product of the normal velocity vector component by concentration of matter on that surface ($w \cdot k$). In this case dimensionality is $g \cdot cm^{-2} \cdot sec^{-1}$. Moreover, because of the coarse differentiation of our observations with respect to depth, we usually do not know the law of concentration variation between the two adjacent horizons, especially at the boundaries of transition layers. Therefore, gradient calculations in a layer are always considerably rougher than concentration determination in a horizon.

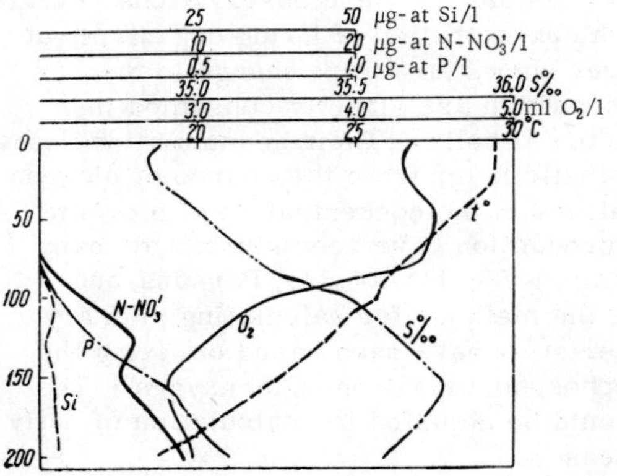

FIGURE 1. Vertical distribution of the concentration of biogens, temperature, and salinity at station 6033-9

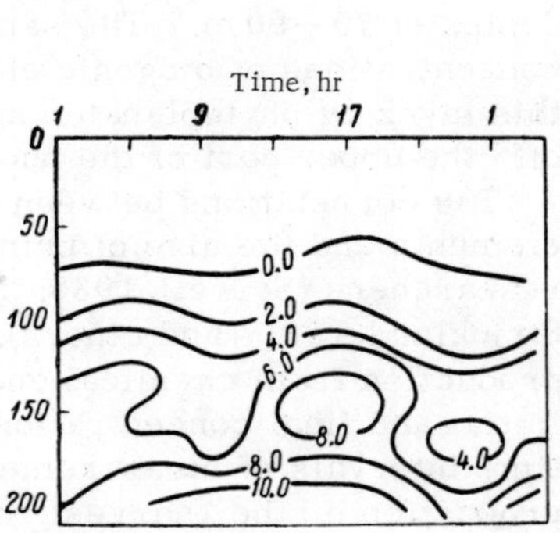

FIGURE 2. Variation of the concentration of nitrates at the diurnal station 6052-13

Several methods are available for calculation of the vertical velocity components. Our choice of method was dictated by the specific conditions of the expedition. Observations in all polygons were conducted in almost absolute calm, and therefore the calculation methods based on taking account of wind conditions could not be used. The vertical velocity components were calculated from the equation of water transport by advection and turbulent diffusion by the method described by Arsen'ev et al. in this collection.

In using the total values of the upward influx of phosphorus and nitrogen for calculation of production expressed in terms of synthetized carbon, we used the well known stoichiometric ratios between these elements and carbon in Richards' (1965) model of organic matter, $C/P=106/1$ for phosphorus, $C/N=106/16$ for nitrate nitrogen. We also assumed the influx of these elements to continue throughout the diurnal period ($8.64 \cdot 10^4$ sec). To convert the carbon production from microgram-atoms to milligrams, the former value must be multiplied by 12 (atomic weight of carbon) and divided by 1,000. Then the final factors for converting the influx of phosphorus and nitrogen to total carbon production are found to be, respectively, $10.97 \cdot 10^5$ and $6.86 \cdot 10^4$, while the conversion formulas are as follows: $C=W \cdot P\ 10.97 \cdot 10^5$ and $C=W \cdot N \cdot 6.86 \cdot 10^4$. The velocities are stated in cm/sec, the nitrogen and phosphorus concentrations in μg-at/l and production in mg/m^2 · day.

The total daily upward influx of phosphorus and nitrogen predetermined the amount of primary production that could be produced by their complete assimilation. If the production were exclusively controlled by the photosynthetic process and completely dependent on the amount of solar energy incident on a unit of surface, it would display a distinct diurnal cycle. At night, a part of the inflowing nutrient salts would not be utilized and would be accumulated in inorganic form. Yet this is not observed (see Figure 2). The zero concentrations of these elements in the upper 50 – 70-m layer throughout the 24-hour period suggest that the conversion of nitrogen and phosphorus to organic matter continues throughout the period and hence that accumulation of inorganic nutrient salts is not entirely limited to photosynthesis.

The results of the calculations are summarized in the table. At station 6052 the thicknesses of the homogeneous layers with zero nitrogen and phosphorus concentrations were 100 and 75 m, respectively, while the vertical velocity in the 75 and 100-m horizons are also different, but the results for calculation of total production from nitrogen and phosphorus for different horizons differ by only 20%. The influx of biogenic elements from below is fairly stable, and is also confirmed by satisfactory agreement between results calculated from one-time determinations at the center of the "cross" and from the mean diurnal values for nitrates and phosphates at the lower boundary of the surface layer.

The calculation for station 6064 should be examined in greater detail. A homogeneous layer devoid of mineral nitrogen and phosphorus was absent at this station throughout the observation period. All the concentrations were significantly different from zero. We calculated the mean weighted concentrations in the 0 – 125-m layer (P_{0-125}). Subtracting them from the concentrations at the lower horizon (P_{125}), we obtained the effective concentration (P_{ef}), thus: $P_{ef}=P_{125}-P_{0-125}$. This quantity was used in calculations of total carbon production. Similar calculations were made for determination of production from the influx of nitrates up into the surface layer. It was assumed that not all the inflowing biogenic elements were converted into primary production but only their excess over their constant concentration in the surface layer.

It is of interest to compare the calculated values of "potential" primary production with direct determinations of production according to C^{14} at the same stations (the last column in the table).

TABLE. Results of calculations of primary production based on the influx of phosphorus and nitrogen into layer

Vertical velocity, cm/sec	Concentrations in the calculation horizon, μg-at/l		Maximum possible production, mg $C/m^2 \cdot$ day	Actual production, mg $C/m^2 \cdot$ day or mg $C/m^2 \cdot$ 24 hours
	phosphorus	nitrogen		
Station 6033 (depth 75 m)				
$1.4 \cdot 10^{-3}$	0.29	—	446	75**
$1.4 \cdot 10^{-3}$	—	4.0	386	—
$1.4 \cdot 10^{-3}$	0.26	—	398*	—
Station 6052 (depth 75 m)				
$3.6 \cdot 10^{-3}$	0.15	—	593	120**
Station 6052 (depth 800 m)				
$2.4 \cdot 10^{-3}$	—	3.0	495	104†
$2.4 \cdot 10^{-3}$	—	3.1	512*	—
$2.4 \cdot 10^{-3}$	—	2.94	493*	—
Station 6064 (depth 125 m)				
$1.2 \cdot 10^{-3}$	0.38	—	503*	175**
$1.2 \cdot 10^{-3}$	—	8.1	667*	224†

* The calculations made use of mean diurnal nitrogen and phosphorus concentrations from diurnal station data.

** After Koblents-Mishke et al. (in this collection).

† After Sorokin (in this collection).

According to Koblents-Mishke et al. (in this collection), photosynthetic production was 75 mg $C/m^2 \cdot$ day at station 6033, 120 mg $C/m^2 \cdot$ day at station 6052 and 175 mg $C/m^2 \cdot$ day at station 6064. According to Sorokin (this collection), photosynthetic production was 104 mg $C/m^2 \cdot$ day at station 6052 and 224 mg $C/m^2 \cdot$ day at station 6064. Both the above authors pointed out that their results were 25—30% too low because the dissolved organic matter was disregarded. An additional and very substantial error is probably introduced by inaccurate and incomplete sampling catches in the layer with the highest concentration of living organisms. This error may also considerably lower the results of determinations.

Our data were obtained by a method that was completely independent of biological determinations and they may therefore be used for a comparative evaluation of productivity in the region.

The proposed method has the following shortcomings. It is well known that calculation of the vertical component of current velocity by all known methods yields an accuracy of 0.5—1.0 of the order of magnitude. Furthermore, some error is introduced by neglecting the horizontal and vertical exchange components which also contribute to the total flow of biogens. An error is similarly introduced by the coarse vertical differentation of observations. The number of horizons in the layer investigated must be considerably increased for calculations of this kind.

An apparent inconvenience of the method proposed is the need to use a large number of different hydrological measurements and analyses. Nevertheless, they are usually included in the standard set of measurements because of the trend toward comprehensive investigation of the region and habitat of all the biological items and communities under study.

The calculations showed that the scheme suggested could be used not only in the case of complete consumption of biogenic elements in the surface layer but also in cases where the concentrations of phosphates and nitrates differ from zero. Therefore, one may assume that analogous calculations need not be limited to oligotrophic regions of the ocean.

Further improvement of the accuracy of direct determinations of primary production will make it possible to solve the converse problem, which is no less interesting, namely, calculation of the vertical influx of the main biogenic elements into the surface layers from below and demarcation of oceanic regions with stronger-than-average vertical advection into the photosynthetic layer. A combination of both methods with detailed hydrological observations may considerably improve the accuracy of estimating the vertical velocities.

Bibliography

Publications in Russian

Arsen'ev, V. S., L. I. Galerkin, E. A. Plakhin, and V. V. Sapozhnikov. Osnovnye cherty gidrologii i gidrokhimii raiona issledovaniya 44-go reisa "Vityazya" (Major Hydrological and Hydrochemical Features of the Region Investigated during the 44th Cruise of the "Vityaz"). (In this collection.)

Finenko, Z. Z. and A. P. Ostapenya. Vertikal'noe raspredelenie vzveshennogo organicheskogo veshchestva v tropicheskikh vodakh Yuzhnogo polushariya Tikhogo okeana (Vertical Distribution of Suspended Organic Matter in the Tropical Pacific Waters of the Southern Hemisphere). (In this collection.)

Koblents-Mishke, O. I. et al. Pervichnaya produktsiya i khlorofill "a" v zapadnoi chasti Tikhogo okeana (Primary Production and Chlorophyll "a" in the West Pacific). (In this collection.)

Smetanin, D. A. Ob otsenke produktsii organicheskogo veshchestva v nekotorykh raionakh Beringova morya (Assessment of Organic Matter Production in Some Regions of the Bering Sea). — Trudy Instituta Okeanologii Akademii Nauk SSSR, No. 17. 1956.

Sorokin, Yu. I. Kolichestvennaya otsenka roli bakterioplanktona v biologicheskoi produktivnosti tropicheskikh vod Tikhogo okeana (Quantitative Evaluation of the Role of Bacterioplankton in the Biological Productivity of Tropical Pacific Waters). (In this collection.)

Vinogradov, M. E., I. I. Gitel'zon, and Yu. I. Sorokin. O prostranstvennoi strukture soobshchestv evfoticheskoi zony tropicheskikh vod okeana (The Spatial Structure of Communities in the Euphotic Zone of the Ocean Tropics). (In this collection.)

Publications in Other Languages

McAllister, C. D., T. R. Parsons, and J. D. H. Strickland. Measurements of Primary Production in Coastal Sea Waters Using a Large-Volume Plastic Sphere. — Limnol. and Oceanogr. Vol. 6, No. 3. 1961.

Richards, F. A. Dissolved Gases Other than Carbon Dioxide. — Chemical Oceanography. No. 1. 1965.

Seiwell, H. R. The Annual Organic Production and Nutrient-Phosphorus Requirement in the Tropical Western North Atlantic. — J. Cons. Internat. Explor. Mer., Vol. 10, No. 1. 1935.

PRIMARY PRODUCTION AND CHLOROPHYLL "A" IN THE WEST PACIFIC

O. I. Koblents-Mishke, A. M. Tsvetkova, M. M. Gromov, and L. I. Paramonova

Shirshov Institute of Oceanology of the Academy of Sciences of the USSR and
Institute of Photosynthesis of the Academy of Sciences of the USSR

A model of a pelagic community, describing the transformation of matter and energy in the trophic chain, can be split into two parts. The first part describes the involvement of mineral components in the biological sphere in the process of primary biosynthesis performed by plants utilizing solar energy. The input of this part of the model consists of the mineral nutrients for plants and solar energy. The second part describes the further transformations of organic matter and the energy accumulated in it in the heterotrophic cycle. Its input is primary production.

The possibility of dividing the model as a whole into two parts depends upon the availability of direct measurement techniques for primary production.

Both parts of the production process in the sea (the primary production and the heterotrophic part) were studied during the 44th cruise of the "Vityaz." Studies of primary production were trained on the meeting point of these two parts, and the results yielded by them can be used in a dual manner. Construction of a primary production model calls for elucidation of the physiological and ecological patterns governing primary biosynthesis, using physical, chemical, and optical data. The results of primary production measurements are in turn utilized for studying the laws governing heterotrophic production.

In view of this we decided to describe the results obtained by the primary production division during the cruise in two papers. This paper, devoted to studies of heterotrophic production, presents data on the magnitude of primary production and the basic photosynthetic pigment chlorophyll "a" (along with its product pheophytin "a") and the results of observations at diurnal stations.

The ecological and physical conclusions are described in the other paper (Koblents-Mishke, in this collection).

METHODS

Work program

Samples for primary production determinations were taken from the sea surface with a bucket and from subsurface horizons with a 17-liter polyethylene sampler. The depth of sampling was chosen in accordance with the results of sounding with a transparency meter. Besides the standard layers, investigations were made of low-transparency layers, where especially high concentrations of phytoplankton could be expected. In parallel, samples for phytoplankton studies and occasionally also for studies of suspended matter were taken from the same samplers. The surface samples were exposed in a transparent aquarium, while samples from the water column were exposed in the same horizons from which they were taken, i.e. in situ.

Water samples for determinations of pigments were taken either from the same samplers as those for in situ experiments or during diurnal stations (also monitored with the transparency meter). Bathythermographic temperature soundings were synchronized with the water sampling as far as possible. The water for chemical analyses was taken from hydrological samplers.

The flasks for in situ production measurements were filled 2 — 3 hours before the beginning of exposure; exposure was either from sunrise to astronomical noon or from noon to sunset.

The surface samples for photosynthesis measurements in the aquarium were taken directly before the beginning of exposure either before sunrise, or before astronomical noon.

Working and calculation procedures

Determination of primary production by radiocarbon techniques. Measurements of photosynthesis of the phytoplankton by the C^{14} method were performed by standard techniques (Vinberg et al., 1960).

Two radiocarbon solutions were used on the cruise. One had been prepared in 1967 in Moscow with twice distilled water, distributed into glass ampules with a volume of 10 — 15 ml and sterilized. The initial activity (or more precisely, the counting rate) of this solution determined in the cruise was $2 \cdot 10^6$ pulses/min · ml. The solution was used in the first three polygons [test areas]. The other, much more concentrated, solution was prepared by Sorokin during the cruise and was used at station 6064 and in Kieta Bay (Bougainville Island).

The stock solution of the isotope was injected in portions of 1 ml with a pipet connected to a syringe into the flask filled with seawater. Flasks of colorless glass with a volume of 650 ml were used for determinations of photosynthesis on the surface, and similar flasks with a volume of about 300 ml were used for in situ measurements. Each determination was carried out in two replications, half of the flasks being exposed in the dark. In this way, every sample was distributed among four flasks (two "light" and two "dark").

After the exposure, the contents of the flasks were filtered through a membrane filter No. 5 in a plexiglass funnel with a filtering area of 2 cm^2. The filters were then processed with a 0.7% solution of HCl, after which they were washed with filtered seawater and dried in a desiccator with silica gel. Their radioactivity was measured in a DP-100 counting device with a BFL-25 end-window counter.

The results were calculated by the formula

$$C_{ph} = \frac{C_k (r_1 - r_d) \cdot \rho_0}{R \cdot \rho_1},$$

where C_{ph} is the magnitude of photosynthesis during the exposure; C_k, total carbon content in all the forms of carbon dioxide in the water calculated from the pH, alkalinity, salinity, and temperature data; r_1, counting rate of the filter used for filtration of the contents of the light flasks; r_d, same, for the dark flasks; ρ_1, counting rate for the standard preparation on the day of determination of initial activity; ρ_0, counting rate for the standard preparation on the day of counting of the filters; R, initial activity of the isotope. The quantity ρ_0/ρ_1 incorporated in the formula reflects changes in the sensitivity of the apparatus. Its introduction is especially important when the counters are replaced.

C_{ph} is calculated only for the surface. The relative variations of this quantity were taken into account for transition to photosynthesis at different depths. In using the results of the primary production determinations provided below it should be borne in mind that they were not corrected for isotopic effect (+ 6%). Moreover, failure to take into account the fraction of organic matter excreted intra vitam lowered the results by 25 — 30%, according to indirect evidence.

These corrections were not introduced for the following reasons. Intravitam excretion of organic matter by phytoplankton in nature has not yet been investigated at all, and the order of magnitude quoted by us was obtained for higher algae (Khailov, 1969). Compared with such large errors of the method aggravated by the considerable variance of the measured values, the 6% correction seemed to be too insignificant to be introduced. Moreover, these corrections are scarcely ever introduced by other authors, and therefore we achieved better comparability for our results by not introducing them.

Determination of the concentration of chlorophyll and pheophytin "a". The pigments were determined according to the original method of Richard and Thompson, which was described in a report of the UNESCO Working Group No. 17 (Determination . . . , 1966), with Lorenzen's modifications (1967) permitting separate measurements of chlorophyll "a" and of its product pheophytin "a".

Seawater samples of 12.5 liters were filtered for 1 — 2 hours through membrane filters No. 5, 65 mm in diameter, coated with a $BaCO_3$ layer thick enough to retain the entire phytoplankton.

In the standard procedure, the filter is not coated with $BaCO_3$ but with a small amount of $MgCO_3$. However, a special methodological investigation revealed that magnesium carbonates used in an amount sufficient for retention of phytoplankton lowered the results of determinations of pigment concentrations. In contrast, the results are not affected in this way by $BaCO_3$ and powdered glass. Of these two adsorbents, we chose $BaCO_3$.

The $BaCO_3$ layer was removed from the filters, after drying and storage for several days in the first three polygons and immediately upon filtration in the last two polygons, and transferred to glass filters No. 4 with addition of 1 — 2 ml of 90% acetone. The extraction was conducted for 15 — 20 min, stirring with a glass rod, and then the extract was transferred to centrifuge test tubes placed in a Bunsen flask under the filters. The procedure was repeated 3 or 4 times. The volume of filtrate was brought to 4 – 5 ml and was then centrifuged at 5,000 – 7,000 rpm for 10 – 15 min.

The extracts thus prepared were usually subjected to immediate spectrophotometric determination. The absorption spectrum of the acetone extract of the pigments was recorded with an SF-10 spectrophotometer in 1-cm cells twice, without modification and on addition of 2 – 4 drops of 0.5% HCl solution in acetone. The results referred to combined absorption by chlorophyll and pheophytin "a" in the first case and by pheophytin in the second case (both the amount originally present in the sample prior to acidulation and that resulting from the pheophytinization). Since it is known that conversion of 1 mg chlorophyll to pheophytin decreases the optical density (D) of the solution in the wavelength 663 nm by a factor of 1.7, it is possible to determine the separate concentrations of chlorophyll "a" and pheophytin. It is true that the method suffers from considerable inaccuracies (Holm-Hansen et al., 1965) because, besides the pheophytins, the samples also contain chlorophyllides and pheophorbides, not only of chlorophyll "a" but also of "b" and "c". Our method is limited to differentiation of the sum of chlorophyll and chlorophyllides from that of pheophytin and pheophorbides (the so-called pheopigments).

The chlorophyll "a" concentration was calculated by the formula

$$C_a = 11.64D_{663} - 2.16D_{645} + 0.1D_{630} \ (\mu g/ml),$$

where D_{630}, D_{645}, and D_{663} are the optical densities in the corresponding wavelengths. The percentage of pheophytin "a" was calculated by the formula

$$\% \, C_{ph} = \frac{D_{663}^{ac} \cdot 1.7 - D_{663}}{0.7 \cdot D_{663}},$$

where D_{663}^{ac} is the optical density in the wavelength 663 nm after acidulation of the extract.

RESULTS

Polygon No. 1 (stations 6025 – 6029) was executed in the lagoon of Butaritari Atoll (Gilbert Islands). Surface samples were taken at five stations on approaches to this atoll for primary production determinations and samples from standard depths down to 100 m for determinations of pigments. The first station was situated 10.6 miles from the entrance to the lagoon, the following stations at distances of 5.0, 2.3, and 0.95 miles from the latter, and the last station was in the lagoon itself. In situ determinations of primary production in the lagoon (28 m deep) were performed in four horizons down to 15 m. The primary production on the surface was 0.14, 0.38, 0.16, 0.66, and 0.19 mg $C/m^3 \cdot$ day, respectively, at stations 6025 (10.6 miles from shore), 6026 (5.0 miles), 6027 (2.3 miles), 6028 (0.95 miles), and 6029 (in the lagoon). Thus phytoplankton production in the lagoon and near the atoll is very low, increasing only in front of the entrance to the lagoon. Production determined in the lagoon in situ was 0.38, 0.74, 1.52 and 0.46mg $C/m^3 \cdot day$, respectively, on the surface and at depths of 5, 10, and 15 m. The primary production for the entire water column from surface to bottom was only 24 mg $C/m^2 \cdot day$.

The data on chlorophyll and pheophytin "a" are given in Table 1. Attention is drawn to the very low concentrations of these pigments at the surface; the concentration is increased only at 75 – 100 m. The concentration of pigments in the waters around the atoll is lower than at the lagoon entrance and in the lagoon. The rise in the concentration of pigments and phytoplankton production near the atoll and in the lagoon is actually very slight. The leading role in organic matter production here is probably played not by phytoplankton but by phytobenthos (probably mainly by microphytobenthos).

TABLE 1. Concentrations of chlorophyll "a" (1st row of figures) and pheophytin (2nd row of figures) in polygon No. 1 (mg/m^3)

Depth, m	Station 6025	6026	6027	6028	6029
0		0.03	0.05	0.12	0.08
		0.0	0.0	0.11	0.05
5					0.07
					0.06
10	0.00	0.00	0.04	0.06	0.09
	0.04	0.04	0.00	0.00	0.10
15					0.07
					0.06
25	0.00	0.00	0.15	0.04	
	0.04	0.07	0.15	0.03	
50	0.04	0.04	0.06	0.01	
	0.00	0.04	0.01	0.04	
75	0.12	0.03	0.03	0.01	
	0.00	0.00	0.03	0.04	
100	0.02	0.06	0.02	0.11	
	0.07	0.06	0.04	0.05	
Total chlorophyll and pheophytin "a", mg/m^3	7.0	7.2	6.9	8.3	2.2

All the evidence indicates that the first polygon was executed in oligotrophic waters.

Polygon No. 2 (station 6033) was executed in the South Equatorial Current. Primary production on the surface from data of 15 determinations varied from 0.04 to 1.6 mg $C/m^3 \cdot day$, i. e. by a factor of 40. The most common production values range from 0.8 to 1.24 mg $C/m^3 \cdot day$ (Figure 1). The most probable value for the polygon is 1 mg $C/m^3 \cdot day$ (mode of the curve). It is somewhat larger than the mean value, which is 0.72 mg $C/m^3 \cdot day$.

Of the three series of primary production determinations at different depths in situ two were performed on days when primary production on the surface was the lowest for the entire ten-day observation period, 0.04 and less than 0.01 mg $C/m^3 \cdot day$. In these series the results were below the experimental error at all the depths. The third series of production determinations in situ was carried out on a day when surface production

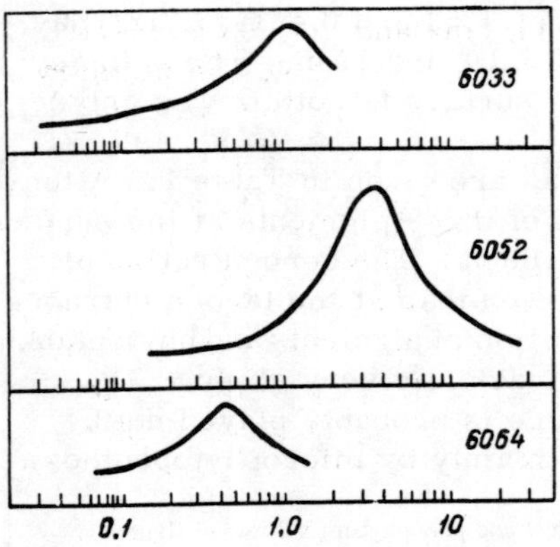

FIGURE 1. Distribution of primary production values on the surface in the oceanic polygons: x-axis — production; y-axis — frequency of occurrence of the values.

was 0.5 mg C/m³·day, i. e. one-half of the modal production for the polygon. The results were 0.14, 0.92, 0.08, 0.0, 0.22, 0.42, 0.32, and 0.10 mg C/m³·day at the respective depths of 0, 10, 25, 35, 50, 75, 95, and 150 m. Thus, the primary production exhibited two maxima, at 10 and at 75 m.

The first maximum was evidently mainly due to the photic maximum of photosynthesis at that depth, while the second maximum was due to a phytoplankton concentration the existence of which could be deduced from the high chlorophyll "a" concentration. Total production for the entire photosynthetic layer was 40 mg C/m³·day.

Since the surface production was below average at the time of observations, the value 40 mg C/m³·day cannot be regarded as characteristic for primary production in the region of polygon No. 2. A more correct approach is to start from the averaged surface production (1 mg C/m³·day) and then to proceed to production for the entire photosynthetic layer by making use of previously derived relationships (Koblents-Mishke, 1967). In this way the primary production in polygon No. 2 can be estimated at 75 mg C/m³·day, which is a characteristic value for tropical oligotrophic waters (Koblents-Mishke, 1967).

The concentrations of chlorophyll and pheophytin "a" determined in two series and once every three hours over a 24-hour period were fully in accordance with the primary production values (Figure 2). Their total concentration did not exceed 0.05 mg/m³ in the greater part of the euphotic layer. The concentration rose to 0.22 mg/m³ at 75—100 m, where there were phytoplankton concentrations, but decreased again farther down. The chlorophyll "a" concentrations determined for the second polygon were in satisfactory agreement with those reported in the literature (Doty and Capurro, 1961; Le Bourhis, Wauthy, and Amadeo, 1966).

Polygon No. 3 (station 6052) was executed in the Coral Sea, a richer region with a relatively high concentration of blue-green algae.

The results were processed as for the preceding polygons. The mode of the distribution curve for production values (see Figure 1) on the surface corresponded to 3 mg C/m³·day, which scaled to the photosynthetic layer yields 120 mg C/m²·day. The vertical distribution of primary production in the third polygon exhibited the same pattern as the second polygon, i. e. two maxima, one at subsurface depths, the other at a depth of about 100 m near the thermocline.

Production in the photosynthetic layer by in situ determinations was below 120 mg C/m². This was because the in situ measurements (similarly as in the preceding polygon) turned out to have been executed on the "poorest" days, when production at the surface was 0.14 — 1.14 mg C/m³·day, whereas it varied from 0.14 to 15.7 mg C/m³·day over the entire working period in the polygons. The concentration of chlorophyll and pheophytin "a" in polygon No. 3 was somewhat higher than in polygon No. 2, with 13.2 mg/m^2 on the average for the 0–150-m layer (Table 2). Accumulation of pigments proceeded more intensively in the discontinuity layer, as is readily perceived from the graph of diurnal variation of the concentration of chlorophyll + pheophytin "a" (see Figure 2). Their combined concentration here reached 0.32 mg/m^3.

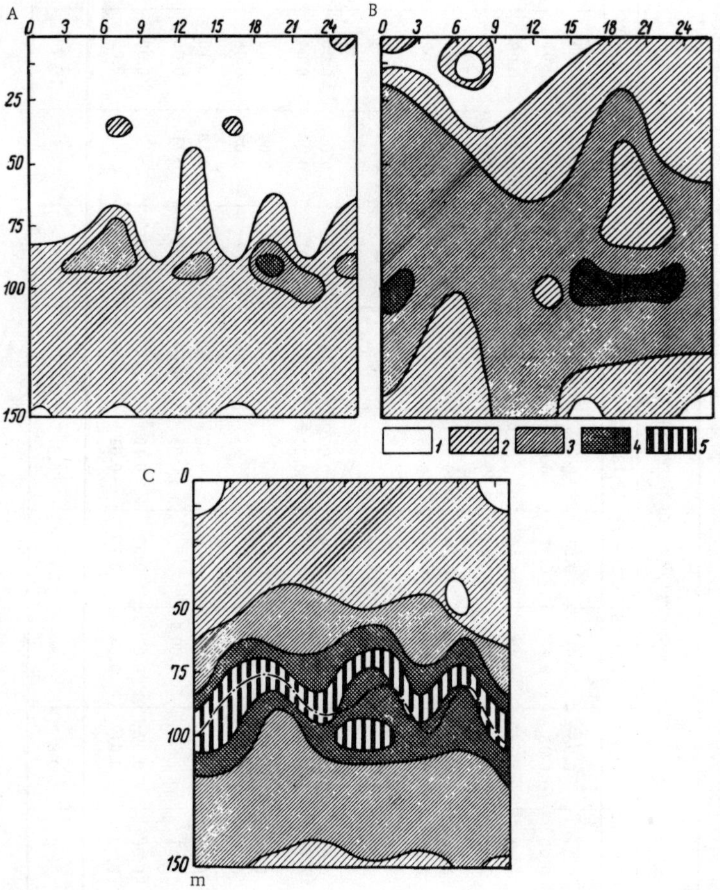

FIGURE 2. Diurnal variations of the combined concentration of chlorophyll "a" and its products in the oceanic polygons:

A – polygon No. 2 (station 6033); B – polygon No. 3 (station 6052); C – polygon No. 4 (station 6064); x-axis – time; y-axis – depth: isolines mark chlorophyll concentrations (mg/m^3): 1 – 0.0–0.05; 2 – 0.05–0.10; 3 – 0.10–0.20; 4 – 0.20–0.30; 5 – > 0.30; the dotted and dashed line represents the upper boundary of the thermocline.

TABLE 2. Concentrations of chlorophyll "a" (1st row of numbers) and pheophytin "a" (2nd row of numbers) in January 1959 in polygon No. 3 (station 6052), mg/m³

Depth, m	Date and hour of observation												Mean combined concentration of chlorophyll and pheophytin "a"
	16;10–12	18;10–11	19;01–02	19;04–05	10;07–08	19;10–11	19;13–14	19;16–17	19;19–20	19;22–23	20;01–02	21;02–04	
0	0.00 0.09	0.08 0.00	0.04 0.05	0.05 0.08	0.01 0.08	0.03 0.08	0.08 0.06	0.08 0.05	0.06 0.03	0.06 0.00	0.08 0.01	0.04 0.10	0.10
10	0.04 0.02	0.11 0.02	0.08 0.05	0.06 0.04	0.02 0.01	0.09 0.04	0.05 0.07	0.07 0.02	0.03 0.00	0.02 0.04	0.06 0.03	0.08 0.03	0.09
25	0.02 0.04	0.10 0.04	0.06 0.06	0.07 0.00	0.08 0.07	0.09 0.03	0.03 0.03	0.02 0.13	0.07 0.03	0.05 0.00	0.07 0.02	0.06 0.05	0.10
35–40	0.08 0.02	0.07 0.01										0.07 0.00	0.08
50–60	0.09 0.00	0.12 0.02	0.02 0.15	0.07 0.04	0.11 0.00	0.04 0.01	0.09 0.00	0.06 0.05	0.07 0.01	0.05 0.06	0.05 0.04	0.09 0.06	0.1
75	0.04 0.11	0.09 0.04	0.12 0.03	0.10 0.00	0.06 0.05	–	–	0.14 0.05	0.05 0.01	0.04 0.03	0.08 0.08	0.07 0.08	0.14
100	0.06 0.11	0.13 0.08	0.21 0.11	– –	0.10 0.00	–	0.06 0.01	0.16 0.10	0.09 0.05	0.18 0.11	0.10 0.05	0.18 0.09	0.21
150	0.00 0.04	0.02 0.02	0.06 0.02	0.07 0.01	0.03 0.03	0.07 0.10	0.10 0.01	0.03 0.00	0.02 0.05	–	0.00 0.02	0.01 0.02	0.07
	11.0*	13.6	18.3	9.6	9.8	10.7	8.8	15.6	11.5	10.4	10.7	14.8	13.2

* This row states the chlorophyll and pheophytin "a" concentrations in mg/m².

The results for polygon No. 3 agree well with those obtained by Jitts (1965) in a nearby region and during the same season. He estimated the total diurnal production in the photosynthetic layer at 150 mg C/m^3·day and the chlorophyll "a" concentration in the 0–150-m layer at about 15 mg/m^2.

The primary production values determined suggest that polygon No. 3 was executed in waters intermediate between the oligotrophic and mesotrophic types. The wide scatter of the data for production at the surface arouses suspicion that a change of waters of different trophic characteristics took place during the work in the polygon.

Polygon No. 4 (station 6064) was executed east of the Solomon Islands. Production on the surface and throughout the water column (determined in situ) proved to be much smaller than in the preceding polygons, whereas the mean combined concentration of chlorophyll and pheophytin "a" was significantly higher, 20.4 mg/m^2 (Table 3). The results of primary production determination are probably erroneous because of some undetected methodological failure related to the transition from the standardized isotope solution to the new, much more concentrated solution (see above). Since the results of primary production determination in polygon No. 4 do not inspire confidence, it would be pointless to dwell upon them in greater detail.

TABLE 3. Concentrations of chlorophyll "a" (1st row of numbers) and pheophytin "a" (2nd row of numbers) in February 1969 in polygon No. 4 (station 6064), mg/m^3

Depth, m	Date and hour of observation										Mean combined concentration of chlorophyll and pheophytin "a"
	7; 01–03	8; 12–13	8; 15–16	8; 18–19	8; 21–22	9; 00–03^{20}	9; 03–04	9; 06–07	9; 10–11	11; 03–04	
0	0.03	0.00	0.00	0.04	0.05	0.04	0.06	0.06	0.08	0.05	0.06
	0.00	0.06	0.07	0.03	0.01	0.00	0.02	0.00	0.00	0.00	
10	0.02	0.04	0.08	0.07	0.04	0.05	0.07	0.04	0.07	0.06	0.06
	0.02	0.02	0.00	0.00	0.00	0.00	0.00	0.03	0.02	0.00	
25	0.00	0.07	0.04	0.04	0.06	0.02	0.06	0.06	0.08	0.06	0.06
	0.04	0.00	0.02	0.04	0.01	0.03	0.00	0.01	0.00	0.00	
50–55	0.04	0.08	0.08	0.10	0.00	0.06	0.08	0.11	0.10	0.09	0.11
	0.19	0.02	0.02	0.03	0.02	0.02	0.01	0.01	0.01	0.00	
70	—	—	0.15	—	0.20	—	0.17	—	—	—	0.35
	—	—	0.41	—	0.05	—	0.08	—	—	—	
75	0.20	0.19	0.13	0.17	0.38	0.03	0.23	0.22	0.16	0.17	0.27
	0.18	0.11	0.03	0.03	0.05	0.14	0.07	0.12	0.05	0.03	
80–95	—	0.24	—	0.30	—	0.21	—	0.22	0.26	0.29	0.36
	—	0.03	—	0.18	—	0.17	—	0.09	0.12	0.04	
100	0.09	0.14	0.22	0.15	0.21	0.24	0.20	0.10	0.14	0.10	0.29
	0.35	0.19	0.16	0.12	0.03	0.12	0.10	0.05	0.09	0.07	
150	0.03	0.05	0.04	0.07	0.06	0.07	0.08	0.06	0.04	0.06	0.10
	0.04	0.05	0.06	0.02	0.08	0.03	0.04	0.03	0.04	0.00	
	22.4*	18.1	17.1	14.5	16.0	18.8	18.2	19.2	15.9	12.8	17.4

* This row states the chlorophyll and pheophytin "a" concentrations in mg/m^2.

In order to estimate the primary production in polygons Nos. 4 and 5 from data for polygons Nos. 2 and 3, we calculated the ratio between the production and the combined chlorophyll and pheophytin "a" concentration in the 0—150-m layer. It was found to be 10 on the average, as against 2.1—2.4 in polygons Nos. 2 and 3. Calculation of production in these two polygons from the chlorophyll concentration by resorting to the conversion factor 10.0 yields fully realistic values (Table 4). The value for polygon No. 4 is 175 mg C/m^3 · day, which is characteristic for mesotrophic waters.

TABLE 4. Production and chlorophyll "a" at different stations

Polygon, station	Primary production on the surface, mg C/m^3 · day	Primary production in the water column, mg C/m^2 · day	Concentration of chlorophyll and pheophytin "a" in the 0—150-m layer, mg/m^2	Ratio of production in the water column to chlorophyll "a" and pheophytin
2; 6033	1.0	75	6.5	11.5
3; 6052	3.0	120	13.2	9.0
4; 6064	0.4*	50*	—	—
	7.0**	175**	17.4	10.0
5; 6065	6.9*	138**	—	—
	55.0**	560**	56	10.0

* Measured production.
** Assumed values.

The diurnal station in polygon No. 4 was executed with especial care. Samples for pigments were taken at the station at intervals of three hours and were monitored by means of the transparency meter with an attached bathythermograph. Sounding with these instruments began half an hour before work was begun with the water sampler. The concentration of the pigments here proved to be appreciably higher than at station 6052. Concentrations below 0.05 mg/m^3 were hardly observed, and the concentration occasionally exceeded 0.5 mg/m^3 in the vicinity of the pycnocline. There was a very distinct maximum of pigment concentration in the lower part of the euphotic layer. This maximum was found to be subject to periodic diurnal vertical shifts along with the thermocline. The amplitude of such shifts was about 30 m and their duration about 6 hours. Apparently, what we observed here were internal waves which shifted the thermocline along with the accumulated phytoplankton and detritus.

Polygon No. 5 (station 6065) was executed in Kieta Bay (Bougainville Island). The maximum photosynthesis and maximum chlorophyll "a" concentration were observed at a depth of 12 m. Kieta Bay should be classified as an eutrophic area by its production and chlorophyll concentrations. Averaged data for this polygon are given in Table 4.

Thus, the 44th cruise included investigations of oceanic regions ranging from oligotrophic (polygons Nos. 1 and 2) to eutrophic (polygon No. 5), with a corresponding increase in the efficiency of utilization of solar energy. This efficiency was calculated by means of the coefficient 9.36 cal/mg C. The numerical results were as follows:

Polygon No.	1	2	3	4	5
Utilization of luminous energy in photosynthesis, %	0.010	0.025	0.035	0.064	0.250

Since the radiocarbon method yields low results (approximately two-thirds of the true values), it must be borne in mind that these data on the efficiency of utilization of luminous energy are also low.

Bibliography

Publications in Russian

Khailov, K. M. Elementy ekologicheskogo metabolizma v pribrezhnoi zone more (Elements of Ecological Metabolism in the Coastal Zone of the Sea). Author's Summary of Doctoral Thesis. Moskva. 1969.

Koblents-Mishke, O. I. Pervichnaya produktsiya (Primary Production). — In: Tikhii okean, Vol. 7. Biologiya Tikhogo okeana, Book 1. Moskva, Izdatel'stvo "Nauka.". 1967.

Vinberg, G. G., Yu. G. Kabanova, O. I. Koblents-Mishke, and N. N. Khmeleva. Metodicheskoe posobie po opredeleniyu pervichnoi produktsii organicheskogo veshchestva v vodoemakh radiouglerodnym metodom (Methodological Manual for Radiocarbon Determination of the Primary Production of Organic Matter in Water Bodies). Minsk. 1960.

Publications in Other Languages

Determination of Photosynthetic Pigments in Sea-water. — Monographs on Oceanographic Methodology, Vol. 1. UNESCO. 1966.

Doty, M. and L. Capurro. Productivity Measurements in the World Ocean. — I. G. Y. Oceanogr. Rept. No. 4. 1961.

Holm-Hansen, O., C. Lorenzen, R. Holmes and J. D. H. Strickland. Fluorometric Determination of Chlorophyll. — J. Cons. perman. int. Explor. mer., Vol. 30, No. 1. 1965.

Jitts, H. R. The Summer Characteristics of Primary Productivity in the Tasman and Coral Seas. — Austral. J. Mar. Freshwat. Res., Vol. 16, N2. 1965.

Le Bourhis, Y., B. Wauthy and M. Amadeo. Mesures de production primaire des croisieres Bora 1, Bora 3 et Bora 4 du N. O. "Coriolis". — ORSTOM, Centre de Noumea. Oceanog. Rapp., No. 14. 1966

Lorenzen, C. J. Determination of Chlorophyll and Phaeopigments. Spectrophotometric Equations. — Limnol. and Oceanogr., Vol. 12, No. 2. 1965.

SOME ECOLOGICAL AND PHYSIOLOGICAL PROPERTIES OF PHYTOPLANKTON

O.I. Koblents-Mishke

Shirshov Institute of Oceanology of the Academy of Sciences of the USSR

The material obtained in the 44th cruise of the "Vityaz" broadened our ideas on the physiological ecology of marine phytoplankton shaped by previous expeditions of the Institute of Oceanology. Research of this kind is conducted by most authors under laboratory conditions (for instance, studies of "light curves" in incubators, etc.). In contrast, the same characteristics are studied in expeditions of the Institute of Oceanology under conditions that are as similar to the natural conditions as possible. Our approach is much more laborious and less resultful than laboratory investigations. Nevertheless, natural experiments are unavoidable in ecological physiology for the following two reasons. On the one hand, such experiments are essential for appraising the validity of extending the laboratory results to natural processes, and on the other, they reveal new phenomena, such as, for instance, increased efficiency of the utilization of solar radiation in photosynthesis in a water layer of high turbidity (Kabanova et al., 1964), and so on.

For elucidation of the ecological and physiological properties of marine phytoplankton, primary production measurements were performed in situ and the concentrations of plant pigments were measured many times over a 24-hour period.

The techniques of these investigations have been described in the preceding paper (Koblents-Mishke et al.). The results were compared against data on light and mineral nutrition. Unfortunately, the measurements of photosynthetic parameters and environmental conditions were not always strictly synchronous. For example, samples for hydrochemical determinations were taken from samplers other than those used for measurements of photosynthesis and the concentrations of pigments, while the amounts of luminous energy penetrating to different horizons were calculated by means of vertical light attenuation factors averaged for each polygon. The data used for this purpose in earlier studies were those obtained on the day of exposure of phytoplankton samples in photosynthetic measurements. For the rest, the same energy calculation techniques were used as in previous investigations (Koblents-Mishke, 1961; Kabanova et al., 1964; Koblents-Mishke and Ochakovskii, 1966).

ASSIMILATION NUMBER

Studies of the assimilation number (or coefficient) of marine phytoplankton are important in modern research. A method for primary production determinations without exposure of samples, based on an assumed constancy of assimilation number under optimal illumination, was earlier proposed by Ryther and Yentsch (1957). Subsequently, however, the assimilation number was found to be dependent not only upon light but also upon many other conditions (mineral nutrients, temperature, etc.). Investigation of these relationships will make it possible to improve the method described by Ryther and Yentsch, which in turn will provide an opportunity of utilizing the large amount of information already in hand on pigments and solar radiation for expanding and improving the accuracy of information concerning the primary production of the World Ocean. Moreover, the assimilation number referred to irradiance in a given horizon will permit estimation of the efficiency of light utilization by the phytoplankton's pigment apparatus under different ecological conditions. Calculation of this unit makes it possible to study the pure effects of such factors as minerals, temperature, etc. (Ryther, 1959).

In the 39th cruise of the "Vityaz" (Koblents-Mishke et al., 1970), in summer 1966 east of the Kurile Islands, it was demonstrated that the efficiency of utilization of solar energy by one milligram of chlorophyll increases with decreasing amounts of light available to the phytoplankton. The same was found during the 44th cruise (Figure 1). The results of these two cruises aptly supplement each other. The efficiency of chlorophyll activity is lower in the tropical zone of the ocean than in the boreal zone, apparently on account of different conditions of mineral nutrition. The concentration of mineral phosphorus may serve as an index of these conditions. The functional efficiency of chlorophyll increases with increasing concentration of phosphates (Figure 2), although the correlation is low (coefficient of correlation, 0.48 ± 0.1). There is a higher coefficient of correlation, -0.76 ± 0.05, between the assimilation number referred to unit light energy and temperature (Figure 3). This means either that the temperature is not only an associated factor bearing an inverse correlation to the mineral nutrition factor but in itself affects the functional efficiency of the pigment apparatus, or else that the concentration of phosphates does not adequately reflect the conditions of mineral nutrition. The first assumption contradicts biochemical concepts.

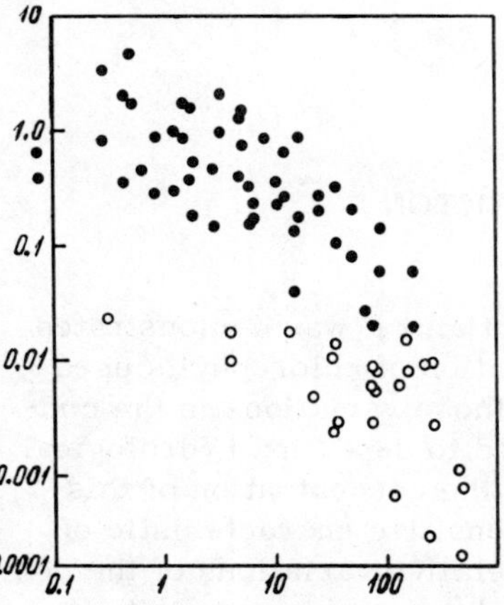

FIGURE 1. Relationship between the assimilation coefficient of chlorophyll "a" referred to light energy (A/E, the y-axis) and underwater irradiance, $cal \cdot cm^{-2} \cdot min^{-1}$ (x-axis).

Dark and light and circles represent, respectively, data from the 39th and 44th cruises of the "Vityaz."

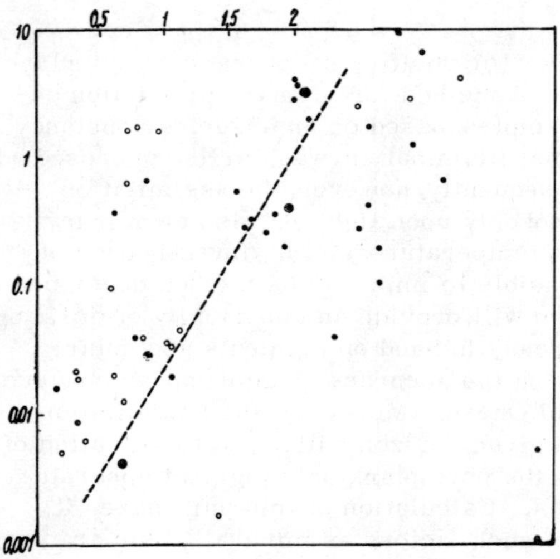

FIGURE 2. Relationship between A/E (y-axis) and the concentration of phosphates, μg-at/l (x-axis). Double circles are the results of averaging within one order of magnitude.

Other notation as in Figure 1.

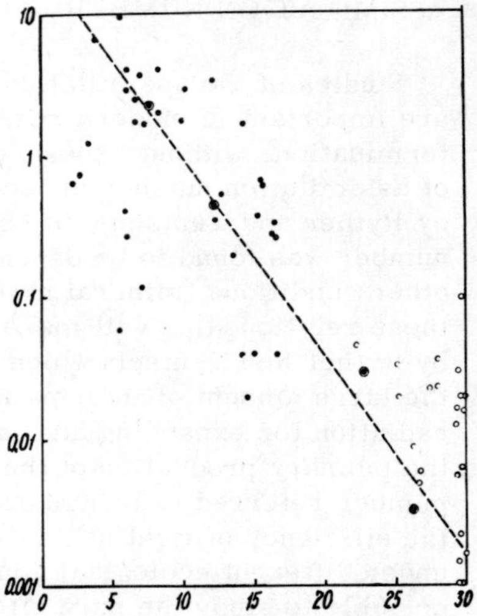

FIGURE 3. Relationship between A/E (y-axis) and temperature (°C, x-axis).

Other notation as in Figures 1 and 2.

DIURNAL VARIATIONS IN THE PHYTOPLANKTON PIGMENT SYSTEM

In the preceding paper (Koblents-Mishke et al.) it was demonstrated that the method for determining the concentration of chlorophylls used in the cruise made it possible to detect only those variations in the concentration of chlorophyll "a" that were related to depth and hydrological conditions. Continuous diurnal variations in the concentration of this pigment were not observed. A much more sensitive characteristic of the pigment composition is provided by the relative variations of the concentrations of different pigments and their light-absorbing capacity. Figure 4B depicts the diurnal variations of the ratio between the combined concentration of chlorophyll "a" and its derivatives (pheophytin, etc.) and the combined concentration of chlorophyll "c" and its derivatives. The ratio exhibits two maxima. The upper maximum is probably due to predominance of the concentration of unaltered chlorophyll "a" over that of chlorophyll "c", while the lower maximum is probably due to a similar relationship of their derivatives, mainly pheophytins. This assumption is based on a fairly large number of papers the authors of which demonstrated a relative increase of pheophytin "a" (in comparison to chlorophyll) with increasing depth (Yentsch, 1965a, 1965b; see also the preceding paper by Koblents-Mishke et al.). Chlorophyll "a" is almost entirely replaced by pheophytin at 50 — 80 m, where the second maximum of the ratio occurs. No method has yet been devised for calculating pheophytin "c", and therefore only assumptions can be drawn.

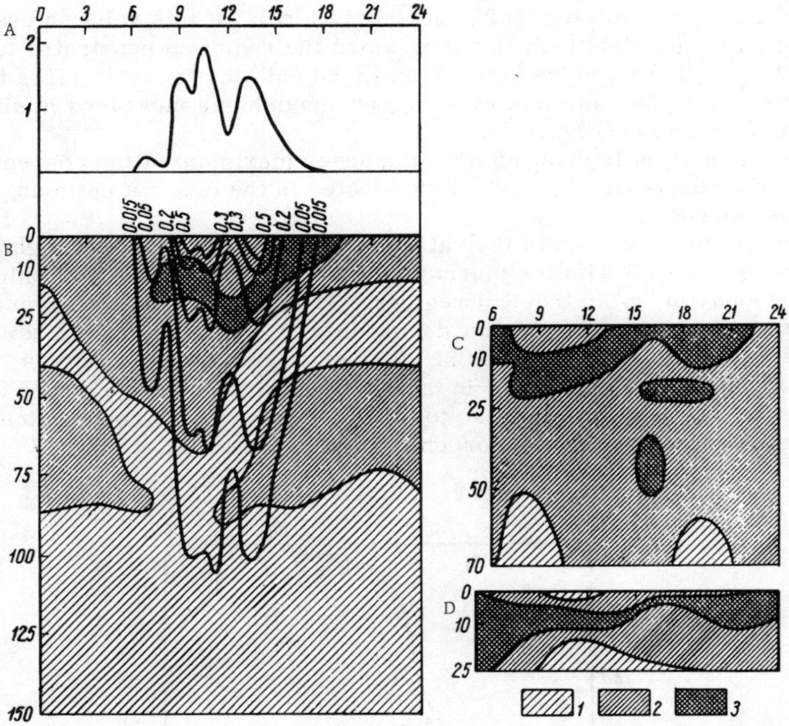

FIGURE 4. Solar radiation and the ratio of the combined concentrations of chlorophyll and pheophytin "a" and "c" (chl. "a"/chl. "c"):

A — variation of incident photosynthetically active radiation (PHAR), $cal \cdot cm^{-2} \cdot min^{-1}$ (y-axis) with respect to time (x-axis) on the day of diurnal observations at station 6064 (February 7, 1969); B — vertical distribution of PHAR and of the ratio chl. "a"/chl. "c" from diurnal observation data at station 6064; C — ratio chl. "a"/chl. "c" at different depths from diurnal observation data at station 5631 during the 39th cruise of the "Vityaz"; D — same, after Yentsch and Scagel (1958) in East Sound, Washington, USA. Limits of the ratio chl. "a"/chl. "c" (shown by the different shadings 1, 2, 3): B — <2; 2 — 10; > 10; C — < 0.5; 0.5 — 1; > 1; D — < 1; 1 — 1.4; > 1.4.

The relative predominance of chlorophyll "a" at the surface at night encompasses a layer a little thicker than 10 m, and a layer as thick as 40 m in the daytime close to the moment of the sun's culmination. Its concentration is relatively lower on the very surface at that time. Maximum predominance occurs at depths of 0 — 25 m, the layer of the maximum being located deeper in the daytime. Apparently, the ratio of concentration chlorophylls "a" and "c" increases at the depth of the photosynthetic optimum, which varies in the course of the light day. This is illustrated in Figure 4B, which shows the amount of solar radiation penetrating to different depths at different times in the 24-hour period. The calculations were performed by means of the curve of the diurnal march of solar radiation (Figure 4A) and the average light-attenuation factor for the polygon. The maximum ratio of chlorophylls "a" to "c" was observed at an irradiance of about 0.25 $cal \cdot cm^{-2} \cdot min^{-1}$. If this value is scaled to the entire daylight period during which the radiation penetrated to water (9 – 17 hours), we arrive at the quantity 60 $cal \cdot cm^{-2} \cdot day^{-1}$. This figure coincides with the optimum value for photosynthesis measured in situ during this cruise (Figure 5).

Hence, it is quite probable that the upper maximum of the concentration ratio of chlorophylls "a" and "c" is located in the layer of optimum of photosynthesis.

The lower maximum of the ratio is especially noticeable at night, which agrees well with the diurnal variation data on the ratio of chlorophyll and pheophytin "a" in tropical regions. The relative concentration of pheophytin was found to decline during the day, with a downward shift of its maximum. This is possibly due not only to transformations of chlorophyll and pheophytin in the light and in the dark described by Yentsch (1965a, 1965b), but also to diurnal migrations of zooplankton, in the intestines of which chlorophyll is pheophytinized.

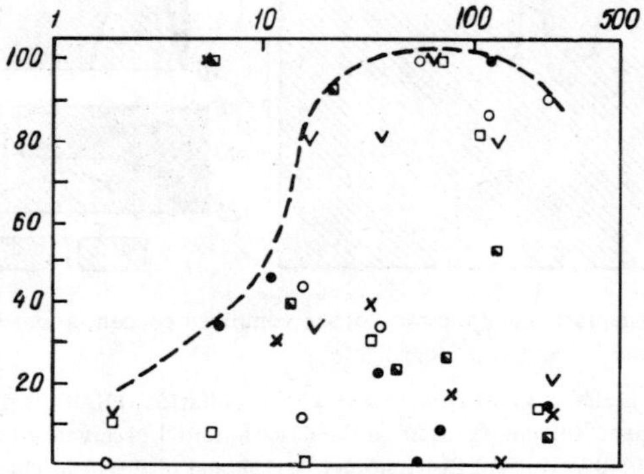

FIGURE 5. Photosynthesis measured in situ (% of the maximum, y-axis) at different irradiances ($cal \cdot cm^{-2} \cdot min^{-1}$, x-axis) according to data from the 44th cruise of the "Vityaz."

The different symbols represent data obtained at different stations.

The diurnal dynamics of the ratio of chlorophylls "a" to "c" had previously been studied by Yentsch and Scagel (1958) in East Sound, Washington, USA (48°40' N) in summer 1954 and by ourselves east of the Kurile Islands in summer 1966 (Figure 4C, D).

Yentsch and Scagel worked in shallow waters during the northern summer with much longer days than in the tropics, and therefore the pattern elucidated by them was less distinct, although in general it follows that which we disclosed. There was a maximum of the ratio in which we are interested, and its upper boundary shifted deeper in the daytime. A chlorophyll "a" minimum occurred on the surface at noon. Apparently the chlorophyll "a" maximum merged with the pheophytin "a" maximum, and therefore the pattern in the lower part of the graph was different from that observed in the 44th cruise of the "Vityaz."

The depth interval 0 — 75 m was investigated in the 39th cruise. The chlorophyll "a" maximum likewise shifted deeper at noon and reached its highest values at depth, where conditions for photosynthesis were apparently optimal.

The other diurnal stations executed during the 44th cruise yielded similar though less distinct results, with the difference that the ratio of chlorophyll "a" and "c" was considerably lower (especially at station 6033) than at station 6064. It was characterized by the following average figures: 0.98 at station 6033, 3.2 at station 6052, and 4.3 at station 6064. The impression is gained that the ratio of chlorophylls "a" and "c" is higher the higher are the primary production values; these averaged 75, 120, and 175 mg $C/m^2 \cdot$ day, respectively.

Unfortunately, the low sensitivity of the method used does not permit a similar analysis of other components of the pigment complex. Analysis of the behavior of individual pigments can be replaced by evaluation of their light-absorbing capacity in different regions of the spectrum as was done for evaluation of the chromatic adaptation of the pigment apparatus (Koblents-Mishke et al., 1970).

The so-called pigment index, i. e. the absorption ratio of acetone extract from phytoplankton in the wavelengths 430 and 655 nm, is very interesting from the ecological standpoint. Absorption in the former wavelength is the total absorption of the chlorophylls and carotenoids, while absorption in the latter wavelength is almost exclusively limited to chlorophyll "a" and its derivatives. The higher the pigment index, the more there is of supplementary pigments in the algal extract and the greater is their variety. According to Margalef (1960), the pigment index grows in parallel with the specific variety of the phytoplankton community. The variety index is an important ecological characteristic of biocenoses. High variety indexes characterize "mature" communities in which each successive stage in the trophic chain exploits to a greater or lesser extent the resources provided by the preceding stage. Margalef demonstrated that the pigment index increases in the course of aging of individual cells and of entire populations of phytoplankton. There apparently exists still another mechanism for diversification of the pigment composition in the phytoplankton, which was not taken into account by Margalef and which consists in active biochemical transformation of pigments in the course of adapatations. All those three mechanisms lead to a complication of the pigment apparatus which manifests itself in higher pigment index values under adverse conditions of habitat.

Studies of the pigment index calculated from data of the 44th cruise of the "Vityaz" confirmed Margalef's principal theses. The mean index for all stations and horizons of each polygon was found to drop with increasing production. The mean index was 48 at station 6033, 11 at station 6052, and 6 at station 6064. Moreover, it changes regularly with depth (Figure 6), mirroring the pattern of change of the chlorophyll "a" concentration, to which it bears an inverse logarithmic correlation (Figure 7). As a result, the diurnal variations of this index at different depths station 6064 (Figure 8) are a mirror image of the diurnal variations of the combined concentration of chlorophyll and pheophytin "a" (Koblents-Mishke et al., in this collection).

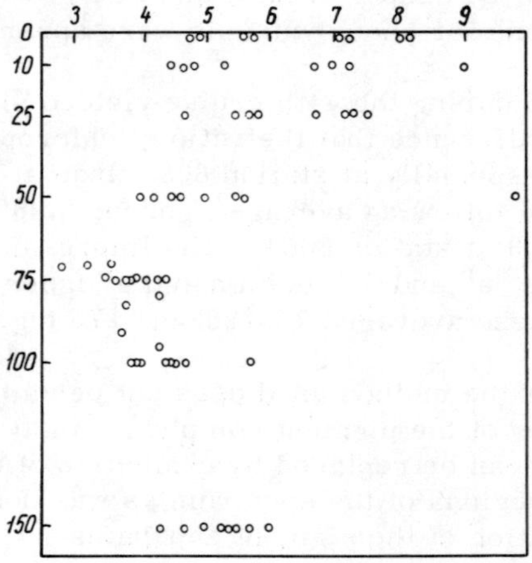

FIGURE 6. Variation of the pigment index D_{430}/D_{665} (x-axis) in relation to depth (y-axis) at station 6064

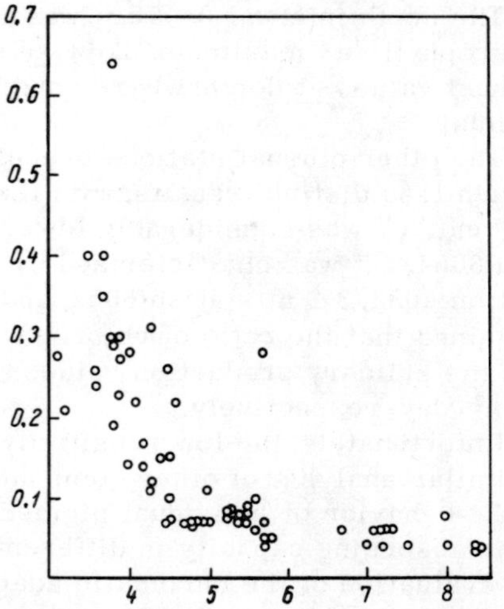

FIGURE 7. Pigment index (x-axis) and combined concentration of chlorophyll and pheophytin (y-axis) at station 6064

The lowest values of the pigment index are found in the layer of phytoplankton accumulation near the thermocline, where the phytoplankton exists under the most favorable nutrition conditions. Death of phytoplankton deeper down beyond the boundary of the euphotic layer is accompanied by accumulation of yellow pigments, which are destroyed more slowly than chlorophyll, and the index increases. The pigment index also increases above the thermocline as the biogenic elements become exhausted, the decrease being less pronounced in the morning and daytime than in the evening and at night. The decrease of the index at the surface in the first half of the day is apparently related to increased photosynthetic activity of the pigment system.

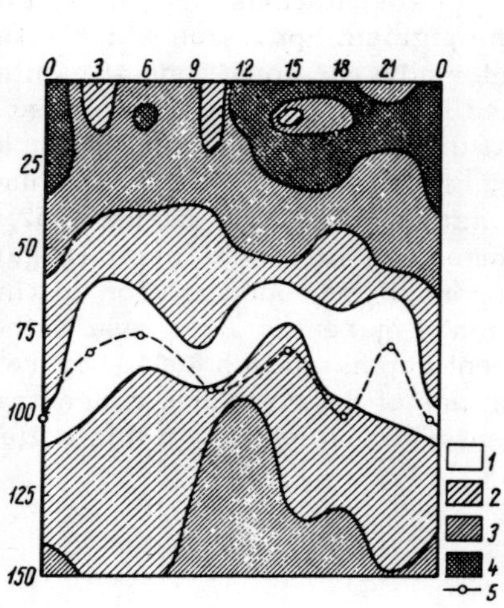

FIGURE 8. Diurnal variations of the pigment index at different depths, station 6064: 1 — < 4; 2 — 4-5; 3 — 5-7; 4 — > 7; 5 — upper boundary of thermocline. The time is marked on the y-axis.

Thus, the data obtained on the 44th cruise of the "Vityaz" revealed the pigment index to be an indicator not only of large-scale (seasonal and geographical) variations in the structure and functions of a phytocenosis but also of their small-scale vertical and diurnal variations.

CONCLUSION

Many hypotheses may be advanced concerning the causes and mechanism of the phenomena described in this paper. For instance, the shifting of the chlorophyll "a" maximum following optimal illuminance may be explained by migration of phytoplankton or by diurnal rearrangements of the pigment apparatus in the same algae. The pigment index appears to be related to the specific variety of phytoplankton.

The author is deeply grateful to his colleagues L. I. Paramonova, M. M. Gromov, and A. M. Tsvetkova, who helped to collect and prepare the data utilized in this paper.

Bibliography

Publications in Russian

Kabanova, Yu. G., O. I. Koblents-Mishke, and V. N. Pelevin. Fotosintez morskogo fitoplanktona na raznykh glubinakh (Photosynthesis of Marine Phytoplankton at Different Depths). — Okeanologiya, Vol. 4, No. 3. 1964.

Koblents-Mishke, O. I. Vidovoi sostav fitoplanktona i pervichnaya produktsiya v severozapadnoi chasti Tikhogo okeana zimoi 1958—1959 gg. (Specific Composition of Phytoplankton and Primary Production in the Northwest Pacific in the Winter of 1958—1959). — Trudy Instituta Okeanologii Akademii Nauk SSSR, Vol. 45. 1961.

Koblents-Mishke, O. I., O. D. Bekasova, V. I. Vedernikov, B. V. Konovalov, V. V. Sapozhnikov, and V. A. Terskikh. Pervichnaya produktsiya i pigmenty v raione Kurilo-Kamchatskogo zheloba letom 1966 g. (Primary Production and Pigments in the Kurile-Kamchatka Trench Region in the Summer of 1966). — Trudy Instituta Okeanologii Akademii Nauk SSSR, No. 86. 1970.

Koblents-Mishke, O. I. and Yu. E. Ochakovskii. Ob izmereniyakh sveta pri izuchenii pervichnoi produktsii v more (Measurements of Light in Studies of Primary Production in the Sea). — Okeanologiya, Vol. 6, No. 3. 1966.

Rabinovich, E. Fotosintez (Photosynthesis), Vol. 3. Moskva, Izdatel'stvo Akademii Nauk SSSR. 1959.

Publications in Other Languages

Margalef, F. Valeur indicatrice de la composition des pigments du phytoplancton sur la productivité, composition taxonomique et propriétés dynamiques des populations. — Rapp. Cons. Internat. Mer. Méditerr., 15, pt. 2. 1960.

Ryther, J. H. Potential Productivity of the Sea. — Science, Vol. 130. No. 3376. 1959.

Ryther, J. H. and C. S. Yentsch. The Estimation of Phytoplankton Production from Chlorophyll and Light Data. — Limnol. and Oceanogr., Vol. 2, No. 3. 1957.

Yentsch, C. S. The Relationship between Chlorophyll and Photosynthetic Carbon Production with Reference to the Measurement of Decomposition Products of Chloroplastic Pigments. — Mem. Ist. Ital. Idrobiol., Vol. 18, Suppl. 1965a.

Yentsch, C. S. Distribution of Chlorophyll and Pheophytin in the Open Sea. — Deep-Sea Res., Vol. 12, No. 5. 1965b.

Yentsch, C. S. and R. F. Scagel. Diurnal Study of Phytoplankton Pigments on in situ Station in East Sound, Washington. — J. Mar. Res., Vol. 17, No. 4. 1958.

60094 1

APPLICATION OF THIN LAYER CHROMATOGRAPHY TO SEPARATION OF PHYTOPLANKTON PIGMENTS ABOARD SHIP

L. I. Paramonova

Institute of Photosynthesis of the Academy of Sciences of the USSR

The standard method used in the 44th cruise of the "Vityaz" for determining the concentration of plant pigments, a modification of the method described by Richards and Thompson (1962), furnishes only a very rough idea of the concentrations of chlorophylls "a", "b", and "c" and the total concentration of carotenoids.

But if we wish to study how the phytoplankton's pigment system adapts itself to vertically variable illuminance conditions more detailed knowledge of its qualitative and quantitative composition is required. An attempt to study this composition by thin layer chromatography was made during the cruise. Most of the analyses were performed on samples from the Coral Sea.

ADSORBENTS

Preliminary testing of adsorbents (sucrose, $BaCO_3$, $CaCO_3$) revealed that $BaCO_3$ mixed with potato starch in a weight ratio of 10:1 was the easiest to work with under tropical conditions. On the one hand, $BaCO_3$ is not hygroscopic and on the other, it hardly affects the pigments because of its low chemical activity (Goodwin, 1965). Addition of starch optimizes its physical properties.

The adsorbent mixture was applied to a plate in the form of a paste prepared with distilled water (in the weight ratio 1:1), 1 ml per 15 cm^2 glass, producing a layer 0.5 — 0.7 mm thick. The plates were first dried in air and then kept in a drying oven for 1.5 — 2 hours at 70 — 80°. The dried plates were stored in a desiccator pending their use.

The aqueous acetone extract of pigments was evaporated to dryness under vacuum before being applied to a plate. The residue was dissolved in a 7:3 mixture of petroleum (b. p. 40 — 70°) and diethyl ethers. The pigments were applied to the plate in a strip.

SOLVENT MIXTURES

The following solvent mixtures were chosen after the testing of various partition systems:

1) 5 — 6% acetone in *n*-hexane;
2) 15% acetone in *n*-hexane + 0.5% *n*-propanol;
3) 3% acetone in *n*-hexane;
4) 1.1% propanol in *n*-hexane.

The choice of a solvent system or a complex of several solvent systems depends upon the particular properties of the material to be chromatographed. The substitution of petroleum ether, which is the usual solvent in such cases, with *n*-hexane was prompted by the former's high volatility, which is a serious hindrance when working in the tropics.

Development of a chormatogram takes 5 — 10 min. Desiccators were used instead of chromatographic chambers. The method permits a fairly distinct separation of chlorophylls, pheophytins, and the principal carotenoid pigments carotene, lutein, violaxanthin, neoxanthin, and zeaxanthin (see figure).

Because the technique was elaborated directly during the cruise, the chromatographic investigation was conducted only at the last two stations, 6064 and 6065, and was of a tentative nature.

Spectra of spectrophotometrically isolated pigments.

x-axis — wavelengths; y-axis — light transmission in relative units.

RESULTS OF THE INVESTIGATION

Station 6064, February 11, 1969, 0020 hr. Analysis of the pigments of phytoplankton sampled from the horizons 0, 20, 70 and 100 m, on $BaCO_3$; extraction with 90% acetone.

Pigments from the 0, 70, and 100-m horizons were separated by double chromatography, first in the system of 6% acetone in *n*-hexane, then in the system of 15% acetone in *n*-hexane + + 0.5% *n*-propanol. The mixture of pigments from the 20-m horizon was quite well separated in the single system of 6% acetone in *n*-hexane.

The following pigments were identified: chlorophylls "a" and "b", carotene, lutein, zeaxanthin, violaxanthin, and heoxanthin. Chlorophyll "a" was differentiated into two zones in the course of chromatography. The distribution of these pigments by horizons is shown in Table 1. The data on the contents of chlorophyll "a" in this and the following tables represent the sum total of chlorophyll "a" concentrations in both zones in the chromatogram.

The chlorophyll "a" concentration increases with depth, reaching its maximum at 100 m. The same applies to α-carotene, but the increase of its concentration with increasing depth is less marked.

Traces of chlorophyll "b" were detected at the 70-m horizon. Its concentration reached 0.695 μg/l at the 100-m horizon, amounting to half the concentration of chlorophyll "a" at the same depth.

TABLE 1. Quantitative distribution of pigments (μg/l) at station 6064, February 11, 1962

Depth, m	Chlorophyll "a"	Chlorophyll "b"	α-Carotene	Lutein	Zeaxanthin	Violaxanthin	Neoxanthyn	Total xanthophylls	Chlorophyll "a": total xanthophylls	Chlorophyll "a": carotene	Total xanthophylls: carotene
0	0.129	—	Trace	0.091	—	Trace	Trace	0.091	1.42	—	—
20	0.149	—	0.056	0.079	—	"	"	0.079	1.89	2.66	1.41
70	0.850	Trace	0.094	0.349	—	0.108	0.119	0.576	1.47	9.05	6.13
100	1.340	0.695	0.189	0.173	0.019	0.386	0.189	0.767	1.74	7.10	4.05

TABLE 2. Quantitative distribution of pigments (μg/l) at station 6064, February 12, 1969

Depth, m	Chlorophyll "a"	α-Carotene	Lutein	Violaxanthin	Neoxanthin	Total xanthophylls	Chlorophyll "a": total xanthophylls	Chlorophyll "a": carotene	Total xanthophylls: carotene
0	0.111	0.07	0.157	0.027	—	0.184	0.60	6.53	10.80
25	0.061	0.036	0.0675	0.071	0.098	0.237	0.26	1.70	6.60
50	0.203	0.085	0.202	0.071	0.046	0.319	0.64	2.39	3.76
60	0.117	0.066	0.054	0.060	0.056	0.170	0.69	1.77	2.58
85	0.122	0.038	0.011	0.064	0.030	0.105	1.16	3.22	2.77
100	0.570	0.118	0.099	0.545	0.106	0.750	0.76	4.83	6.35

TABLE 3. Quantitative distribution of pigments (μg/l) at station 6065, February 15, 1969

Depth, m	Chlorophyll "a"	α-Carotene	Lutein	Violaxanthin	Unidentified yellow pigment	Chlorophyllide from start	Total xanthophylls	Chlorophyll "a": total xanthophylls	Chlorophyll "a": carotene	Total xanthophylls: carotene	Chlorophyll "a": chlorophyllide "a"
0	1.535	0.310	0.110	0.470	0.490	0,304	1.070	1.44	4.95	3.45	5.10
12	2.610	0.417	0.806	0.437	0.603	0.885	1.346	1.94	6.27	3.24	2.95
40	1.844	0.443	0.236	0.315	0.320	2.410	0.871	2.11	4.16	1.97	0.77

Lutein, violaxanthin, and neoxanthin are present in all the horizons. The maximum concentration of lutein occurs in the 70-m horizon. It decreases at 100 m, in contrast to increased concentrations of violaxanthin and neoxanthin and the presence of zeaxanthin, which was found only in that horizon.

The combined concentration of xanthophylls increases with increasing depth, similar to that of chlorophyll. The ratio of chlorophyll "a": total xanthophylls scarcely changes with depth.

Station 6064, February 12, 1969, 0730 hr. The samples were taken from the horizons 0, 25, 50, 60, 85, and 100 m; the techniques for the filtration of phytoplankton, and extraction and separation of pigments were the same.

The chromatogram displayed two chlorophyll "a" zones, α-carotene, lutein, violaxanthin, neoxanthin (Table 2). As with the preceding station, the concentrations of chlorophyll "a" and α-carotene increase appreciably with depth, reaching a maximum at 100 m. Lutein has two maxima, on the surface and at 50 m. Its concentration drops at the 85-m horizon, increasing again at 100 m. The concentration of violaxanthin minimal on the surface, becomes 2.6 times higher at 25 m, then remains stable down to 85 m, and again increases by a factor of 8.5 at 100 m. The highest concentration of neoxanthin occurs at 25 and 100 m. This station lacked the distinct upward trend of the concentrations of individual pigments and their ratios with increasing depth that had been observed at the preceding station.

Station 6065 (Kieta Bay), February 15, 1969, 1140 hr. The samples were obtained from the depths 0, 12, and 40 m. In contrast to the preceding stations, an unidentified yellow pigment, with high adsorption in the spectrum region 420 — 500 nm without characteristic maxima, occurs at this station in fairly large amounts at all depths. Its spectrum has a plateau extending over 15 — 20 nm in the region 440 — 460 nm.

Chromatography showed this unidentified pigment to be closely associated with the lower chlorophyll "a" zone, and the solvent mixtures used in the preceding analyses proved inadequate for their separation.

After some preliminary experimentation, a mixture of 1.1% *n*-propanol in *n*-hexane was used. Chromatography in this mixture permitted distinct separation of the zones of these pigments. Apart from the unidentified yellow pigment, the other pigments were the same as in the preceding stations (Table 3).

At this station an attempt was made to establish the composition and concentration of pigments that do not lend themselves to chromatography and remain at the start. Spectrophotometric analysis showed that chlorophyllide "a" remained at the start. Moreover, the somewhat diffuse absorption band in the blue region of the spectrum suggests the presence of irreversibly altered carotenoid pigments.

It is of interest to note that one of the maxima (615 nm), the nearest to the main red peak of chlorophyll "a", becomes extended in the direction of the longer wave lengths up to 635 nm in chlorophyllide solution, suggesting the presence of chlorophyll "c" or its analogue. Possibly our chromatographic technique does not permit isolation of chlorophyll "c" from the overall mass of pigments.

The maximum of concentrations of chlorophyll "a," lutein, unidentified pigment, and total xanthophylls was found at 12 m.

The carotene concentration was fairly high at this station. Violaxanthin concentration was somewhat higher than that of lutein. An important contribution to total xanthophylls belonged to the unidentified xanthophyll. None of the analyzed pigments except for chlorophyll "a" displayed marked variations of concentration at different horizons.

The concentration of chlorophyllide "a" increases with depth. The chlorophyll "a": chlorophyllide "a" ratio was 5.1 on the surface but less than unity at 40 m. Therefore, the chlorophyll "a" concentration determined by the standard method consists of the sum total of concentrations of the true and modified chlorophylls, the latter's concentration increasing with depth much more rapidly than the concentration of unmodified chloropyll.

In addition to the principal stations, a qualitative analysis of the pigments in the phytoplankton was performed for the surface water layer in the vicinity of Heron Island.

Chromatographic determinations were carried out in two solvent systems, 1) 2% acetone in *n*-hexane and 2) 15% acetone in *n*-hexane + 0.5% *n*-propanol.

The following pigments were isolated: carotene, zeaxanthin, chlorophyll "a", and lutein; zeaxanthin was quantitatively dominant. We also analyzed pigments isolated from zooxanthellae from mantles of Tridacna. Chromatography revealed the presence of carotene, chlorophyll "a", some violaxanthin, and a large amount of unidentified xanthophyll that was also encountered at station 6065.

Bibliography

Goodwin, T. W. Chemistry and Biochemistry of Plant Pigments. London — New York, Acad. Press. 1965.

Richards, P. A. and T. G. Thompson. The Estimation and Characterization of Plankton Populations by Pigment Analysis. II. A Spectrophotometric Method for the Estimation of Plankton Pigments. — J. Mar. Res., Vol. 11, No. 2. 1952.

QUANTITATIVE EVALUATION OF THE ROLE OF BACTERIOPLANKTON IN THE BIOLOGICAL PRODUCTIVITY OF TROPICAL PACIFIC WATERS

Yu. I. Sorokin

Institute of Biology of Inland Waters of the Academy of Sciences of the USSR

INTRODUCTION

Synthesis of the biomass of bacteria is one of the major large-scale processes of transformation of organic matter in the aquatic environment. Its importance in the formation of food resources for the animal population of fresh waters has been demonstrated by many investigators (Gorbunov, 1946; Rodina, 1951; Kuznetsov, 1952; Sorokin, 1964; Romanenko, 1965). On the other hand, the trophic role of the microflora in seas and oceans has been little studied (Jørgensen, 1962; Wood, 1967; Vinogradov, 1968). The observation made by some authors that phytoplankton production in the sea is often patently inadequate for the nutrition of zooplankton speaks in favor of the participation of microflora in the food chain (Riley, 1956). Barber 1966). Various aggregated forms of organic matter including bacteria have been suggested as a supplementary source of food for the zooplankton (Sieburth, 1968). Bacteria were shown to be capable of utilizing dissolved organic matter for biosynthesis even at very low concentrations (ZoBell and Grant, 1943; Parsons and Strickland, 1961). But these investigations did not provide any quantitative estimates of the alimentary role of natural microflora in communities of marine organisms. Other factors hindering recognition of the trophic role of marine microflora in plankton communities are the sparsity of the populations of marine microorganisms and the prevalence in the pelagic waters of seas and oceans of coarse filter feeders for which bacteria of $1-2\,\mu$ are supposedly inaccessible.

In order to clarify the part played by microflora in the food chains of marine communities it was necessary to obtain the following marine microflora characteristics: a) the biomass and production of bacterioplankton in the water column; b) the functional activity of natural bacterial populations in the water column; c) the concentrations of the major sources of organic substances for bacterial biomass production; d) the degree of aggregation of bacterioplankton; e) the capacity of different species of aquatic invertebrates for consuming and assimilating bacterioplankton at its natural concentration.

In the course of experimental investigations during the 44th cruise of the "Vityaz", multidiurnal polygons yielded some information on the above points. A considerable part of this material was new for the ocean.

The characteristics obtained certainly need to be checked and their accuracy improved, and similarly the conclusions derived from them; the major importance of the latter appears to be that they pose new problems for further research.

The methods applied to these investigations were developed for studies of fresh water microflora and adapted to marine work in the 43rd cruise of the "Vityaz" and in investigations carried out by the author in 1967 in the Karadag Branch of the Institute of Biology of Southern Seas of the Academy of Sciences of the Ukrainian SSR in collaboration with T. S. Petipa and E. V. Pavlova.

METHODS OF INVESTIGATION

Sampling

The samples for determination of total abundance, activity, and production of bacterioplankton were taken with plastic samplers with a capacity of 4 — 7 liters. The samplers were thoroughly rinsed with hydrochloric acid (1:3) before casting. A special check carried out during the 43rd cruise showed that samplers thus sterilized were practically uncontaminated with microflora.

Life activity of natural populations of heterotrophic microflora

The life activity of the bacterioplankton was measured using a radiocarbon tag. The samples were taken from 10 to 15 horizons in the water column (two samples per horizon) in 300-ml flasks sterilized with HCl (1:4) mixed with Lugol's solution (iodine in potassium iodide solution). In filling, the flasks were washed clean of the sterilizing solution by six rinsings with the water to be tested.

A solution of algal protein hydrolysate tagged with C^{14} was added to a series of samples. A portion of the added solution contained about 30 μg organic carbon with activity about $1 \cdot 10^5$ pulses/min. After exposure of 10 — 12 hours in the dark at 27 — 29°, 50 — 100 ml water from the samples were filtered through No. 3 membrane filters. The filters were washed free of residual tagged organic matter by additional filtration of 50 ml seawater and dried. This was followed by determination of the radioactivity acquired by bacteria by assimilation of the tagged organic matter. The experiment was suspended at the beginning of the logarithmic growth phase, which occurred 8 — 10 hours after addition of the tagged organic matter, and therefore the radioactivity of the bacterioplankton was proportional to the abundance and activity of the initial population. The magnitude of radioactivity (% of radioactivity of bacterioplankton in the surface sample) served as the criterion of the relative functional activity of natural bacterioplankton (Sorokin, 1969).

Total abundance and biomass of bacterioplankton

The total number of bacteria was measured on VUFS membrane filters (pore diameter $0.1-0.3\,\mu$, filter area $0.9\ cm^2$) by direct microscopic examination. The portion of water used for filtration was usually 20 – 40 ml. The filters were stained with carbolic erythrosin and examined under an oil immersion microscope. The biomass of bacteria (B) was calculated by multiplying the total bacterial count (N) by the average volume of bacterial cells (V). The volume was measured on stained membrane filters, introducing the correction factor 2.5 for converting the volume of dried bacterial cells to that of live cells (Troitskii and Sorokin, 1967). All biomasses are stated in wet weight (mg/m^3).

Production of bacterioplankton

Production of bacteria was measured by the radiocarbon method from heterotrophic assimilation of CO_2 (Romanenko, 1964; Sorokin, 1971). Determinations were made of the total dark CO_2 assimilation by the microflora (Sorokin, 1964), assuming the virtual absence of chemosynthesis in the ocean waters.

Solution of a radioactive carbonate with known radioactivity $8-12\cdot10^6$ pulses/min filtered through a No. 2 membrane filter was added to the water samples in the dark. The flasks were exposed in the dark at a temperature close to the natural. The time of exposure was 1 – 1.5 days for samples obtained from depths of 0 – 200 m, 2 – 3 days for those obtained from 200 – 800 m and up to 4 days for those obtained from below 800 m. The magnitude of potential production was determined by exposing the flasks for 4 – 5 days at an even temperature (28 – 30°). The index of potential productivity of bacteria was used to characterize the reserve of organic matter in the water that was available to the bacteria.

Calculation of destruction according to assimilation of CO_2*

An empirical ratio of CO_2 assimilation by the microflora and oxygen demand due to bacterial destruction for marine bacterioplankton, determined in special experiments, was found to be 150 on the average (Sorokin, 1971). By using this ratio, the experimentally determined bacterial assimilation of CO_2 (A, $mg\ C/m^3$) could be used for bacterial destruction (D) calculations: $D = 150\cdot A\ mgO_2/m^3\cdot day$.

Aggregation of natural bacterioplankton

The percentage of bacteria in aggregates was calculated by dividing the tagged natural bacterioplankton by filtration. The bacterioplankton was

* According to Romanenko (1965).

tagged by adding C^{14}-tagged dissolved organic matter or a carbonate to seawater samples filtered through No. 76 plankton bolting silk. The amount of dissolved matter added was very small, about 20 $\mu g/l$. By assimilating this organic matter, the bacteria acquired the tag with practically no effect on the bacterial count. In another case, in water samples with a tagged carbonate exposed in the dark for 1 — 2 days the bacteria acquired the tag by assimilation of C^{14} in the course of heterotrophic assimilation of CO_2.

The degree of aggregation was measured by filtering one part of a water sample with tagged bacterioplankton through a membrane filter No. 3 with pores of 0.4 μ and measuring the total radioactivity (R_1), while another part of the same sample was filtered through a No. 6 membrane filter with pores of 4 μ, for measuring the radioactivity of aggregated bacterioplankton (R_2). The ratio $R_2 : R_1$ expressed the fraction of aggregated bacterioplankton.

Microflora as food for aquatic animals

The capacity of aquatic invertebrates for feeding on bacterioplankton was determined by the assimilation index C_{assim}/C used as the criterion of feeding intensity and efficiency (Sorokin, 1966).

Primary photosynthetic production of phytoplankton

Primary production was calculated by the radiocarbon method (Sorokin, 1959). Aimed samples were taken for determination of the distribution coefficients K_d with a bioluminescence transducer. The coefficients K_t were determined by using previously published curves for similar regions in tropical waters (Sorokin, 1959). For more complete measurements of the total primary production, we also determined the loss by phytoplankton in a separate sample of dissolved photosynthetic assimilates in the course of a flask experiment. For this purpose, a solution of a C^{14} carbonate with radioactivity $3-4 \cdot 10^7$ pulses/min was added to water samples in 300-ml flasks. Some of the samples were kept in the dark. The samples were exposed in the sea at depths of 2 m for a daylight period and filtered through a No. 5 membrane filter; the filter was processed with 1% HCl solution and measurements were made of the radioactivity of CO_2 assimilated in the course of photosynthesis. The filtrate received an addition of 5 ml of 10% soda solution, acidified with 10% HCl, and heated to boiling, followed by another addition of soda, acidulation, and another boiling for complete removal of tagged CO_2. The filtrate was neutralized with 5% KOH, received an addition of 5 ml of 5% KOH + 0.1 ml of 5% soda solution + traces of $AgSO_4$ + 2 ml 0.1 $KMnO_4$ solution, and boiled for 15 min in a small flask, after which the flask was connected to the device for distilling off CO_2 (Sorokin, 1966). One milliliter H_2O_2 and 5 ml 10 H_2SO_4 were added to the flask. CO_2 produced by oxidation with permanganate of dissolved organic matter liberated by the algae was driven off by the boiling. It was absorbed in alkali, precipitated as $BaCO_3$, and filtered on a membrane filter for determination of its radioactivity. The radioactivity (corrected for self-absorption of radiation within the precipitate) was multiplied by 1.25, this being the empirical correction for incomplete oxidation of the dissolved organic matter by the permanganate.

Consumption of cobalt by the bacterioplankton in the water

Consumption of cobalt by the bacterioplankton was measured by an isotope technique with Co^{57}; 0.5-ml portions of the working isotope solution (radioactivity $2.5 \cdot 10^5$ pulses/ml, cobalt content 2 μg/ml) were added to the flasks containing the water samples. There were three experimental variants, a light flask, a dark flask, and a formalin control. Parallel determinations of bacterial production in the same water were performed by the radiocarbon method. The samples were exposed for a daylight period, and filtered through membrane filter No. 3; the bacterioplankton was washed free of residual dissolved isotopes with an acid solution of $CoCl_2$ and measurements were made of the radioactivity due to Co^{57} absorbed by the bacteria in the course of the experiment (r). The absolute amount of cobalt assimilated by the bacteria (K) was calculated by multiplying r by A (reciprocal specific radioactivity of cobalt in the working solution, μg/pulse). The average cobalt concentration in seawater being $^1/_{20}$ its concentration in the working solution added to the sample, its specific activity remained practically unaffected upon addition of this solution to the seawater sample.

RESULTS

Activity, biomass and production of bacterioplankton in the open ocean

Data on the quantitative characteristics of bacterioplankton in the water column were obtained at three stations: 6033, 6052, and 6064 (Figure 1). Aimed samples for hydrological and hydrochemical analyses were taken at stations 6052 and 6064 (Vinogradov, Gitel'zon, and Sorokin, 1971).

Analysis of the aimed samples showed that luminescent zooplankton accumulates in the layers with the highest concentration of food (algae and bacteria). These layers are rather thin and as a rule remain unnoticed in sampling from standard horizons (Figure 2, Table 1). The bacterioplankton biomass in the layers of high concentration rises to 100 — 300 mg/m^3 as against its average biomass 15 — 30 mg/m^3 in the surface layer of oligotrophic tropical waters. Such values are characteristic of mesotrophic shelf waters (Sorokin, 1964). The relative activity of microflora in the layer of maximum luminescence was several times higher than in the adjacent horizons and bacterial production was 3 — 4 times higher than the average diurnal photosynthetic production in the surface layer (see also Vinogradov, Gitel'zon, and Sorokin, 1971).

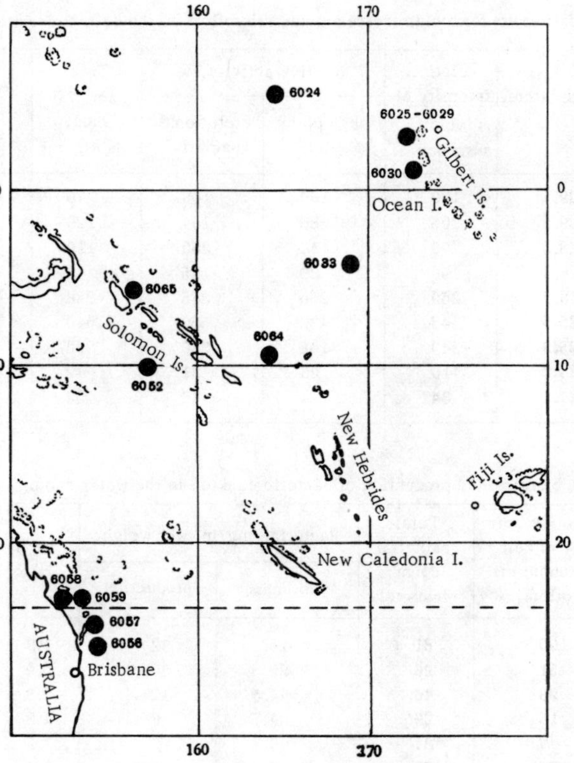

FIGURE 1. Location of the stations.

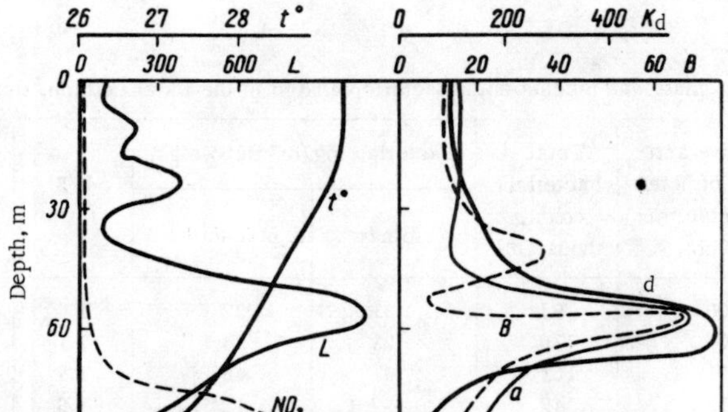

FIGURE 2. Analytical results for aimed samples, station 6064:

$t°$ — temperature; L — relative intensity of bioluminescence, %; NO_3 — nitrate nitrogen, μg-at/l; K_d — relative distribution of active phytoplankton, %; a — relative activity of bacterioplankton, %; B — bacterial biomass, mg/m^3.

TABLE 1. Analytical results for bioluminescence and microflora in aimed samples, station 6052, January 22, 1969

Depth, m	Temperature, °C	Relative intensity of biolumi-nescence, %	Relative activity, %		Total bacterial count, thous. /ml	Bacterial biomass, mg/m^3	Bacterial production, mg/m^3
			phytoplank-ton	heterotrophic bacteria			
5	29.5	100	100	100	76	23	26
36	29.5	205	86	194	126	38	36
50	28.5	270	114	360	110	32	46
70	26.4	190	39	115	43	13	42
78	26.2	250	250	345	286	86	44
84	25.5	545	432	550	1040	380	72
88	25.3	340	158	286	290	88	54
111	24.2	110	23	31	42	12	12
114	24.1	34	14	46	57	18	14

TABLE 2. Activity, biomass, and production of bacterioplankton in the water column, station 6033

Depth, m	Relative activity of heterotrophic microflora, %	Total bacterial count, thous. ml	Bacteria, mg/m^3 wet weight		P/B ratio	Bacterial destruction, mg $O_2/m^3 \cdot$ day
			biomass	production		
10	100	81	25.	32	1.3	24.0
150	51	96	29	30	0.9	22.5
220	96	46	14.5	12	2.8	9.0
300	15	29	4.7	6	1.3	4.5
400	7	37	6.7	9	1.3	6.9
500	82	72	17.4	17	1.0	12.8
600	3	19	3.4	1.6	0.5	1.2
750	3	23	2.2	0.16	0.07	0.12
1000	1	5.2	0.4	0.02	0.2	0.06
2000	2	3.1	0.2	0.04	0.2	0.03
3000	1	6.0	0.4	0.06	0.15	0.04

TABLE 3. Activity, biomass, and production of bacterioplankton in the water column, station 6052

Depth, m	Relative activity of heterotrophic microflora, %	Total bacterial count, thous. /ml	Bacteria, mg/m^3 wet weight		P/B ratio	Bacterial destruction, mg $O_2/m^3 \cdot$ day
			biomass	production		
0	100	95	31	32	1.0	24.0
40	106	110	33	34	1.1	25.6
70	71	151	41	46	1.7	33.5
150	15	42	7.6	12	1.6	9.3
220	35	34	15.4	22	1.4	17.3
300	38	33	6.1	7	1.1	5.0
400	270	61	10.6	5	0.5	3.9
500	193	123	26.0	33	1.3	24.8
750	11	18	4.5	3	0.7	2.1
1000	6	8	1.6	0.8	0.5	0.6
1500	9	5	1.0	0.5	0.5	0.4
2000	17	12	2.4	0.3	0.1	0.2

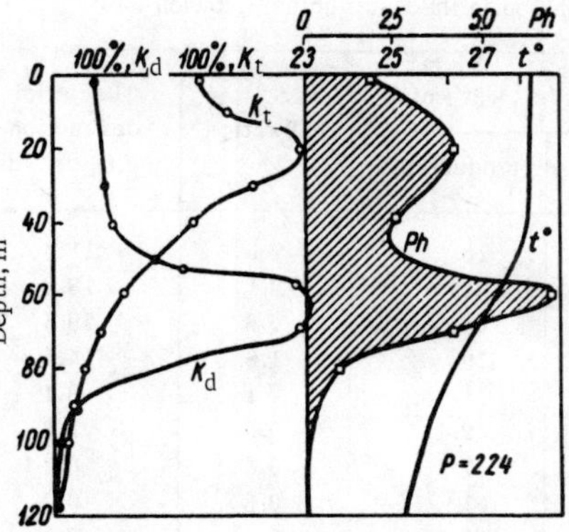

FIGURE 3. Characteristics of photosynthesis in the water column, station 6052:

K_d — relative distribution of active phytoplankton; K_t — vertical variation of photosynthesis intensity as a function of underwater illuminance, %; *Ph* — intensity of photosynthesis in the water column, mg C/m³ · day; *P* — diurnal phytoplankton production, mg C/m².

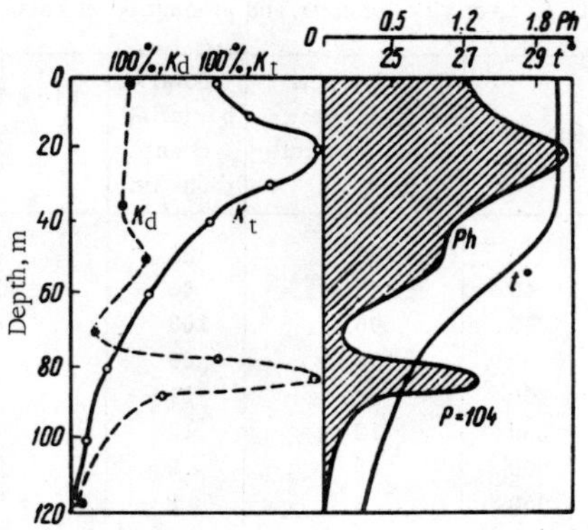

FIGURE 4. Characteristics of photosynthesis in the water column, station 6064.

Notation as in Figures 2 and 3.

The presence of a high concentration of phytoplankton at 70 – 80 m leads to a considerable intensity of the photosynthesis in this layer, although it is reached by only a small fraction (5 – 10°) of physiologically active solar radiation (Figures 3 and 4). Primary production at stations 6052 and 6064 was 104 and 224 mg C/m² · day, which is usual for tropical waters in the equatorial currents.

Minimum concentrations of phytoplankton and bacterioplankton occur directly above and below the main maximum. The concentration and activity of bacterioplankton at 90 – 100 m, i.e. 10 – 15 m below the main maximum, were so low that they were sometimes no higher than the corresponding values at 600 – 700 m. The data on abundance, activity, and production of bacterioplankton in the pelagic zone (Tables 2 – 4) reveal the presence of two main maxima of these quantities within the main thermocline, both located within the upper boundaries of water masses characterized by an inflection in the TS-curve (Arsen'ev et al., 1971). The upper, very narrow maximum, detected with aimed samples, is situated in the layer with the maximum phytoplankton, which develops here on account of the biogens brought by turbulent exchange from underlying layers (Sorokin, 1959; Vinogradov, Gitel'zon, and Sorokin, 1971). The lower, considerably more extended maximum is situated in the lower part of the main thermocline at depths of 400 – 500 m. The activity, biomass, and production of bacterioplankton in the layer of the second maximum are commensurable with those in the surface layer.

The abundance and the activity of the bacterioplankton decline sharply below the second maximum at 700 – 800 m. The abundance and the biomass of the bacteria throughout the column of deep waters are $^1/_{20}$ – $^1/_{40}$ of those in the layers of the maximum, i.e. 2 – 6 thous./ml and 0.2 – 1 mg/m³.

TABLE 4. Activity, biomass, and production of bacterioplankton in the water column, station 6064

Depth, m	Relative activity of heterotrophic microflora, %	Total bacterial count, thous/ml	Bacteria, mg/m^3 wet weight		P/B ratio	Bacterial destruction, $mg\ O_2/m^3 \cdot day$
			biomass	production		
5	100	21	6	15	2.5	11.2
40	142	46	12	25	2.1	18.8
60	570	103	46	39	0.8	29.3
90	76	26	12	19	1.6	14.4
160	23	21	5.2	11	2.1	8.4
250	12	12	3.0	2.4	0.8	1.8
300	10	9	2.0	1.2	0.6	0.9
350	6	8	1.9	1	0.5	0.7
400	3	5	1.3	0.8	0.6	0.6
460	7	14	5.1	5.4	1.1	4.1
550	21	31	14	7.4	0.5	5.5
600	14	13	4.7	0.2	0.04	0.15
730	2	5	1.0	0.15	0.15	0.1
900	1	5	1.0	0.2	0.2	0.15
1300	1	4	0.8	0.16	0.2	0.12
1800	0.4	2	0.4	0.08	0.2	0.06
2700	0.2	2	0.4	0.1	0.2	0.15

Diurnal bacterial production under 1 m^2 ocean surface down to 1,000 m was highest at station 6052 in the Solomon Sea near the islands, about 1,300 $mg\ C/m^2 \cdot day$. Bacterioplankton production at that station was over 10 times the plankton production. At station 6064, bacterial production under 1 m^2 was only one-third and primary production was three times that at station 6052, so that the bacterial production was 1.6 times the phytoplankton production (Table 5). Knowing that "assimilation of CO_2:consumption of O_2" = 150, it is possible to calculate the diurnal oxygen demand in bacterial destruction (see Tables 2 – 4). The position of its maximum, 70 – 90 m, corresponded to the maximum bioluminescence and phytoplankton concentration. However, the diurnal oxygen demand even in this layer was only 20 – 30 mg O_2/m^3, i. e. it was at the sensitivity limit of oxygen determination according to Winkler. The diurnal oxygen demand in surface tropical waters outside the maximum layer was 8 – 10 mg O_2/m^3. The second oxygen demand maximum occurs at depths of 450 – 600 m, where it reaches 12 – 17 mg $O_2/m^3 \cdot day$. Oxygen demand in deep waters is extremely small, 0.1 – 0.2 mg $O_2/m^3 \cdot day$. With such a low demand, the reserve of oxygen in the deep waters should suffice for 50 – 100 years.

The P/B ratio of bacterioplankton is usually 1 – 2 in the main thermocline layer, 0.2 – 0.4 in deep waters. The mean P/B ratio for the 0 – 1,000-m layer is 1 – 1.2. The production and biomass of bacterioplankton in the top 0 – 600-m layer account for more than 90% of their total through the entire water column.

Activity and production of bacterioplankton in coastal waters and lagoons

The level of primary production in lagoon and shelf waters is comparatively high (Doty and Oguri, 1956), and moreover, these waters receive a

TABLE 5. Biomass (B) and diurnal production (P) of bacterioplankton, diurnal bacterial destruction (D), and primary phytoplankton production (Ph) in the water column, g C/m^3

Location of stations	Station	Total depth, m	Thickness of layer under calculation, m	B	P	P/B	D	D, % of total content of organic matter in the water	Ph	Production of primary food, PP*	$\frac{P}{Ph}$	$\frac{D}{Ph}$
Lagoon	6029	24	22	1.49	0.41	0.3	1.15	1.60	0.15	0.52	2.7	7.7
Shelf	6058	70	70	1.20	0.81	0.7	2.27	1.62	0.62	1.27	1.3	3.7
	6056	210	200	0.85	0.56	0.7	1.57	0.39	0.35	0.82	1.6	4.5
	6028	670	300	1.38	0.93	0.7	2.60	0.43	0.24	1.12	3.9	11.0
Neritic zone	6057	2900	300	0.53	0.58	1.1	1.63	0.27	0.12	0.67	4.8	13.6
Pelagic waters	6052	3500	1000	1.31	1.34	1.0	3.85	0.25	0.10	1.42	13.4	38.5
	6064	4500	1000	0.395	0.48	1.2	1.34	0.09	0.29	0.70	1.6	4.6
	6033	5000	1000	0.98	0.99	1.0	2.80	0.19	0.20**	1.14	5.0	14.0

* Production of primary food was calculated as the sum of primary bacterial production from "external" organic matter: $PF = Ph + P - 0.25\ Ph = P + 0.75\ Ph$ (0.25 Ph is the bacterial production due to consumption of primary phytoplankton production).

** This is the mean production for the region (Koblents-Mishke, 1967).

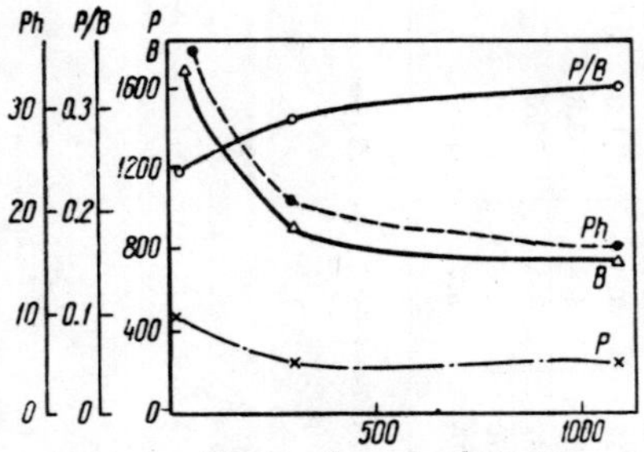

FIGURE 5. Primary production (Ph, mg $C/m^3 \cdot day$), bacterial biomass (*B*, mg/m^3), bacterial production (*P*, mg/m^3), and the *P/B* ratio in surface water samples over a section extending from the shore to the center of the lagoon of Butaritari Atoll

large influx of organic matter from land and coral reefs (Marshall, 1965, 1968). Quantitative data on bacterioplankton in coastal tropical waters are almost totally lacking. In the 44th cruise of the "Vityaz" special attention was paid to studies of bacterioplankton in lagoons and on the shelf.

In the lagoon of Butaritari Atoll (Gilbert Islands) the bacterioplankton biomass near the low, muddy shore opposite sparse mangrove thickets reached 1,600 mg/m^3 (Figure 5), nearly 100 times its biomass in the open ocean. Such a high bacterioplankton biomass is characteristic of eutrophic waters. The bacterial biomass was only one-half of that value, 800 mg/m^3, in the center of the lagoon. Bacterioplankton production was here commensurable with phytoplankton primary production, being equivalent to 20 — 40 mg $C/m^3 \cdot day$.

TABLE 6. Primary production and bacterioplankton in the surface layer off Heron Island

Distance from the shore, m	Primary production, mg $C/m^3 \cdot day$	Bacterioplankton			P/B ratio	Percentage of aggregates in the bacterioplankton
		count, thous./ml	biomass, mg/m^3, wet weight	production, mg/m^3		
5	21	980	680	236	0.35	14
100	67	365	410	202	0.49	17
300	50	610	670	158	0.24	23
600	69	1040	740	408	0.55	25

The P/B ratios in the lagoon water were $\frac{1}{2}$ — $\frac{1}{3}$ those in the surface waters of the open ocean. One obvious cause for the difference was the abundance of benthic microflora, washed out of the bottom sediments by waves and the tidal currents. The vertical distribution of bacterioplankton in the water column of the lagoon was fairly uniform. The bacterial count increased somewhat in the near-bottom layer, evidently due to transport of roiled sediments by the currents. The bacterioplankton biomass in the near-bottom layer was 1 g/m^3. Such a concentration of bacteria is sufficient for the normal nutrition of invertebrate filter feeders, both planktonic (see below) and benthic (Sorokin, 1966).

A similarly high level of bacterial production and biomass was found off Heron Island (Great Barrier Reef, Australia) in the center of the inner reef lagoon (Table 6). High bacterial production in this lagoon was ensured by phytoplankton productivity, which was 30 — 60 times that in the open ocean, up to 70 mg $C/m^3 \cdot day$. Nevertheless, even in this lagoon bacterioplankton production was 30 — 50% of the primary production.

The effect of the influx of the highly productive lagoon and shelf waters on the productivity of the open waters is of considerable interest. Sections were made from the shallows toward the open ocean and several stations were executed on the shelf of the east coast of Australia.

On the section extending from the lee side of the ring reef to Butaritari Atoll, four vertical profiles were executed with determinations of the quantitative characteristics of bacterioplankton at the four horizons, 10, 60—80 (beginning of the thermocline), 200, and 300 m. The atoll has very steep slopes and oceanic depths begin 3—4 miles from the outer reef.

The analyses showed that the productivity of waters even above depths of 1,500—4,000 m in direct proximity to the large atoll with its highly productive lagoon water was as low as in the open ocean (Table 7). A bacterioplankton maximum occurs at the upper boundary of the thermocline, the bacterial count and biomass decreasing abruptly below this level. The influence exerted by the atoll is confined to a distance of 1.5 miles over depths 600—1,500 m. It manifests itself in a higher bacterial biomass and production, first at 200—300 m and then also in the surface layer closer to the atoll.

TABLE 7. Quantitative characteristics of bacterioplankton in a series of vertical profiles near Butaritari Atoll

Station	Depth, m	Primary production, mg C/m^3·day	Sampling depth, m	Bacteria			P/B ratio	Bacterial destruction, mg O_2/m^3·day
				total count, thous. /ml	biomass, mg/m^3	production, mg/m^3		
6025	4000	0.43	10	88	26	38	1.4	28.5
			78	136	41	52	1.3	39.0
			200	21	6.2	13	2.1	10.2
			300	15	4.5	3.2	0.7	2.4
6026	1900	0.41	10	102	33	52	1.6	39.0
			76	240	72	94	1.3	70.1
			200	28	8.3	16	1.9	12.3
			300	16	4.8	4.8	1.0	3.6
6027	1500	0.60	10	164	48	64	1.3	51.0
			56	181	54	46	0.8	34.5
			200	129	39	38	1.0	28.5
			300	151	45	56	1.2	42.0
6028	670	5.7	10	221	100	73	0.7	110.0
			56	115	52	31	0.60	46.6
			200	81	36	26	0.7	38.0
			300	94	42	26	0.6	38.0

The data obtained on the Australian shelf opposite the Great Barrier reef (stations 6056—6058) showed average bacterioplankton biomass and production to be here 4—6 times that in the euphotic zone of the open ocean. The water column in this area was markedly stratified, as in the open ocean. A layer in which the concentration and functional activity of phytoplankton and bacterioplankton are 3—6 times those in the adjacent layers is consistently present in the upper part of the thermocline (Figure 6). In the immediate vicinity of the Great Barrier Reef above depths of 70 m (station 6058) the maximum bacterial biomass and production occurs in the near-bottom layer, in which detritus washed out of the reef apparently settles (Figure 7). The P/B ratio for the water column on the shelf is lower than in the open ocean, probably because these waters are reached by part of the microflora washed off the land and from bottom sediments in shallow waters, and this microflora has a low activity in the oligotrophic oceanic waters. Bacterial production under 1 m^2 in the water column is 2 — 4 times the primary production. Average bacterial destruction is 0.04 — 0.06 mg O_2/l · day in the upper water layers, 0.15 mg/l in the lagoon. With this level of destruction the existing reserves of organic matter should last approximately 250 — 300 days in the upper layers of shelf waters, and 40 — 60 days in the lagoon waters.

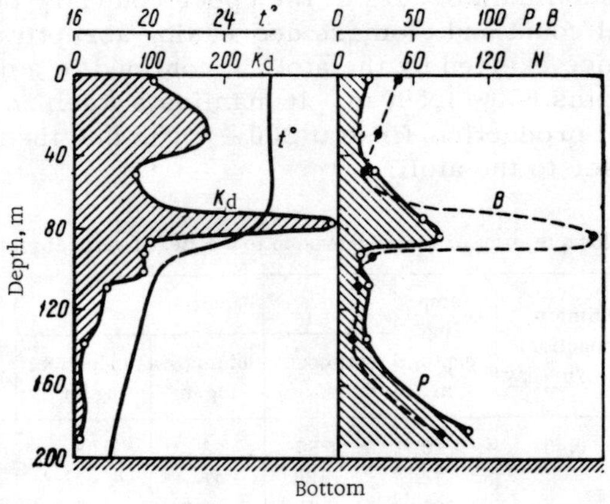

FIGURE 6. Vertical distribution of temperature ($t°$), active phytoplankton (K_d, %), bacterial biomass (B, mg/m^3), and bacterial production (P, mg/m^3) on the Australian shelf, station 6057, depth 2,900 m.

The zone of influence of the eutrophic waters of the atoll lagoons and shelf on the productivity of the surrounding oligotrophic oceanic waters is very limited (Figures 8 and 9). At a distance of 5 — 7 km productivity already drops to the level characteristic for the tropical waters in the open ocean, the primary production and biomass of the bacterioplankton decreasing by a factor of 10 — 20 and bacterioproduction by a factor of 5 — 10. Hence, the "insular effect" (Daty and Oguri, 1956) on the productivity of the waters around atolls with steep slopes is confined to a very narrow zone. The P/B ratio increases regularly in the waters of the open ocean.

This zone is somewhat wider at the Great Barrier Reef, where the influence of the coastal waters and highly productive lagoon waters on the productivity of the surrounding oceanic waters extends for up to 12 — 20 km. The broader extent of the high production zone is probably related to gentler slopes (Figure 10), a stronger influence of land runoff, and the washing out of organic detritus from the extensive reef massifs (Figure 11).

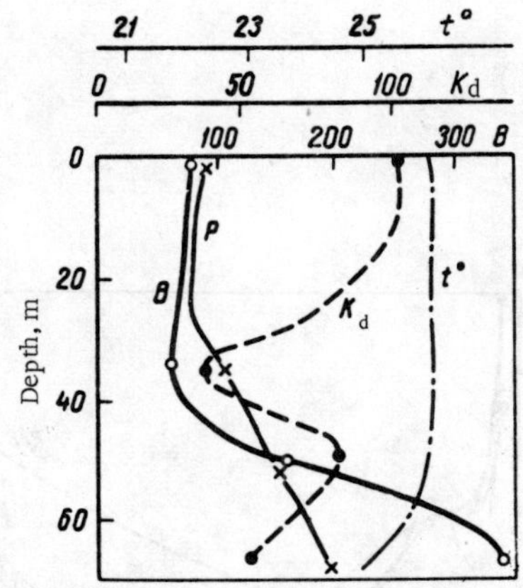

FIGURE 7. Vertical distribution of the same parameters as in Figure 6 at station 6058, depth 70 m.

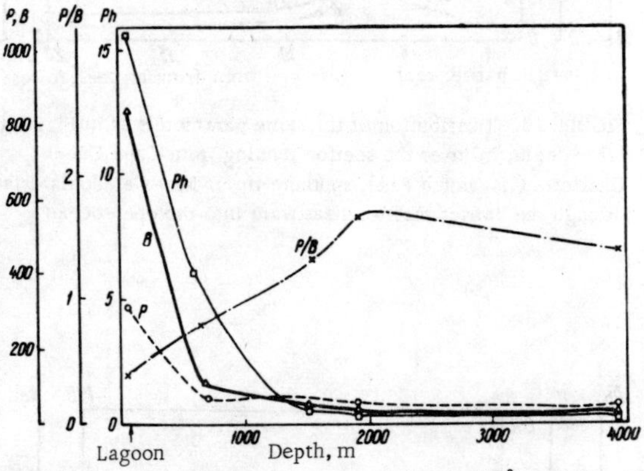

FIGURE 8. Distribution of the biomass (B, mg/m^3) and diurnal production (P, mg/m^3) of bacterioplankton, the P/B ratios, and primary production (Ph, mg C/m$^3 \cdot$ day) on the horizontal section from the center of the lagoon of Butaritari Atoll southwestward into the open ocean.

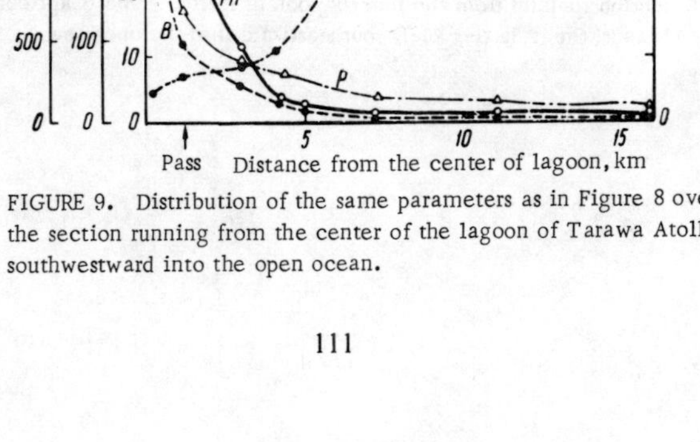

FIGURE 9. Distribution of the same parameters as in Figure 8 over the section running from the center of the lagoon of Tarawa Atoll southwestward into the open ocean.

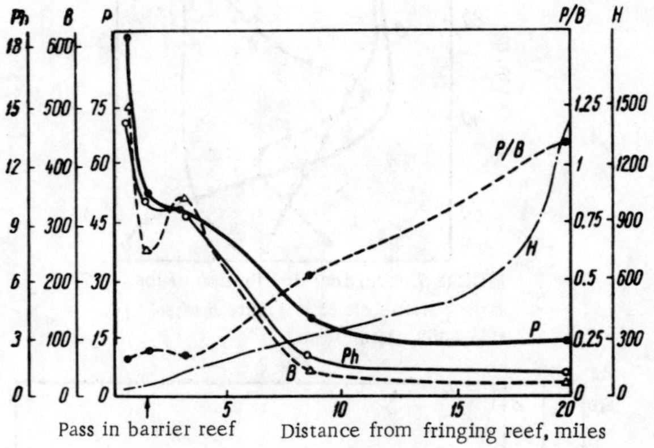

FIGURE 10. Distribution of the same parameters as in Figure 8 (H — depth, m) over the section running from Cape Queen Charlotte (Havannah Pass), southern tip of New Caledonia Island, through the barrier reef southeastward into the open ocean

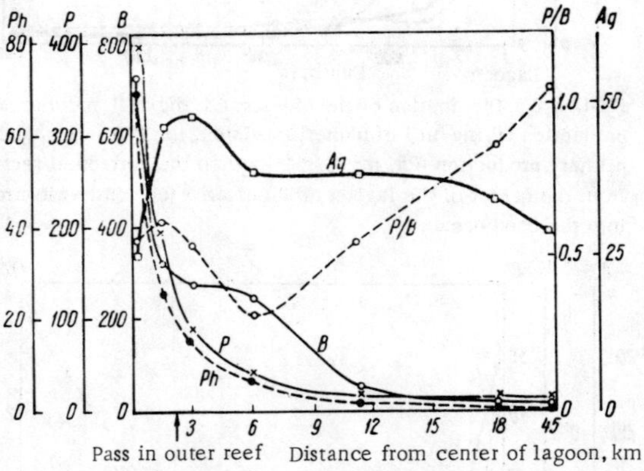

FIGURE 11. Distribution of the same parameters as in Figure 8 (Ag — percentage of aggregated bacterioplankton) over the section running from the inner lagoon of Heron Island (Capricorn Islands, Great Barrier Reef) southeastward into the open sea

Potential bacterioplankton productivity

The results of determinations of the potential productivity in the water column of the open ocean and distribution curves for actual diurnal bacterial production in flasks exposed in situ are shown in Figures 12 and 13. The potential bacterioplankton productivity in the surface layers is 3—4 times the actual production at the same temperature, in accordance with the more prolonged exposure during its determination. In the deep waters, where actual bacterial production is small, to the point of being nonexistent, the potential productivity of bacteria is not lower, and is often even higher, than in the surface water layer, although it may appear to be self-evident that the surface water layers should be richer in readily assimilable organic matter.

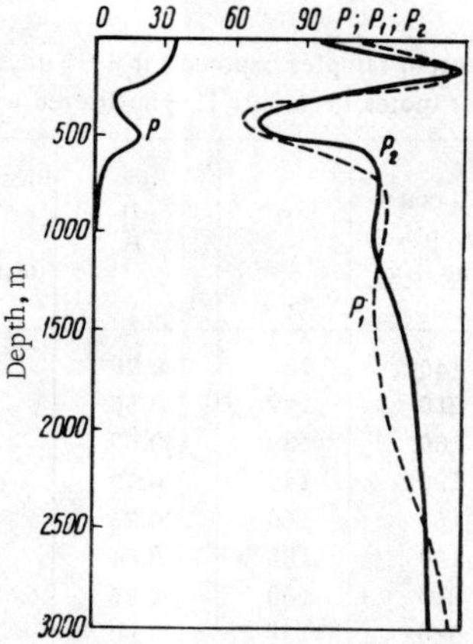

FIGURE 12. Potential bacterial productivity in samples taken from the water column in the open ocean, station 6033:

P — actual diurnal bacterial production, mg/m^3; P_1 — potential bacterial productivity in water filtered through a No. 5 membrane filter with pore diameter 1μ, mg/m^3; P_2, same in unfiltered water.

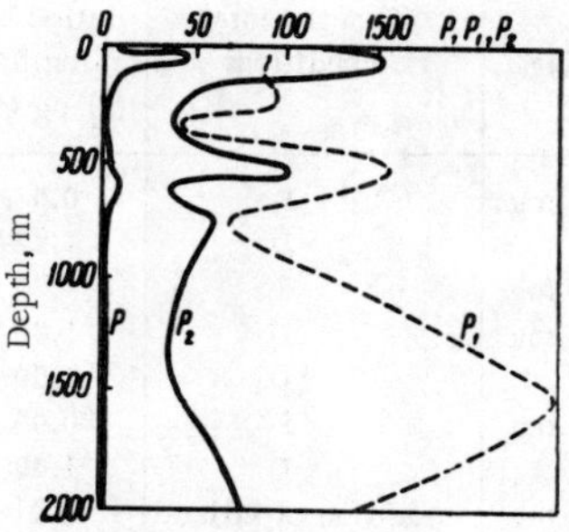

FIGURE 13. Potential bacterial productivity in water samples taken over vertical profiles at station 6052 (P_1) and station 6064 (P_2), mg/m^3:

P — actual diurnal bacterial production at station 6064.

It is important to note that the potential bacterial productivity is little affected by preliminary filtration of the samples through a membrane filter which removes particles coarser than 1μ. This means that the chief source of readily assimilable organic matter in the water column of the open ocean is dissolved organic matter, its mass exceeding many times that of suspended organic matter (Skopintsev, 1966). The potential bacterial productivity is appreciably reduced by filtration only in coastal waters, where the contribution of detritus to the total organic matter in seawater is larger. The difference disappears rapidly at increasing distances from the shore. The potential bacterioplankton productivity in deepwater samples is 120 — 150 mg/m³, according to comparatively brief experiments, corresponding to a bacterial destruction of approximately 90 mg O_2/m³, or 34 mg C/m³, so that the diurnal loss amounts to more than 2% of the total reserve of organic matter in seawater.

Biosynthetic efficiency of the bacterial biomass

The maximum radioactivity of bacteria which consumed tagged dissolved organic matter in the experiments (stated in percentages of the initial radioactivity of tagged organic matter added to the water) corresponds to the maximum efficiency of utilization of readily assimilable organic matter by aquatic microflora for biosynthesis. The efficiency determined was fairly high, 45 — 46% (Figure 14).

TABLE 8. Biosynthetic efficiency of bacterioplankton in seawater, in samples exposed for 3—5 days at 30°. F — water subjected to preliminary filtration through a No. 5 filter (pores 1—1.5μ); U — unfiltered water

Station and sampling depth	Experimental conditions	CO_2 assimilation by microflora (a), μg C/l	Bacterial production (P), μg/l, wet weight	O_2 consumption, μg/l	$\frac{D}{a}$	$\frac{D}{P}$	Biosynthetic efficiency $(K_1)\ \frac{P\cdot 100\%*}{D+P}$
6052, 65 mg	F	0.9	180	140	154	0.78	25
	U	1.22	244	210	172	0.86	24
6958, 60 mg	F	1.20	240	160	132	0.67	26
6064, 64 mg	U	1.30	260	190	146	0.73	26
	U	1.20	240	180	150	0.75	25
6065, 10 mg	F	0.95	190	160	168	0.84	24
	U	1.40	280	210	150	0.75	20
6064, 10 mg	Addition of killed	6.2	1240	980	158	0.80	25
	culture of algae	3.9	740	590	151	0.80	25

* P and D are stated in μg C/l; 10 mg wet weight = 1 mg C.

The biosynthetic efficiency of the bacterioplankton was determined by simultaneous measurements of bacterial production by heterotrophic assimilation (of tagged CO_2) and O_2 consumption in the same samples.

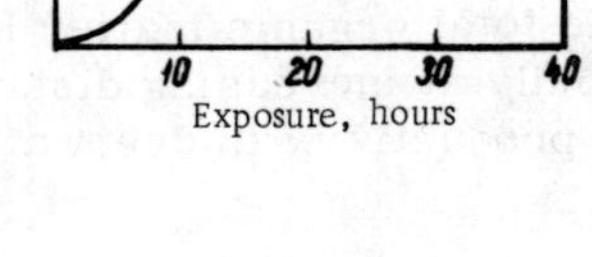

FIGURE 14. Efficiency of utilization of C^{14}-tagged protein hydrolysate by natural bacterioplankton of the surface layer of seawater.

R — radioactivity of the microflora, percent of initial activity of the hydrolysate added to the sample.

The experimental results (Table 8) confirm the previously demonstrated (Romanenko, 1965) high stability of the ratio "assimilated carbon dioxide/consumed oxygen." These findings validate the possibility of using the magnitude of this ratio (150 on the average) for measuring the intensity of destruction and BOD in seawater. The relative stability of the ratio "consumed oxygen/bacterial biomass production" (0.75 on the average) permits the use of this ratio for approximate calculations of bacterioplankton production from measurements of destruction and BOD.

The mean biosynthetic efficiency (Ivlev's coefficient K_1) for utilization of organic matter in seawater by natural bacterioplankton populations is 20—25%, below the value found for utilization of readily assimilable organic matter of algal protein hydrolysate.

Aggregation of marine bacterioplankton

The alimentary role of marine bacterioplankton necessarily depends upon the degree of its aggregation which controls its availability to the coarse filter feeders and predators which are predominant among the marine zooplankton. The degree of aggregation of natural bacterioplankton (percentage content of aggregates larger than 4—5μ) was determined in the open ocean, lagoons, and coastal waters of atolls. The last were especially important because of a report that the formation of aggregates was intensified in the vicinity of coasts and especially near coral reefs, which enrich the water with patches of slime and detritus (Marshall, 1968; Sieburth, 1965, 1968). The content of bacterial aggregates in the water column of the open ocean varies little with depth, increasing to some degree only in the lower part of the main thermocline (Figure 15). It averages 20—40% but increases tens of times in the vicinity of coral reefs, in the lagoons of atolls. However, the microflora incorporated in the aggregates constitutes at most 15—20% of the total amount of microorganisms (see Figure 11). In experiments with C^{14}-tagged dissolved organic matter it was demonstrated that the aggregates were formed in the process of bacterial growth, the presence of suspended particles assumed to be the nuclei of formation of aggregates not being a prerequisite. The formation of aggregates proceeds at the same speed in natural seawater with suspended matter and in water that has been filtered through a membrane filter No. 5, which retains not only suspended matter but also about 90% of the entire microflora (Figure 16). The effect of aggregating bacterioplankton on very dilute dissolved organic matter added to the matter is such that after only a few hours (10—15) some 15% of it is retained by the coarse preliminary membrane filter No. 6, which lets through even some of the small phytoplankton.

It is natural to assume that aggregates are available not only to fine but also to coarse filter feeders of the type of large calanids and euphausiids.

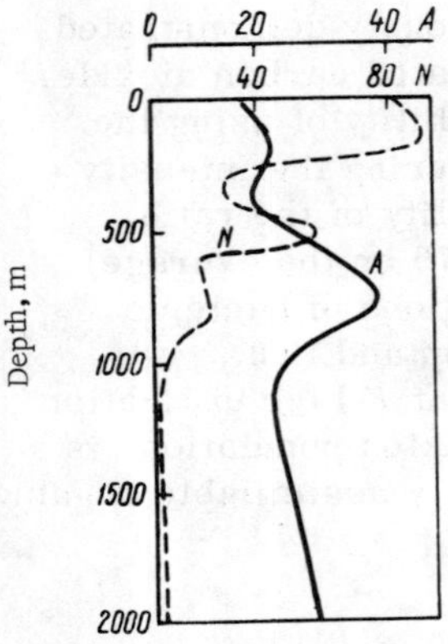

FIGURE 15. Relative content of bacterioplankton aggregates (A, %) in the water column at station 6033:

N — total bacterial count, thous. /ml. The value of A was determined by analyzing the assimilation of $C^{14}O_2$ by the bacterioplankton in experimentally exposed flasks at temperature close to that in situ.

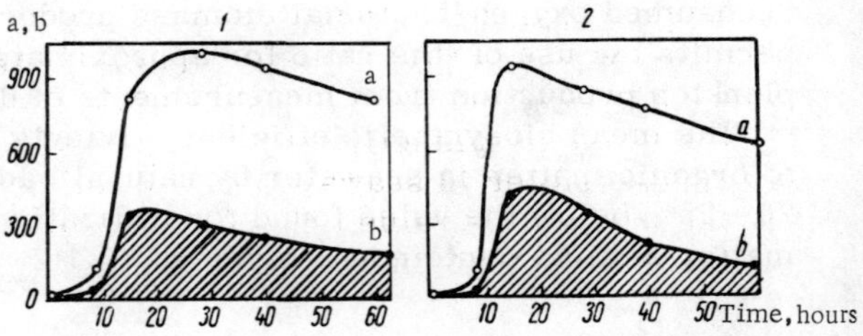

FIGURE 16. Increase of the total radioactivity of bacterioplankton (a) and that of aggregated bacteria (b), pulses/120 ml, in exposed samples of unfiltered surface water (1) and water filtered through a membrane filter No. 5 with pore diameter 1μ (2), after addition of C^{14}-tagged algal hydrolysate

Bacterioplankton as food for aquatic invertebrates

In order to obtain quantitative data of the role of bacterioplankton in the diet of planktonic filter-feeders, experiments were carried out with C^{14} making a comparison possible of the rate of nutrition of invertebrates on phytoplankton and on bacterioplankton. The investigations were carried out on common tropical copepods and some other animals (Petipa, Pavlova, and Sorokin, 1971).

As was to be expected, bacterioplankton was consumed the most vigorously by the fine filter feeders, namely appendicularians in the open ocean and water fleas (Penilia) in neritic waters. As far as water fleas are concerned, bacterioplankton is even better food than algae. At the optimum concentration (0.2 — 0.3 g/m³) the daily ration of Penilia reaches 50 — 100% of their weight and assimilability is 40 — 60%. Similar assimilability was observed when the water fleas fed on algae.

Coarse filter feeders — large copepods and euphausiids — also feed actively on the densely aggregated natural bacterioplankton. The intensity of their feeding on bacterioplankton approaches the intensity of their feeding on algae (Pavlova, Petipa, and Sorokin, 1971; Ponomareva, Tsikhon-Lukanina, and Sorokin, 1971). The optimum bacterioplankton concentration for the nutrition of Calanoida is very high, 2 — 3 g/m³ on the average, 2 — 3 times higher than the optimum concentration for their feeding on algae. The difference is readily explained by the fact that only the aggregated fraction of the bacterioplankton (20 — 25%) is available to the coarse filter feeders.

Removal of bacterial aggregates by filtering through a No. 6 filter is apt to reduce markedly the nutritive qualities of bacterioplankton. The fraction of bacterioplankton passing through the filter, consisting of individual cells and small aggregates, is not utilized by the coarse filter feeders even in the presence of a fairly high concentration of bacteria.

Experiments showed that even predaceous animals and larvae of fish are capable of consuming aggregated bacterioplankton, utilizing it as a supplementary source of food and of vitamins that are in short supply (B_{12}).

The assimilability of bacterial food consumed by the filter-feeding calanids is fairly high (40—60%), approaching that of algae.

Measurement of intravitam liberation of organic matter by phytoplankton

As already shown, bacterioplankton intensively and efficiently utilizes the dissolved organic matter in seawater even if the concentration of the latter is very low. We previously showed that even if the concentration of available organic matter is very low, 3—5 μ g/l, it is utilized by microflora with an efficiency (K_1) of up to 30%. In view of the reported considerable intravitam liberation of dissolved organic matter by phytoplankton (Fogg et al., 1965), it may be assumed that the life activities of marine bacterioplankton are supported not only by the organic matter of dead algae but to a considerable extent also by organic metabolites liberated by living phytoplankton. It was therefore important to measure the diurnal liberation of assimilates by phytoplankton in tropical waters, its characteristic feature being a high level of metabolism due to the high water temperature. Even in phytoplankton of temperate latitudes, the intravitam liberation of assimilated organic matter amounted to 10—50% of the total weight (Watt, 1966).

TABLE 9. Diurnal liberation of dissolved organic matter by phytoplankton in flasks exposed in the sea at 3 m

Sampling site	Exposure conditions	Radioactivity of algae on filters, pulses/l		Radioactivity of dissolved organic matter, pulses/m	
		total	in light flask minus the control (R_1)	total	in light flask minus the control (R_2)
Open ocean, station 6064	Light flask	14,220	12,310	5,240	3,380
		12,730		4,710	
	Dark flask	1,170	—	530	—
		1,060		1,250	
Kieta Bay (Bougainville Island)	Light flask	5,140	5,310	1,820	1,620
		6,370		1,990	
	Dark flask	420	—	300	—
		490		280	

The amount of liberated metabolites increases with increasing illuminance, which is extremely high in the tropics. Quantitative estimation of the liberation of photosynthetized organic matter from the cells into water also makes it possible to determine the error of standard measurements of photosynthetic production, which is especially important in the oligotrophic tropical waters. The results of these determinations (Table 9) show that about one-third of the daily assimilated amount of carbon from $C^{14}O_2$ is liberated by phytoplankton cells into the water and escapes primary production determinations by the radiocarbon method. The daily influx of assimilates into the zone of photosynthesis in tropical waters amounts to 50—100 mg/m^2 · day on the average. Their assimilation by microflora supports the latter's production of 100—300 mg/m^2 wet biomass in the euphotic zone.

Assimilation of cobalt from seawater by bacterioplankton for synthesis of vitamin B_{12}

Microorganisms are known to be capable of synthetizing the major vitamins. Consumption of bacterioplankton supplies aquatic animals not only with protein but also with essential vitamins. The marine phytoplankton also needs the vitamins produced by bacteria — biotin, thiamine, and vitamin B_{12} (Provasoli, 1963), especially the last, which is synthetized exclusively by bacteria (Table 10).

TABLE 10. Consumption of cobalt by bacterioplankton and phytoplankton in seawater (exposure 1 day; *L* — light flask; *D* — dark flask)

Sampling site	Analyzed substance	Reciprocal specific radioactivity of the substance (C_r)10^{-6} μg/pulse	Experimental conditions	Radioactivity of organisms on filters at end of experiment minus control (r)	Biogenic consumption of cobalt (K) and carbon (A), μg/l (K, A = r C_r)	Biosynthesis by bacterioplankton (P) μg/l dry weight, P = 40 · A	$\frac{K}{P}$, %	Ratio "vitamin B_{12} synthesis: P", % $\left(\frac{K \cdot 18 \cdot 100}{P}\right)$
Lagoon of Tarawa Atoll	Co^{57}	0.8	L	2,710	0.0022	—	—	—
			D	3,040	0.0024	—	—	—
	C^{14}	950	L	23,380	22.2	—	—	—
			D	740	0.70	28	0.0086	0.15
Brisbane Bay	Co^{57}	0.8	L	3,170	0.0025	—	—	—
			D	3,680	0.0029	—	—	—
	C^{14}	950	L	26,500	25.1	—	—	—
			D	810	0.77	31	0.0094	0.17

The rate of vitamin B_{12} synthesis by marine bacterioplankton was estimated in a series of experiments involving determination of the rate at which the bacterioplankton consumes cobalt, which is the chief characteristic component of the vitamin (Table 10). The experiments were carried out in the coastal zone, where bacterial production was fairly high. The experiments established a distinct inhibition of Co^{57} assimilation on addition of extraneous vitamin, thus confirming the consumption of Co^{57} by microflora and its predominant utilization for biosynthesis of vitamin B_{12} (Table 10, Figure 17).

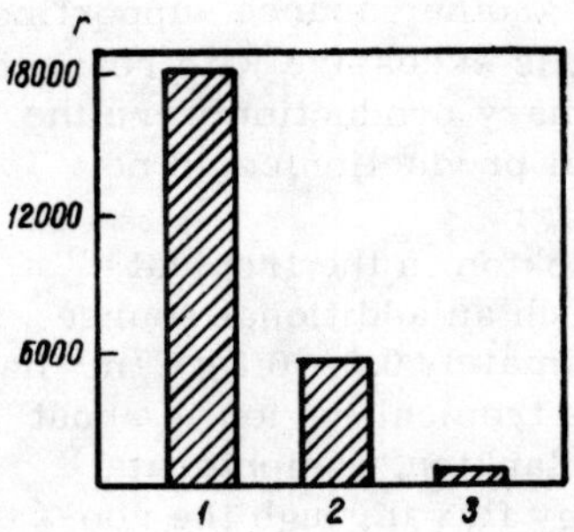

FIGURE 17. Effect of vitamin B_{12} addition on assimilation of cobalt (Co^{57}) by aquatic microflora; 20 mg/l glucose + Co^{57} with activity 40,000 pulses/200 ml added to the water. Exposure time 2 days:

1 — without addition; 2 — with addition of 10 μg/200 ml vitamin B_{12}; 3 — control (+ formalin); r — Co^{57} radioactivity in bacterial cells, pulses/200 ml.

Bacterioplankton assimilates about 0.1 μg Co per mg dry weight, corresponding to a production of about 1.7μg vitamin B_{12}. Hence the diurnal consumption of cobalt in the water column of the tropical regions of the ocean (in the 0—1,000-m layer) averages 0.1 mg/m^2 and the diurnal vitamin B_{12} production, 1 mg/m^2. This production of vitamin B_{12} is quite sufficient to satisfy the requirement of all the remaining animal and plant population of the water column. The vitamin B_{12} concentration in Pacific waters is approximately 1.5 μg/m^3, varying slightly with depth (Holm-Hansen et al., 1966). Its content in the 0 — 1,000-m layer is therefore about 1.5 mg/m^2 and consequently, the microflora ensures its diurnal replenishment.

Physicochemical sedimentation of cobalt is impeded because the element is present in ionic form in seawater and is poorly adsorbed by suspended particles. Bacterial flora, assimilating cobalt, incorporates it in vitamin B_{12} and draws it into the food chain, thereby facilitating its concentration in organic detritus. The latter settles and is transmitted along the ladder of food migrations, carrying the cobalt to the bottom.

This biological mechanism probably governs the supply of cobalt from seawater to bottom sediments.

DISCUSSION OF RESULTS

The above data show that a considerable biomass of bacterioplankton is produced in the upper layer of heated tropical waters. Bacterial production in the water column is often considerably higher than primary production per m^2 in areas with sinking surface waters in the West Pacific. Similar results were obtained during the 43rd cruise of the "Vityaz" (Sorokin, 1970). The relationship between phytoplankton and bacterial production in tropical waters clearly calls for further experimental research. New efforts must be made to eliminate methodological errors in determination of the primary production itself, as the results are usually much too low. The effect of isolation of the samples in flasks upon the rate of reproduction of the microflora in them should also be verified. The most recent evidence indicates that this effect is limited (Laevastu et al., 1965). It may be that after all the methodological errors have been taken into account, the ratios determined between the primary and bacterial productions will prove to be somewhat too high. Nevertheless, incorporation of the above-mentioned corrections will hardly change the conclusion concerning the predominance of bacterial over primary production. It follows that there are two sources, not one, of external energy in the tropical waters, supplying the ecosystem of this pelagic region. One of them is solar radiation, utilized by phytoplankton in photosynthesis, and in addition, there has to be another source supporting the production of a bacterioplankton biomass that is excessive with respect to the primary production. If autochthonous primary production were the only source of energy for the ecosystem, bacterial production could not exceed one-third of it.

Calculation of the food requirements of zooplankton in the tropical waters also speaks in favor of the existence of such an additional source of energy. This requirement amounts to approximately 0.6 — 0.8 g $C/m^2 \cdot$ day.

In fact, the average zooplankton biomass in the tropical region is about 12 g/m^2, i. e. 1.2 g C/m^2 (without counting microplankton, Coelenterata, Thaliacea, and fishes). Assuming a diurnal energy flux through the zooplankton population to be 0.15 cal/g, i. e. only one-half of the average (Sushchenya, 1969), we find that with 60% assimilability of food, the zooplankton must consume 3 cal energy, or about 0.6 g food carbon in order to maintain even such a low energy flux. Remembering the above-mentioned unaccounted for component of the ecosystem — the benthos and bacterial flora — the real energy requirements of heterotrophic stages in the pelagic ecosystem of tropical waters may be expected to increase to 1 — 1.5 g $C/m^2 \cdot$ day. By average data, the primary production in tropical waters is only about 0.2 g $C/m^2 \cdot$ day (Koblents-Mishke, 1967). A similar discrepancy was detected by Sushchenya (1969) and several others (Marshall et al., 1955; Baylor and Sutcliffe, 1963; Riley, 1963). On the other hand, calculation of primary food production actually measured in our experiments (phytoplankton + bacterioplankton produced at the expense of "external" organic matter) yields values that are commensurate with the ecosystem's calculated requirements, 0.7 — 1.4 g $C/cm^2 \cdot$ day (see Table 5). The only possible additional source of "external" energy for bacterial biosynthesis in the tropical waters is organic matter in the seawater brought by the global circulation of oceanic waters from highly productive upwelling regions and from the subarctic and subantarctic regions of formation of deep and intermediate waters. Destruction in these waters is inhibited by the low water temperature (—1 to +3°) in these regions, while primary production is fairly high (Fedosov, 1965; Horne et al., 1969). The calculated destruction in the surface layer in these regions is only 0.3 — 0.6 mg $C/l \cdot$ year (Table 11), while production is nearly 500 times higher. The deep and intermediate waters forming here, enriched in assimilable organic matter, move meridionally. The reserve of organic matter in these waters is expended very slowly on account of the low temperature. Measurements of the potential bacterial production (see Figure 12) and BOD show that even in the "oldest" deep waters reaching the equator, the reserve of readily available organic matter is often commensurate with that in the surface water of the euphotic zone. A similar conclusion is drawn from experiments on determination of BOD in seawater samples obtained from different depths (Novoselov, 1962). The total reserve of organic matter, consisting of the solute fraction to the extent of 90 — 95%, amounts to about 10 kg C/m^2 in oceanic waters. This quantity is sufficient to support additional production of bacterioplankton that will meet the energy requirements of the ecosystem in the tropical waters for 10 — 20 years. Transport of deep oceanic waters to the surface takes on the average 15 years, while the horizontal transport of water from high latitudes to the equator takes 15 — 20 years (Chekotillo, 1965). These periods are fully sufficient to ensure a constant influx of adequate amounts of organic matter from the productive regions of subpolar and boreal waters into the tropical latitudes. Advection of cold water toward the low latitudes is so rapid that the waters carry even subarctic zooplankton (Vinogradov, 1968).

TABLE 11. Tentative calculation of annual destruction in the Pacific Ocean

Latitude	Mean surface temperature, °C	Diurnal destruction in the layer, * mg O_2/m^3 **	Diurnal destruction in the layer 0–600m, mg O_2/m^2 †	Diurnal destruction in the layer 0–4,000 m, g O_2/m^3 ††	Zonal area, 10^{12} m^2	Annual destruction for the entire zone, 10^6 tons O_2	Annual destruction, 10^9 tons O_2
Southern							
60° – 70°	−1.3	1.1	200	0.37	11.9	4.4	—
50° – 60°	5	2.2	395	0.56	13.1	7.3	—
40° – 50°	11	4.0	715	0.89	12.1	10.8	—
20° – 40°	20	9.5	1710	1.88	26.2	50.1	—
20° S – 20° N	26	16	2880	3.05	63.2	194.0	—
Northern							
20° – 40°	21	10	1800	1.97	21.8	43.2	—
40° – 60°	7	3	540	0.71	16.9	12.1	—
All the latitudinal zones	—	—	—	—	165.2	322	43

* [The depths are not given in the original.]

** Calculated from Krogh's curve for Q_{10} = 2.5 (Sorokin, 1969) and diurnal destruction in the surface layer at 23° = 18 mg $O_2/m^3 \cdot$ day.

† Calculated from the vertical distribution curve for intensity of destruction at station 6064.

†† Diurnal destruction in the layer 600 – 4,000 m was calculated from measurements performed at station 6033.

Evidently, organic matter transported by the intermediate and deep water masses can be a real source of energy for additional bacterial production. Rising to the surface in the regions of upwellings and divergences, these waters are heated, and this heating stimulates the development of bacterioplankton which utilizes the dissolved organic matter for biosynthesis of its own mass, replenishing the stock of primary food in the tropical waters. Since this biomass of bacterioplankton is produced on account of a source of energy that is external with respect to the oceanic area in question, the production thus created should be equated to the primary production and added to the latter for calculation of the energy flux in the ecosystem of the water column (Table 5).

As already mentioned, the calculated values proved to be closely similar to the ecosystem's daily energy requirements.

The new idea expounded here, concerning the influx of energy into the ecosystem of surface tropical waters from underlying areas with allochthonous dissolved organic matter, puts previous notions concerning the energy fluxes in the pelagic ecosystem of tropical waters in a new light. Thus, the surface layer of heated waters, the volume of which does not exceed 1% of the total volume of the water column, functions as a peculiar reactor, "burning" the bulk of dissolved organic matter brought from other oceanic regions. Calculations (see Table 11) show that about 50% of all the organic matter undergoing destruction in the Pacific water column is oxidized within this thin layer.

Determinations of potential bacterial production in deep waters (see Figures 12 and 13) revealed that a rise of the incubation temperature from 2° to 30° accelerates the process of bacterioproduction a hundredfold, corresponding to the temperature coefficient $Q_{10^\circ} = 3.4$. The BOD value in seawater decreases practically to zero as the temperature drops to 2° (Figure 18). Determinations of Q_{10° values for bacterioplankton in deep water samples enriched with readily available matter yielded values of 2.4–2.6 (Sorokin, 1969). This means that the microflora's utilization of dissolved organic matter that is present in a low concentration in the water is inhibited by the temperature drop much more strongly than could be expected from the Q_{10° coefficient obtained from Krogh's curve or determined in bacterial cultures. On the other hand, heating of the waters correspondingly stimulates the utilization of dissolved organic matter by microflora.

Data on high potential production of bacteria and high BOD rate constant in samples of deep waters do not bear out the view that organic matter in deep waters consists of fractions of aquatic humus that are mineralized with difficulty (Skopintsev, 1966, 1967). This view has resulted from a misunderstanding of the ways by which organic matter is supplied to the deep waters. According to this mistaken opinion, the entire ocean water throughout its depth is regarded as a column which is "enclosed" in a cylinder. Organic matter is produced in the euphotic zone by phytoplankton, the cells of which die or are consumed, while the readily assimilable metabolites undergo destruction in the upper layer. The intermediate and deep waters, on the other hand, receive only detritus settling from the upper layers and aquatic humus. Production of these stable components of organic matter is calculated from the results of model experiments in which decomposition of phytoplankton took place at a considerable concentration of matter and a fairly high temperature (Skopintsev, 1967). If, however, we take into account the real structure of the water column, the real origin of the deep waters, and the real transport velocities of water masses, we have to give up any attempt to consider the dynamics of organic matter and energy balance of the ecosystem only over a vertical section in an isolated column of water. The latter contains strong currents which flow in different directions (Arsen'ev et al., 1971) and transport huge masses of water with dissolved organic matter at fairly high velocities. Therefore, the energy dependence upon overlying layers of the biological processes taking place in them and especially in their boundary layers, is very relative. It can be maintained to any considerable degree only by vertical transport of organic matter along the food chain by migrating zooplankton or rapidly settling coarse detritus. However, as already shown, the chief role in the supply of bacterial biosynthesis with energy is played by solute organic matter accounting for 90–95% of the total reserve of

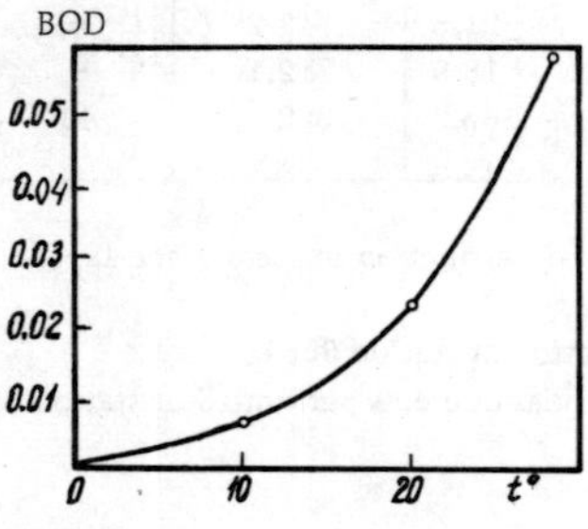

FIGURE 18. BOD, mg-at O_2/l in seawater, as a function of temperature (after Leavastu et al., 1965)

organic matter in seawater (Krey, 1961). On the other hand, vertical transport of organic solutes in the tropical region proceeds extremely slowly. Its main influx is quite different, proceeding via the intermediate and deep waters that are formed from surface subarctic and subantarctic waters and then, sinking, proceed to flow meridionally. In this way, the low latitudes receive organic matter accumulated in the surface waters of high latitudes and preserved at the low temperature. This conclusion is confirmed by results of analyses of the distribution of organic matter and its components in the water column (Figures 19 and 20). The findings show that the concentrations of total, suspended, and dissolved organic matter undergoes little vertical variation down to great depths (Skopintsev and Timofeeva, 1962; Menzel and Goering, 1966; Starikova, 1966). The concentration of nitrogenous organic matter, sugars and vitamins is also little subject to vertical variation (Holm-Hansen et al., 1966; Menzel and Goering, 1966; Park et al., 1962; Walsh and Douglas, 1966).

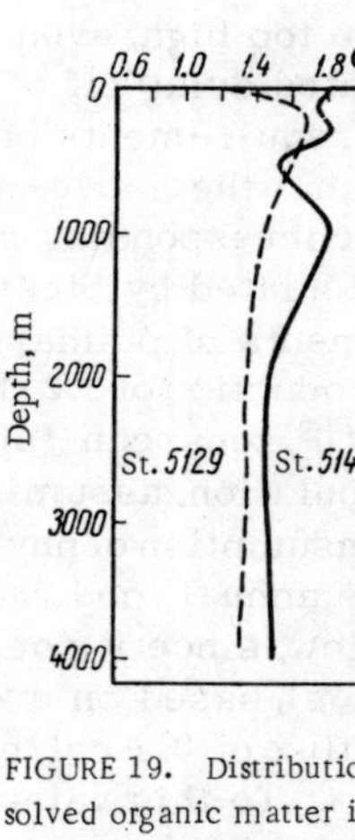

FIGURE 19. Distribution of dissolved organic matter in the water column of the Central Pacific (after Starikova, 1966)

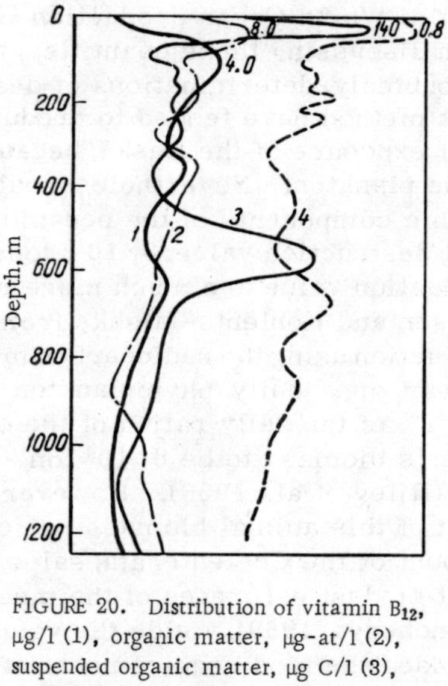

FIGURE 20. Distribution of vitamin B_{12}, μg/l (1), organic matter, μg-at/l (2), suspended organic matter, μg C/l (3), and dissolved organic matter, mg C/l (4) in the East Pacific off California

The figures designate the concentrations of the respective substances at the points of maxima.

The conclusion concerning the important role of bacterial production in the formation of food resources in tropical waters cannot be validated unless the diurnal primary production of phytoplankton in other zones of the ocean exceeds the destruction so that the production and destruction balance each other on the scale of the entire ocean. Unfortunately, no direct data are available concerning the magnitude of destruction in the oceanic water column because sensitive methods for its measurement are still lacking.

Information currently available on the average destruction was obtained by calculation from the total primary production measured by the radiocarbon method (Skopintsev, 1966). In such a calculation, the production and destruction "balance out" automatically, everything depending upon correct determination of the primary production. For the future, it is extremely important to obtain destruction values measured directly in different regions of the ocean so as to be able to compare the value calculated for the entire ocean with the magnitude of primary production.

We attempted a tentative calculation of the total annual destruction in the Pacific from results of direct measurements in the pelagic tropical waters (see Tables 2—4). The results (Table 11) estimated the total annual destruction at $4.3 \cdot 10^{10}$ tons C or an average of 0.18 mg O_2/l · year (0.0005 mg O_2/l · day). The total value for the entire World Ocean should then be $8.4 \cdot 10^{10}$ ton C/year. This destruction value is 6 times the total primary production calculated by Steemann Nielsen (1960), 4 times the same value calculated by Koblents-Mishke (1967), twice the value adopted by Skopintsev in his calculations (1964), and is close to Vinberg's upper limit of World Ocean production (1960).

In discussing these quantities, the following must be taken into account. Undoubtedly, determinations of destruction and bacterial production by the flask method have tended to produce results that are too high, even with brief exposure of the flasks, because of the accelerated dying off of a part of the plankton. Nevertheless, calculations of food requirements of heterotrophic components of the oceanic ecosystem show that the above-indicated total destruction value, $8 \cdot 10^{10}$ tons C/year, and the corresponding primary production value are much more real than those calculated by Steemann Nielsen and Koblents-Mishke from direct measurements of primary production using the radiocarbon method. In fact, a production of $2.2 \cdot 10^{10}$ tons C/year or a daily phytoplankton biomass of $0.6 \cdot 10^{9}$ tons can furnish only 2% of the daily ration of the ocean's animal population, assuming the latter's biomass to be $8 \cdot 10^{9}$ tons and the rate of consumption of phytoplankton 25% (Riley et al., 1956). However, calculation of the annual food requirement of this animal biomass (which is patently too low, since it does not take account of the Coelenterata, salps, nekton, and benthos), based on even the most modest estimates of the mean diurnal energy flux of 0.2 cal/g biomass (Sushchenya, 1969), yields the value $6.3 \cdot 10^{10}$ tons C.* To this value should be added the food requirements of the microplankton and bacteria. Direct measurements of their rate of respiration showed that it consumes most of the primary production. The latter, as determined by the radiocarbon method, scarcely compensates for the metabolic expenditures of only the microplankton and bacteria (Pomeroy and Johannes, 1968).

In the light of these calculations the value $8 \cdot 10^{10}$ tons C for the annual production of the ocean appears not only real, but obviously the minimum value required to meet the ecosystem's energy requirements. Hence, the gap between the food requirements of the heterotrophic components and primary production values measured from C^{14} is real not only for the tropical waters but for the entire ocean. Underestimation of primary

* $\frac{8.3 \cdot 10^{9} \cdot 0.2 \cdot 365}{9.3} = 6.3 \cdot 10^{10}$

production measured by the radiocarbon flask method and that calculated in the water column under 1 m^2 is quite real, since all the errors of the method in question lower the results of measurements. The following errors may be mentioned: 1) a lowered reproduction activity of phytoplankton when isolated in the flasks and failure to take full account of the diurnal rhythm of photosynthesis (Steemann Nielsen, 1952; Vollenweider and Nauwerck, 1961; Odum et. al., 1963); 2) losses of C^{14} with dissolved assimilates reaching 30—50% (our data; Watt, 1966); 3) losses of C^{14} when the phytoplankton on the filters is washed free of carbonates (Strickland, 1960) and during the storage of filters (Wallen and Geen, 1968); 4) incorrect determination of total radioactivity because of failure to take full account of backscattering (Sorokin, 1962); 5) failure to take full account of the population of photosynthetizing phytoplankton in taking samples from the standard horizons or in relation to the degree of penetration of light resulting in failure to detect the phytoplankton concentration maxima (Vinogradov, Gitel'zon, and Sorokin, 1971); 6) inadequate information on the seasonal fluctuations of phytoplankton abundance related to vigorous, but short-term blooming*; 7) failure to take full account of the production of the neritic zone (data on its productivity are extremely scant), as well as the production of phytobenthos and symbiotic algae of the type of zooxanthellae.

The sum of these errors in primary production calculations can easily produce an error factor of 3—4 in estimations of the total annual production. If, as expected, these errors are minimized in the next stage of investigations of the biological productivity of oceans, we shall receive truer values of photosynthetic production that will agree with the real requirements of the ecosystem's heterotrophic components.

In considering the structure of the pelagic ecosystem and the place of bacterioplankton in it, special attention must be paid to the stratified distribution of organisms in the water column. Layers are formed by zooplankton and nekton (Bernard, 1955; Moore, 1950; Vinogradov, 1968; Chindonova, 1966), phytoplankton (Semina, 1967; Sorokin, 1959), and bacterioplankton (Sorokin, 1964). The stratified distribution of phytoplankton, bacterioplankton, and zooplankton in the euphotic zone of tropical waters has been clearly traced (Vinogradov, Gitel'zon, and Sorokin, 1971). One more layer of maximum abundance and activity of bacterioplankton was discovered under the euphotic zone at depths of 400—600 m (see Figures 5—6). This layer was traced over an immense area of the Central and West Pacific (Sorokin, 1971).

The concentration of food particles in the layers of maxima far surpasses that in the adjacent layers. Detection and quantitative investigation of the layers of maxima are extremely important for appraisal of the availability of food to the consumers, the filter-feeding zooplankton. The average

* Even in the subtropical and tropical zones of the ocean, where seasonal phenomena are largely obliterated, there occur outbursts of phytoplankton abundance that are easily missed if seasonal observations are not performed. For instance, total production in the Sargasso Sea, which is one of the poorest regions of the ocean, reached 1.5—2 g $C/m^2 \cdot$ day in the spring period of bloom (Menzel and Ryther, 1959), i.e. higher than the average values observed in eutrophic waters, although it was 1/10 of the latter during the rest of the year. Disregard of this maximum would reduce the total annual production by more than 50%. Outbursts of blooming in boreal waters are even more vigorous and brief.

primary food concentrations in oligotrophic tropical waters, 5 — 15 mg/m^3 phytoplankton + 4 — 6 mg/m^3 bacterioplankton available to the coarse filter feeders are as a rule insufficient for the requirements of filter feeders. Energetically profitable filtration calls for a 10 — 20 times higher food concentration, as follows from calculations of the food requirements and filtration rate of zooplankton (Sushchenya, 1963; Jørgensen 1962) and from measurements of the dependence of the feeding intensity of common species of filter feeders in tropical waters upon the concentration of food (Petipa, Pavlova, and Sorokin, 1971). Only the presence of layers with high primary food concentrations even in the poorest tropical waters can explain the normal existence of filter-feeding zooplankton in these regions; this zooplankton actively concentrates in such layers. It seems that the actual trophic level of waters in a certain region is controlled not so much by the concentration of food material in the layers as by the number and thickness of the layers themselves. These layers are more numerous and thicker in eutrophic than in oligotrophic waters, and consequently the mean food concentrations in the water column are higher in the former.

An important aspect of the trophic dynamics of pelagic waters is the origin of the formation and stability of concentration layers. The formation of the main layer, which was constantly observed at 70 — 90 m in the euphotic zone (see Figure 2), was evidently caused by the presence of a horizon with dynamic equilibrium between the outflow of biogenic elements with turbulent flows from below and the limit of luminous energy arriving from above that is available to the phytoplankton (Sorokin, 1959; Vinogradov, Gitel'zon, and Sorokin, 1971). The position of the layer in the water column is governed by the rate of ascent of biogens in the process of turbulent exchange through the thermocline. If the exchange is intensified, the layer of the maximum rises to the zone of higher illuminance sufficient for the consumption of the excess of incoming biogens by the phytoplankton. In this case, total production per m^2 increases.

The stability of a layer is also conditioned by the high stability of the waters. Layers are usually formed at depths where there is a considerable density gradient that is conducive to the retention and accumulation of settling detritus particles and aggregates carrying a considerable part of the microflora (Paramonov, 1965; Holm-Hansen et al., 1966; Oppenheimer and Vance, 1960; Rakestraw, 1958). The bacteria consume feces and excreta of zooplankton and produce several essential vitamins, the concentration of which in the water depends upon the intensity of bacterial life (Daisley and Fisher, 1958; Droop, 1957; Strickland, 1965; Khailov, 1965). Hence, the layers of maxima are peculiar communities the organisms of which are drawn together by close metabolic ties.

The layer of high bacterial activity discovered in the Pacific Ocean at depths of 400 — 600 m lies in the lower part of the thermocline at the upper boundary of the intermediate water mass at the inflection in the *TS*-curve at arbitrary density (σ_t) 27.0 — 27.3. Characteristically, accumulations of detritus in the Atlantic (Seiwell and Fage, 1948), and of detritus and bacteria in the Pacific (Holm-Hansen et al., 1966) are discovered at the same values. The considerable local density gradients seem to be essential for the formation of this layer (Stommel and Fedorov, 1967).

Heating of the intermediate waters is one of the main causes of enhanced bacterial activity in this layer.

The development of concepts on the aggregation of marine microflora (Jones and Jannasch, 1959; Kane 1967; Sieburth, 1965; Barber, 1966; Carlucci and Williams, 1965) and establishment of the fact that bacteria are the chief nutritive component of detritus (Rodina, 1963) constituted an important stage in studies of the trophic role of marine microflora. Reports by several authors on the possibility of aggregating the molecules of solute organic matter by blowing air through seawater (Baylor and Sutcliffe, 1963; Riley, 1963) were not confirmed by further research (Menzel, 1966; Sheldon et al., 1967). Under sterile conditions, blowing of air does not produce aggregations even of large organic molecules with a molecular weight of up to 10^5. In contrast, aggregation of organic matter does occur in the presence of microflora (Barber, 1966). Similar results were obtained in our experiments with C^{14}-tagged algal hydrolysate as the organic solute. The water from the reservoir was sterilized, 100 — 200μg C/l hydrolysate were added, and air was blown through. Formation of aggregates was detected from the appearance of radioactive organic matter retained by filters with 1-μ pores. The blowing of air did not stimulate accumulation of organic particles. Aggregations of bacterial cells that are readily filtered by water fleas appear en mass only in nonsterile water (Figure 21). In contrast, the tagged algal hydrolysate was not absorbed by the water fleas when air was blown through sterile water.

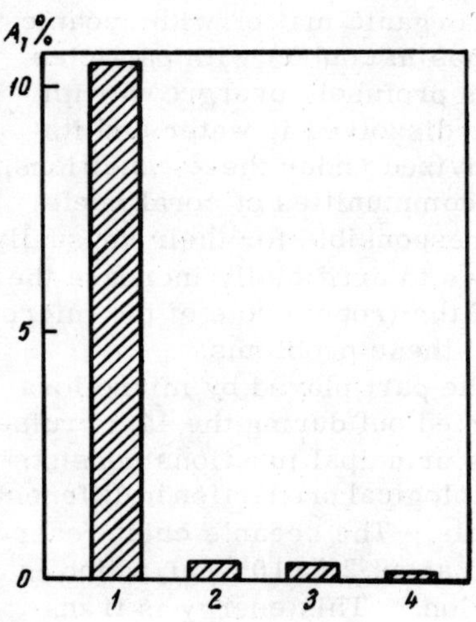

FIGURE 21. Rate of incorporation of substances from C^{14}-tagged algal hydrolysate by water fleas (D. pulex) (A, % of initial radioactivity of hydrolysate):

1 — in a nonsterile medium; 2 — in a sterile medium without blowing of air; 3 — in a sterile medium with blowing of air; 4 — in the control, in which the animals had been killed with formalin (water fleas sterilized with mycetin were placed in a sterile medium).

Hence, bacterioplankton plays the main role in the aggregation of dissolved organic matter. Experiments with natural bacterioplankton from tropical waters showed that 20 — 40% of it was included in aggregates of a size that permitted their consumption by coarse filter feeders (Figure 16). It is important to note that their formation does not require the presence of any particulate matter. The bacterial growth in itself proceeds largely in the form of microcolonies forming the aggregates.

Estimation of the degree of aggregation of natural bacterioplankton is important for analyzing the relationship between the rate of consumption of bacterioplankton by filter feeders and its concentration in the water, and for defining which concentrations of bacterioplankton are apt to satisfy the food requirements of the zooplankton. Estimates of optimal concentration should not be based on the total bacterial count but only on the amount of aggregated bacteria. The lower boundary of optimal concentrations of aggregated bacterioplankton for the feeding of coarse filter feeders such as Calanoida may be fixed at 0.3 — 0.5 g/m^3, or 50 — 100 μg dry weight per liter.

TABLE 12. Effect of blowing air through sterile water from a reservoir on the formation of organic aggregates, from C^{14}-tagged dissolved organic matter (concentration 0.12 mg/l)

Exposure	Radioactivity of organic matter on filters, pulses	% of initial radioactivity of dissolved organic matter
0	70	0.4
1 hr 15 min	50	0.3
5 hr 20 min	40	0.23
8 hr	25	0.15

Also to be considered is the great importance of bacterial biosynthesis in the nutrition of highly productive coastal and lagoon communities, where the production of the bacterial population is uncommonly high, up to 1 $g/m^3 \cdot$ day. It is often not below that in ponds and eutrophic lakes, although the coral reefs and atolls are surrounded by oligotrophic surface waters. The total production of the bacterial population in the water and bottom sediments of lagoons and atolls is as a rule commensurate with or exceeds the total photosynthetic production by phytoplankton and phytobenthos on sandy bottoms.

For the ecosystem to function, there must probably be an influx of additional (with respect to primary production) energy also here. It may be assumed, by analogy with the pelagic community, that here as well the additional energy arrives in the form of dissolved organic matter with oceanic water which is heated here to 30 — 35° and comes in contact with particles of sand and detritus, the surface of dead corals profusely overgrown with microflora, etc. Adsorption of organic matter dissolved in water and its utilization by microflora must be strongly activized under these conditions.

Investigations of the energy balance of the communities of coral reefs and elucidation of the biological mechanisms responsible for their unusually high productivity are important for finding ways to artificially increase the productivity of marine ecosystems. Studies of the trophic role of the microflora will be of especial value in the solving of these problems.

Summarizing the results of research into the part played by microflora in the biological productivity of the ocean carried out during the 44th cruise of the "Vityaz," it is to be noted that one of its principal functions consists in implementing the processes of formation of biological production in different regions with different primary production levels. The ocean's chief reserve of energy is its organic matter and amounts to about $2.5 \cdot 10^{19}$ cal, which is 30 — 40 times the annual phytoplankton production.* This energy is transported by organic matter which is carried by the water masses and is incorporated in the food chain with the participation of aquatic microflora. With the efficiency of 25 — 30%, the latter utilizes the disperse organic matter in seawater to synthetize the protein particles of its own bodies that are available to filter-feeding zooplankton and filter-feeding and mud-eating benthic and sessile animals. The capacities of local ecosystems

* Probably the size of this reserve is itself largely controlled by the velocity of meridional transport of the intermediate and deep waters from the regions of maximum production to those of the most intensive destruction of organic matter. The probable average meridional transport velocity thus calculated is 2 — 3 cm/sec.

for production cannot be correctly estimated without taking into account the extra biosynthesis of bacterial food from "outside" organic matter brought from other regions of the ocean or land.

Bibliography

Publications in Russian

Arsen'ev, V. S., L. I. Galerkin, E. A. Plakhin, and V. V. Sapozhnikov. Osnovnye cherty gidrologii i gidrokhimii raiona issledovanii 44-go reisa "Vityazya" (Major Hydrological and Hydrochemical Features of the Region Investigated during the 44th Cruise of the "Vityaz"). (In this collection.)

Chekotillo, K. A. Vremya vertikal'nogo perenosa vod v okeane (The Time Involved in the Vertical Transport of Waters in the Ocean). Okeanologicheskie Issledovaniya, No. 13. 1965.

Chindonova, Yu. G. O skopleniyakh pelagicheskikh ryb i drugikh zhivotnykh, registriruemykh ekholotom v Atlanticheskom okeane (Accumulations of Pelagic Fishes and Other Animals Recorded with an Echo Sounder in the Atlantic). — Tezisy dokladov II Mezhdunarodnogo Okeanograficheskogo kongressa. Moskva, "Nauka." 1966.

Fedosov, M. V. Usloviya formirovaniya pervichnykh kormovykh resursov okeana (Conditions for the Formation of Primary Food Resources in the Ocean). — Trudy Vsesoyuznogo Nauchno-issledovatel'skogo Instituta Rybnogo Khozyaistva i Okeanografii, Vol. 57. 1965.

Gorbunov, K. V. Tsellyuloznye bakterii kak zveno v pishchevoi tsepi presnykh vodoemov (Cellulose Bacteria as a Link in the Food Chain of Bodies of Fresh Water). — Mikrobiologiya, Vol. 15, No. 2. 1946.

Khailov, K. M. Perspektivy dinamicheskoi biokhimii morya (Prospects of Biochemistry of Marine Dynamics). — Okeanologiya, Vol. 5, No. 1. 1965.

Koblents-Mishke, O. I. Pervichnaya produktsiya (Primary Production).— In: Tikhii okean, Vol. 7. Biologiya Tikhogo okeana, Book 1. Plankton. Moskva, "Nauka." 1967.

Kuznetsov, S. I. Rol' mikroorganizmov v krugovorote veshchestv v ozerakh (Role of Microorganisms in the Cycle of Matter in Lakes). Moskva, Izdatel'stvo Akademii Nauk SSSR. 1952.

Novoselov, A. A. Izuchenie potrebleniya kisloroda v vodakh severnoi chasti Atlanticheskogo okeana (Studies of Oxygen Demand in the North Atlantic). — Okeanologiya, Vol. 2, No. 1. 1962.

Pavlova, E. V., T. S. Petipa, and Yu. I. Sorokin. Rol' bakterioplanktona v pitanii morskikh pelagicheskikh organizmov (Role of Bacterioplankton in the Nutrition of Pelagic Marine Organisms). (In this collection.)

Paramonov, A. N. Nekotorye izmereniya kharaktera raspredeleniya vzvesi v Chernom more (Some Measurements of the Distribution of Suspended Particles in the Black Sea). — Okeanologiya, Vol. 5, No. 1. 1965.

Petipa, T. S., E. V. Pavlova, and Yu. I. Sorokin. Izuchenie pitaniya massovykh form planktona tropicheskoi oblasti Tikhogo okeana radiouglerodnym metodom (Radiocarbon Studies of the Feeding of Mass Plankton Forms in the Tropical Zone of the Pacific). (In this collection.)

Ponomareva, L. A., E. A. Tsikhon-Lukanina, and Yu. I. Sorokin. O potreblenii fitoplanktona i bakterii tropicheskimi evfauziidami (Consumption of phytoplankton and Bacteria by Tropical Euphausiidae). (In this collection.)

Rodina, A. G. O roli otdel'nykh grupp bakterii v produktivnosti vodoemov (Role of Some Groups of Bacteria in the Productivity of Water Bodies). Trudy Problemnogo tematicheskogo soveshchaniya, No. 1. Moskva — Leningrad, Izdatel'stvo Akademii Nauk SSSR. 1951.

Rodina, A. G. Kormovoe znachenie i stroenie detrita (Nutritive Importance and Structure of Detritus). — In Sbornik: "Biologicheskie resursy vodoemov, puti ikh rekonstruktsii i ispol'zovaniya." Moskva, "Nauka." 1966.

Romanenko, V. I. Geterotrofnaya assimilyatsiya CO_2 bakterial'noi floroi vody (Heterotrophic Assimilation of CO_2 by Aquatic Bacterial Flora). — Mikrobiologiya, Vol. 33, No. 4. 1964.

Romanenko, V. I. Sootnoshenie mezhdu potrebleniem kisloroda i uglekisloty u geterotrofnykh bakterii pri roste na peptone (Ratio of Oxygen and Carbon Dioxide Consumption of Heterotrophic Bacteria Cultured on Peptone). — Mikrobiologiya, Vol. 34, No. 3. 1965.

Semina, G. I. Fitoplankton (Phytoplankton). — In: Tikhii okean, Vol. 7. Biologiya Tikhogo okeana, Book 1. Plankton. Moskva, "Nauka." 1967.

Skopintsev, B. A. Raschet obrazovaniya i okisleniya organicheskogo veshchestva v morskikh vodakh (Calculation of the Formation and Oxidation of Organic Matter in Seawater). — Okeanologicheskie issledovaniya, No. 13. 1964.

Skopintsev, B. A. Nekotorye soobrazheniya o sostoyanii i raspredelenii organicheskogo veshchestva v vodakh okeana (Some Considerations on the State and Distribution of Organic Matter in Oceanic Waters). — Okeanologiya, Vol. 6, No. 3. 1966.

Skopintsev, B. A. O balanse ugleroda v vodakh Mirovogo okeana (The Carbon Balance in the Waters of the World Ocean). — Doklady Akademii Nauk SSSR, Vol. 174, No. 6. 1967.

Skopintsev, B. A. and S. N. Timofeeva. Soderzhanie organicheskogo ugleroda v vodakh Baltiiskogo i Severnogo morei, v subtropicheskoi i tropicheskoi oblastyakh severnoi chasti Atlanticheskogo okeana (Content of Organic Carbon in the Baltic and North Seas and in the Subtropical and Tropical Regions of the North Atlantic). — Trudy Morskogo Gidrofizicheskogo Instituta Akademii Nauk SSSR, Vol. 25. 1962.

Sorokin, Yu. I. O vliyanii stratifikatsii vodnykh mass na pervichnuyu produktsiyu fotosinteza v more (Influence of the Stratification of Water Masses on the Primary Production of Photosynthesis in the Sea). — Zhurnal Obshchei Biologii, Vol. 20, No. 6. 1959.

Sorokin, Yu. I. Opredelenie popravochnykh koeffitsientov na samopogloshchenie izlucheniya C^{14} pri opredelenii produktsii fotosinteza i khemosinteza v vodoemakh (Determination of Correction Factors for Self-absorption of C^{14} Radiation in Determinations of Photosynthetic and Chemosynthetic Production in Water Bodies). — Mikrobiologiya, Vol. 31, No. 1. 1962.

Sorokin, Yu. I. Rol' temnovoi bakterial'noi assimilyatsii uglekisloty v trofike vodoemov (The Part Played by Dark Bacterial Assimilation of Carbon Dioxide in the Trophics of Water Bodies). — Mikrobiologiya, Vol. 33, No. 5. 1964.

Sorokin, Yu. I. Primenenie radiouglerodnogo metoda dlya izucheniya pitaniya i pishchevykh svyazei vodnykh zhivotnykh (Application of the Radiocarbon Method to Studies of Feeding and Trophic Relationships among Aquatic Animals). — Trudy Instituta Biologii Vnutrennikh Vod, Vol. 12(15). 1966.

Sorokin, Yu. I. Eksperimental'naya otsenka vliyaniya davleniya i temperatury na metabolizm mikroflory glubinnykh vod okeana (Experimental Evaluation of the Effect of Pressure and Temperature on the Metabolism of Microflora in Deep Oceanic Waters). — Mikrobiologiya, Vol. 38. 1969.

Sorokin, Yu. I. O chislennosti i produktsii bakterii v tolshche vody tsentral'noi chasti Tikhogo okeana (Abundance and Production of Bacteria in the Water Column of the Central Pacific). — Okeanologiya, Vol. 11, No. 1. 1971.

Starikova, N. D. Rastvorennoe organicheskoe veshchestvo (Dissolved Organic Matter). — In: Tikhii okean, Vol. 3, Khimiya Tikhogo okeana. Moskva, "Nauka." 1966.

Sushchenya, L. M. Ekologo-fiziologicheskie osobennosti fil'tratsionnogo pitaniya planktonnykh rakoobraznykh (Ecological and Physiological Features of Filter Feeding among Planktonic Crustaceans). — Trudy Sevastopol'skoi biologicheskoi stantsii, Vol. 16. 1963.

Sushchenya, L. M. Kolichestvennye zakonomernosti pitaniya i obmena planktonnykh rakoobraznykh (Quantitative Patterns of Feeding and Metabolism in Planktonic Crustaceans). Doctoral Thesis. Moskva. 1969.

Troitskii, A. S. and Yu. I. Sorokin. K metodike rascheta biomassy bakterii v vodoemakh (Methods for Calculating the Bacterial Biomass in Water Bodies). — Trudy Instituta Biologii Vnutrennikh Vod Akademii Nauk SSSR, Vol. 19. 1967.

Vinberg, G. G. Pervichnaya produktsiya vodoemov (Primary Production in Water Bodies). Minsk, Izdatel'stvo Akademii Nauk SSSR. 1960.

Vinogradov, M. E. Vertikal'noe raspredelenie okeanicheskogo zooplanktona (Vertical Distribution of Oceanic Zooplankton). Moskva, "Nauka." 1968.

Vinogradov, M. E., I. I. Gitel'zon, and Yu. I. Sorokin. O prostranstvennoi strukture soobshchestv evfoticheskoi zony tropicheskikh vod okeana (The Spatial Structure of Communities in the Euphotic Zone of the Ocean Tropics). (In this collection.)

Publications in Other Languages

Barber, R. T. Interaction of Bubbles and Bacteria in the Formation of Organic Aggregates in Sea Water. — Nature (Lond.), Vol. 211, No. 5046. 1966.

Baylor, E. R. and W. H. Sutcliffe. Dissolved Organic Matter in Sea Water as a Source of Particulated Food. — Limnol. and Oceanogr., Vol. 8, No. 4. 1963.

Bernard, F. Densité du plancton vu au large de Tulon en bathyscaphe. — Bull. Inst. océanogr. Monaco, Vol. 52. No. 1063. 1955.

Carlucci, A. F. and P. M. Williams. Concentration of Bacteria from the Sea Water by Bubble Scavenging. — J. Cons. internat. Expl. Mer., Vol. 30, No. 1. 1965.

Daisley, K. N. and L. R. Fisher. Vertical Distribution of Vitamin B_{12} in the Sea. — J. mar. biol. Ass. U. K., Vol. 37, No. 3. 1958.

Doty, M. and M. Oguri. The Island Mass Effect. — J. Cons. internat. Expl. Mer., Vol. 22, No. 1. 1956.

Droop, N. R. Autotrophy and Organic Compounds in the Nutrition of Marine Phytoplankton. —J. Gen. Microbiol., Vol. 16. 1957.

Fogg, G. E., C. Nalewajko and W. D. Watt. Extracellular Products of Phytoplankton Photosynthesis. —Proc. Roy. Soc. B. Vol. 162. 1965.

Holm-Hansen, O., J. D. Strickland and P. M. Williams. A Detailed Analysis of Biologically Important Substances in a Profile off Southern California. — Limnol. and Oceanogr., Vol. 11, No. 4. 1966.

Horne, A. J., G. Fogg and D. J. Eagle. Studies in situ of the Primary Production of an Area of Inshore Antarctic Sea. — J. mar. biol. Assoc. U. K., Vol. 49, No. 2. 1969.

Johannes, R. E. Ecology of Organic Aggregates in the Vicinity of a Coral Reef. — Limnol. and Oceanogr. Vol. 12, No. 2. 1967.

Jones, G. E. and H. W. Jannasch. Aggregates of Bacteria in Sea Water as Determined by Treatment with Surface Active Agents. — Limnol. and Oceanogr., Vol. 4, No. 3. 1959.

Jørgensen, C. B. The Food of Filter Feeding Organisms. — Rapp. Cons. Internat. Expl. Mer., Vol. 153. 1962.

Kane, J. Organic Aggregates in Surface Waters of the Ligurian Sea. — Limnol. and Oceanogr., Vol. 12, No. 2. 1967.

Krey, J. Der Detritus im Meere. — J. Cons. internat. Expl. Mer., Vol. 26, No. 3. 1961.

Laevastu, T., H. Zeitlin and K. Song Moon. Notice on the Oxygen Consumption in Sea Water. — Limnol. and Oceanogr., Vol. 10, No. 1. 1965.

Marshall, N. B. Detritus over the Reef and its Potential Contribution to Adjacent Waters of Eniwetok Atoll. — Ecology, Vol. 46, No. 3. 1965.

Marshall, N. B. Observations on Organic Aggregates in the Vicinity of Coral Reefs. — Marine Biology, Vol. 2, No. 1. 1968.

Marshall, S. M. and A. P. Orr. The Biology of a Marine Copepod, Calanus finmarchicus (Gunnerus). Edinburgh, London (Oliver a. Boyd). 1955.

Menzel, D. W. Bubbling of Sea Water and the Production of Organic Particles: A Re-evaluation. — Deep-Sea Res., Vol. 13, No. 5. 1966.

Menzel, D. W. and J. J. Goering. The Distribution of Organic Detritus in the Ocean. — Limnol. and Oceanog., Vol. 11, No. 3. 1966.

Menzel, D. W. and J. H. Ryther. Annual Cycle of Primary Production in the Sargasso Sea off Bermuda. — Reprints 1st Internat. Oceanogr. Congr. Washington, D. C. 1959.

Moore, H. B. The Relation between the Scattering Layer and the Euphausiacea. — Biol. Bull. (Woods Hole), Vol. 99, No. 2. 1950.

Odum, H. T., R. Beyers and N. Armstrong. Consequences of Small Storage Capacity in Nannoplankton, Pertinent to Measurement of Primary Production in Tropical Waters. — J. Mar. Res., Vol. 21, No. 3. 1963.

Oppenheimer, C. H. and M. H. Vance. Attachment of Bacteria to the Surfaces of Living and Dead Microorganisms in Marine Sediments. — Ztschr. Allgem. Microbiol. Vol. 1, No. 1. 1960.

Park, K., W. T. Williams, J. M. Prescott, and D. W. Hood. — Aminoacids in Deep Sea Water. — Science, Vol. 138, No. 3539. 1962.

Parsons, T. R. and J. D. H. Strickland. On the Production of Particulate Organic Carbon by Heterotrophic Process in Sea Water. — Deep-Sea Res., Vol. 8, No. 3/4. 1961.

Pomeroy, L. R. and R. E. Johannes. Occurrence and Respiration of Ultraplankton in the upper 500 meters of the Ocean. — Deep-Sea Res., Vol. 15, No. 3. 1968.

Provasoli, L. Organic Regulation of Fhytoplankton Fertility. In: The Sea. Ed. by Hill., Vol. 2. 1963.

Rakestraw, N. W. Particulate Matter at the Oxygen Minimum Level. — J. Mar. Res., Vol. 17, 1958.

Riley, G. A., S. A. Conover, G. B. Deevey, R. J. Conover, S. B. Wheatland, E. Harris, and H. L. Sanders. Oceanography of Long Island Sound, 1952 — 1954. — Bull. Bingham. Oceanogr. Coll., Vol. 15. 1956.

Riley, G. A. Organic Aggregates in Sea Water and the Dynamics of their Formation and Utilization. — Limnol. and Oceanogr., Vol. 8, No. 4. 1963.

Seiwell, R. B. and L. Fage. Minimum Oxygen Layer in the Ocean. — Nature, Vol. 162, No. 4129. 1948.

Sheldon, R. W., T. P. Evelin, and T. R. Parsons. On the Occurrence and Formation of Small Particles in Sea Water. — Limnol. and Oceanogr. Vol. 12, No. 3. 1967.

Sieburth, J. M. Organic Aggregation in Sea Water by Inorganic Nuclei during the Formation of Ammonia by Bacteria. — J. Gen. Microbiol., Vol. 41. 1965.

Sieburth, J. M. Observations on Planktonic Bacteria. In: Proc. U. S. — Japan seminar on Marine Microbiol. Kyoto. 1968.

Sorokin, J. I. On the Primary Production and Bacterial Activities in the Black Sea. — J. Cons. internat. Expl. Mer., Vol. 29, No. 1. 1964.

Steemann, Nielsen, C. E. The Use of Radioactive Carbon (C^{14}) for Measurement Organic Production in the Sea. — J. Cons. Internat. Expl. Mer., Vol. 18, No. 1. 1952.

Steemann Nielsen, C. E. Productivity of the Oceans. — Ann. Rev. Plant. Physiol., Vol. 11. 1960.

Stommel, H. and K. N. Fedorov. Small Scale Structure in Temperature and Salinity near Timor and Mindanao. — Tellus, Vol. 19, No. 2. 1967.

Strickland, J. D. H. Measuring the Production of Marine Phytoplankton. — Bull. Fish. Res. Board Canada, No. 122. 1960.

Strickland, J. D. H. Production of Organic Matter in the Primary Stages of the Food Chain. In: Chemical Oceanography, Ed. Riley J. and G. Scirrow, Vol. 1. 1965.

Vollenweider, R. A. and A. Nauwerck. Some Observations on the C^{14} Method for Measuring Primary Production. — Verh. Intern. Verein. Limnol., Vol. 10. 1961.

Wallen, D. G. and G. H. Geen. Loss of Radioactivity during Storage of C^{14} — Labelled Phytoplankton. — J. Fish Board Canada, Vol. 25, No. 10. 1968.

Walsh, G. E., and J. Dooglas. Vertical Distribution of Dissolved Carbohydrate in the Sargasso Sea off Bermuda. — Limnol. and Oceanogr., Vol. 11, No. 3. 1966.

Watt, W. D. Release of Dissolved Organic Material from the Cells of Phytoplankton Population. — Proc.Roy. Soc., Vol. 164. 1966.

Wood, E. J. F. Microbiology of Oceans and Estuaries. Amsterdam — London — N. Y. 1967.

ZoBell, C. E. and C. W. Grant. Bacterial Utilisation of Low Concentrations of Organic Matter. — J. Bacteriol., Vol. 45. 1943.

RADIOCARBON STUDIES OF THE FEEDING OF MASS PLANKTON FORMS IN THE TROPICAL ZONE OF THE PACIFIC

T.S. Petipa, E.V. Pavlova, and Yu.I. Sorokin
Institute of Biology of the Southern Seas of the Academy of Sciences of the Ukrainian SSR and the Institute of Biology of Inland Waters of the Academy of Sciences of the USSR

In the 44th cruise of the "Vityaz," devoted to studies of the structure and productivity of tropical planktonic communities in the Pacific Ocean, considerable attention was paid to the feeding of pelagic animals.

Very little is known about the feeding of tropical zooplankton. Geinrikh (1958) investigated the contents of the intestine and structure of the mouth parts in several species of Copepoda from tropical regions of the Pacific and differentiated three groups of species differing to some degree with respect to their intestinal contents. She concluded that free phytoplankton plays an important part in the food chains of pelagic tropical waters, Coccolithophoridae and Silicoflagellata being extensively consumed in addition to diatoms. The fraction of copepods (with respect to the biomass) belonging to predatory families is considerably larger than in cold-water oceanic regions. Arashkevich (1969) analyzed the morphological features of the mouth parts and determined the mode of feeding of 13 families of Calanoida, classifying them as phytophages, mixed feeders, and predators. Experimental studies of the feeding of several copepods were made by Mullin (1965, 1966), who concluded that the animals have a rather broad spectrum. The size of the food is more important to them than its systematics.

These investigations show contradictions and confusion with respect to the composition of food and the mode of its capture, but their main shortcoming is the absence of clear-cut quantitative feeding indexes.

We aimed to determine the quantitative characteristics of the consumption and utilization of food by the chief representatives of the community inhabiting the upper layers of tropical pelagic waters. This paper presents the results of three series of experiments:

1) studies of the food spectrum of mass species;

2) studies of the dependence of the intensity of feeding on the concentration of food;

3) determination of the principal elements of the food energy balance.

MATERIALS AND METHODS

Studies of the feeding of planktonic animals in the tropical Pacific were carried out by the radiocarbon method elaborated by Sorokin (1966). Its application to marine organisms has been described by Pavlova and Sorokin (1970) and Petipa, Sorokin, and Lanskaya (1970). Summarizing the work of these authors, we shall provide a fuller description of the methods used in all three series of experiments in the marine environment.

Preparation of food

The food consisted of bacteria, small and large diatoms and peridinians, as well as several species of planktonic Copepoda.

The C^{14}-tagged bacterial food was prepared as a suspension of cells or as natural bacterioplankton. In the former case, a strain of bacteria from the genus Pseudomonas isolated from seawater was cultured on a mineral medium with addition of 0.5 g/l tagged glucose. The bacterial cells were washed from the medium by centrifuging and the sediment was made into a thick suspension. The suspension was measured for reciprocal specific activity of carbon in the bacterial cells (C_r) and their abundance by direct microscopic examination on membrane filters. This homogeneous suspension was used for estimates of the consumption of dispersely distributed bacteria. The desired concentration was achieved by introducing the required amount of tagged bacterial suspension into seawater that had been freed of bacteria by filtration through a membrane filter No. 3.

Tagged natural bacterioplankton was obtained by adding a portion of C^{14}-tagged dissolved organic matter (algal hydrolysate) to seawater freed of plankton by filtration through double bolting silk No. 76. The hydrolysate was prepared by acid hydrolysis of C^{14}-tagged chlorella. The hydrolysate contained about 1 mg C_{org}/ml and its radioactivity was about $2 \cdot 10^6$ pulses/min. Addition of 0.1 – 0.5 ml/l of this solution to seawater for the experiment had little effect on the natural content of organic matter in the seawater (2 mg/l). The natural bacterioplankton absorbed the tagged organic matter, acquiring a tag of 300 – 500 pulses/ml in 1 or 2 days. The water containing the tagged bacterioplankton was used in the experiments after preliminary determination of its bacterial count and after its dilution with seawater that had been filtered through a membrane filter No. 3.

The reciprocal specific activity of tagged natural bacterioplankton was calculated by the formula $C_r = \frac{B}{10 \cdot r}$ μg C/pulse, where C_r is the reciprocal specific activity of bacterial matter; B, wet bacterial biomass (mg/l) determined from the bacterial count on the membrane filters and the size of the bacteria, with a correction of 2.5 in consideration of shrinkage upon drying (Troitskii and Sorokin, 1967); r, radioactivity of bacteria in the same volume.

The average size of individual cells of Pseudomonas was $0.5 - 0.6 \times 1.5 - 2\mu$, and the size of bacterial clusters in the natural bacterioplankton greater than 4μ.

The experiments included cultures of Amphidinium klebsii from the Red Sea, Gymnodinium lanskoi from the Caribbean, and Asterionella japonica and Biddulphia catenata from the Mediterranean; all cultures were kindly supplied by Lanskaya.* The cultures were maintained under scattered daylight during the expedition. They were transferred every 5 — 7 days at the longest, or else the cultures received an addition of nutrient solution. If the growth of algae became weaker, vitamin B_{12} was added (4 — 8 μg/l) in the transfers. The nutrient solution was prepared according to Goldberg's formula modified by Kabanova (1958).

The algal cultures used in feeding studies were subjected to preliminary tagging with C^{14} carbonate and grown on a nutrient solution prepared with seawater from which the HCO_3^- ion had been removed. For this purpose the water was acidulated with 0.4 ml conc. HCl, boiled for 5 — 10 min, cooled, and neutralized to pH 8 with alkali free of CO_2. The algae were separated from the active medium by filtration through membrane filters Nos. 5 — 6 with several portions of seawater. The sediment was washed off with a small volume of seawater. The same culture was used simultaneously to prepare a suspension with distilled water for determination of C_r. Radioactivity of the food suspension was analyzed on glass targets to which a precisely measured volume of the suspension (0.05 ml) was applied, uniformly distributed over a circular area of 3 cm^2 and dried. The C_{org} content in 0.05 ml of the suspension was determined by wet digestion followed by titration with dichromate mixture. The reciprocal specific activities of the algae and bacteria ranged from $0.9 \cdot 10^{-6}$ to $14 \cdot 10^{-6}$ mg C/pulse.

The animal food was prepared in the following manner. Various Copepoda, mainly small Calanoida, were picked from freshly gathered plankton samples; 10 — 50 crustaceans were placed in vessels with a capacity of 1 — 2 liters containing water that had been filtered through a No. 5 membrane filter, adding 1 — 2 ml of the thick suspension of C^{14}-tagged algae so that the concentration of algae in the vessel should be optimal for the feeding of crustaceans (1—2 mg/l). The animals were kept in the vessels for 1 — 2 days, then washed free of the tagged algae with 2 — 3 liters of seawater, and a few individuals were sampled for radioactivity determinations performed from dried homogenates on glass targets. The reciprocal specific activity of animal food (C_r) was calculated by dividing the amount of carbon in the body of a single animal by its radioactivity. The carbon content in the bodies of all the animals used in the experiment was determined by wet digestion. Large specimens underwent preliminary conversion to homogenate. The radioactivity of small animal food (Acartia, Paracalanus, Scolecithrix, Temora) was usually 60 — 200 pulses/specimen, that of large animal food (Eucalanus, Rhincalanus, Undinula) 400 — 1700 pulses/specimen. The lowest radioactivity, about 40 pulses/specimen, was found in dead crustaceans used as food. The reciprocal specific activity of the animal food varied within a range of $10 — 15 \cdot 10^{-6}$ mg C/pulse.

* The authors are deeply grateful to L. A. Lanskaya for her help.

Preparation of experimental animals

The animals were taken from samples of freshly caught net plankton.* The most active animals were placed in 0.5 — 2 liters of pure, unfiltered water, in which they were kept for 2 — 3 hours while the experiment was being prepared. The briefness of the animals' stay in the laboratory prior to the experiments was conducive to preservation of the natural level of their motor activity, energy metabolism, and consequently food requirements.

In one experiment, 1 — 70 animals (depending on their size) were placed in vessels of 1 — 2 liters' capacity containing seawater that had been filtered through a No. 5 membrane filter, with addition of the tagged food. The concentration of forage organisms in the experimental vessels was calculated beforehand from their abundance in the initial food suspension, size, and weight. The weight of forage and experimental animals was estimated from the carbon content of their bodies. The water temperature in all experiments was the same as that on the sea surface.

Studies of the food spectrum and feeding intensity as a function of food concentration

The nutritive value of forage organisms and the extent of their consumption by animals were evaluated by the quantity C_{assim}/C (index of assimilation), defined as the ratio of carbon assimilated and incorporated in the animal's body to the animal's total carbon content. It provides the most complete characterization of the nutritive value of forage organisms, being a combined index of their consumption and assimilation.

The C_{assim}/C values were determined by maintaining the animals on tagged food for 4 — 6 or 24 hours, followed by washing with 2 — 3 liters of seawater and a few milliliters of distilled water and determinations of the radioactivity of their bodies. The radioactivity values were multiplied by C_r of the food to calculate the amount of food assimilated in the course of the experiment, in mg of carbon. All the results were scaled to unit time (24 hours) and stated in percentages.

The food spectrum was investigated in 32 species of planktonic animals including Copepoda (27 species), Ostracoda, Decapoda, Appendicularia, and Salpae. The food consisted of bacteria, small peridinians (*Gymnodinium lanskoi*, *Amphidinium klebsii*), large diatoms (*Biddulphia catenata*), blue-green algae (*Oscillatoria* sp.), and animals: a mixture of Calanoida, young *Euchaeta*, *Eucalanus*, *Scolecithrix*, *Rhincalanus*, *Temora*, *Undinula*. Animal detritus was also used (dead crustaceans). In all, 150 experiments were carried out, each lasting 5 — 10 hours. The experiments were conducted in those hours during which the animals feed intensively.

The dependence of feeding intensity on food concentration was estimated from the C_{assim}/C values determined at different concentrations of the forage organisms. The volume of the experimental vessels and the concentrations of food in them were chosen in such a way that the total amount of food in a

* The authors are grateful to members of the plankton division headed by Yu. A. Rudyakov, for supplying zooplankton at all hours.

vessel should exceed the total ration of the consumer organisms even at the lowest food concentration. The food concentrations in the experiments varied within the same limits as the natural food concentrations, 0.02 – 2 g/m³.

Determination of the principal balance equation components

The first series of experiments was concerned with the initial components of the food balance equation. The animals were maintained on tagged food of a known concentration for a time that was sufficient for a single filling of their intestine (from 15 – 40 min to 3 hours, depending upon the species of animal and type of food). The animals were then washed clean of the tagged food with seawater and placed in a vessel containing 40 – 100 ml seawater freed of CO_2/HCO_3^- (see above) and with pH 8.0 – 8.2. The water received an addition of some untagged food of the same kind as the tagged food in order that its concentration in the experimental vessel be approximately equal to that of the tagged food in the preceding stage. The exposure time for animals on untagged food must be equal to the digestion time for all the tagged food consumed until the intestines are fully free of its remains. The digestion time should preferably be determined in advance. In order to obtain correct information on the size of rations, the exposure of animals on tagged food should continue to the beginning of defecation. An animal consumes a new portion of food while the preceding portion is being digested and excreted, and therefore the exposure time of the animals on tagged and untagged food should be approximately the same.

At the end of the experiment the animals were killed by heating, leaving their outer covers intact, removed from the vessel with a pipet, and washed free of salts with distilled water. The washed animals were crushed in a small mortar. The homogenate was applied to glass targets for determination of the radioactivity of the bodies (R_2) i. e. the fraction of assimilated food that had been incorporated in the substance of the body.

The remaining water was investigated for the radioactivity of carbon dioxide liberated by the animals (r_c), radioactivity of formed feces (r_f), and radioactivity of liquid excreta present in water in the form of dissolved organic matter (r_d). For this purpose, the water was alkalized with 0.2 ml of 0.2 *N* KOH after the removal of the animals, mixed and filtered through a No. 4 membrane filter, on which the radioactivity of feces was determined. The results were corrected (correction factor 1.25) for counting conditions of the preparations on glass targets for C_r determination. For determination of r_c, the filtrate was acidified in the apparatus (Sorokin, 1966), and CO_2 was driven off, absorbed in alkali, and precipitated as $BaCO_3$ on addition of 1 ml of 10% $BaCl_2$ solution. The radioactivity of the $BaCO_3$ precipitate was determined on a membrane filter No. 2, correcting for self-absorption and counting conditions (correction factor 1.25). For determination of r_d, the filtrate remaining after CO_2 had been driven off was alkalized (5 ml of 5% KOH per 150 ml), followed by addition of soda as carrier (0.1 ml of a 2% solution) and 2 ml 0.1 N $KMnO_4$ + 1 ml of 1% Ag_2SO_4. The filtrate was then boiled in a small flask with a reflux condenser for 15 min. After the boiling, the flask was connected to the apparatus for the driving off of CO_2 with addition of 5 ml of 20% H_2SO_4 and 1 ml hydrogen peroxide solution. The solution in the flask was decolorized and was brought to boiling, followed by driving off of CO_2, precipitation of $BaCO_3$, and determination of the radioactivity of the precipitate on a No. 2 filter, with the same corrections as those used in determination of r_c. The empirical correction factor for completeness of oxidation (1.25) was additionally included. When determining radioactivity one must always bear in mind that the counting conditions under the counter should be the same as those involved in determination of the C_r of the food.

The radioactivity of food consumed by the animals during the experiment (R_1) was determined from the sum of the initial balance component $R_1 = R_2 + r_c + r_f + r_d$. The conversion from radioactivities to absolute quantity of organic matter stated in carbon or energy units was performed by multiplying these quantities by the reciprocal specific activity of the consumed food (C_r) and by the conversion factor from carbon to calories (1 mg C = 9.361 cal).

Calculation of the daily supply and expenditure of matter and energy necessitates a knowledge of the diurnal rhythm of the filling of intestines and digestion time. These data were taken from the following authors: Delalo, 1964; Petipa, 1966; Petipa, Sorokin, and Lanskaya, 1970. In our opinion, the diurnal feeding rhythm of tropical copepods is similar to that observed in the same and related species in other warm seas.

The information on the diurnal feeding rhythm was used to calculate the conversion factors for converting the initial balance component to mean diurnal values. This was followed by determination of the mean diurnal digestion time and correspondingly the second factor based on the dependence of the rate of digestion upon feeding intensity at the same diurnal rhythm (Petipa, 1966). The rate of digestion was assumed to be doubled as the temperature rose to 10°, similar to the rates of other physiological processes. The mean diurnal digestion time for plant food was assumed to be 20 min for hyponeuston (Pontillidae), 30 min in nonmigrant and weakly migrating inhabitants of the upper layers (Acartia, Temora, Eucalanus), and 3 hours in migrating interzonal animals (Euchaeta, Pleuromamma). Application of both factors to all the food balance components yields diurnal indexes. The diurnal food balance was expressed as the equation $R = U + F$, where R stands for the daily ration, U for the food assimilated during the diurnal period, and F, the food which remained unassimilated during the same period.

Assimilated food consists of matter incorporated in the body and respiration losses, while unassimilated food consists of solid and liquid excreta. The balance equation components were determined for 5 animal species in 29 experiments.

RESULTS

Food spectrum

The qualitative composition of the food was studied in 32 plankters, showing which kinds of food these animals consumed and to what extent they were utilized. Comparison of indexes of assimilation in feeding different kinds of food to the animals showed that only Appendicularia were capable of consuming bacteria and small algae at a fairly intensive rate, $C_{assim}/C = 10-50\%$ (Table 1). Small Calanoida, filter feeders, and "graspers," consumers of mixed food (*Undinula, Acartia, Temora*), utilize bacteria and small algae much less thoroughly, $C_{assim}/C = 2-3\%$. Consumption of these foods by all the animals investigated proved insignificant, the index of assimilation not exceeding 0.5 — 1% and as a rule 0.01 — 0.2% (Table 1).

Large filter feeders and "graspers," consumers of mixed food, and predators — Pontellidae, *Eucalanus, Scolecithrix, Rhincalanus, Pleuromamma, Euchaeta* — can consume considerable numbers of large diatoms and blue-green algae, such as *Biddulphia* and *Oscillatoria*; the index of assimilation varies from 0.5 to 13%. In the case of small animals feeding on large algae the index of assimilation did not exceed 1 — 2.5%.

The highest values of the index of assimilation (C_{assim}/C) were observed in the case of predatory copepods feeding on animals, 6 — 54 for *Oncaea*, 8 — 123 in *Candacia*, 1 — 28 in *Corycaeus*, 5 — 230 in *Euchaeta*, and 13 — 140% in *Pleuromamma*.

Larvae and young of the flying fish had indexes of assimilation of 0.2 — 33% in consumption of crustacean plankton.* Their consumption of algae and bacteria was extremely limited, $C_{assim}/C = 0.002-0.02\%$.

Detritus of animal origin (killed crustaceans) was vigorously eaten by sucking predators *Oncaea* and *Corycaeus*, $C_{assim}/C = 12-54\%$.

Feeding intensity as a function of food concentration

In studying the dependence of feeding intensity upon food concentration in the case of several species of copepods feeding on algae and bacteria, we used small and large filter feeders and "graspers" — *Temora, Eucalanus, Undinula, Pleuromamma*, mixture of Calanoida. Algal food consisted of small *Amphidinium* and large *Biddulphia* and *Oscillatoria*.

All the species investigated showed (Table 2, Figure 1) the same trend, their index of assimilation increasing with increasing food concentration to an optimum at 0.5 — 3.0 g/m^3. At a still further increase of concentration, the rise of C_{assim}/C values became markedly slower and the curve formed a plateau. The concentration optimum for animals feeding on bacterioplankton (0.5 — 1.0 g/m^3) was lower than in the case of algal food (1 — 3 g/m^3) (Sorokin, Petipa, and Pavlova, 1970). These results were in good agreement with previously established relationships between feeding intensity and food concentration (Petipa, Sorokin, and Lanskaya, 1970; Pavlova and Sorokin 1970).

* The authors are grateful to N. N. Gorbunova for raising the larvae from eggs and selecting the fish larvae and young.

TABLE 1. Characteristics of the food spectrum and feeding intensity for different planktonic animals in the tropical Pacific

Animal	Size of specimen, mm	Weight of specimen, mg C	Kind of food	Concentration of food, g/m³ wet biomass	C_r of food, mg C·10^{-6}/pulse	Ultimate radioactivity of animals' bodies, pulses/specimen·day	C_{assim}/C, %
Calanoida	2.50	0.03	Bacteria	1.4–2	1.13	294	1.08
Pontellidae	3.00	0.038	Animals	1.2	6.00	60	9.47
Calanidae (Undinula etc.)	1.95	0.021	Bacteria	1.4–2	0.91	232	1.16
	2.50	0.050	Amphidinium	1.0–2.0	1.50	670	1.60
	3.00	0.077	Biddulphia	0.58	1.55	596	1.37
	2.90	0.069	Green-blue algae	2–3	13.94	55	1.11
Acartia sp.	1.28	0.004	Bacteria	1.4–2	0.91	109	2.75
Euchaeta marina	3.20	0.066	Bacteria	1.4–2	0.90	85	0.11
	3.10	0.060	Biddulphia	5.4	1.33	234	0.55
Prestandrea	3.05	0.058	Green-blue algae	2–3	13.94	13	0.31
	3.26	0.075	Animals	5.41	558.66	58	52.03
Euchaeta wolfendeni Scott	2.25	0.028	Bacteria	1.4–2	0.90	20	0.07
	2.55	0.032	Gymnodinium	1.25	0.88	5	0.01
	3.05	0.059	Biddulphia	0.67	1.50	68	0.95
Eucalanus attenuatus Dana	3.84	0.045	Bacteria	1.4–3	5.25	340	2.48
	3.78	0.043	Biddulphia	1.0–2.0	1.42	313	1.02
	3.73	0.041	Amphidinium, Gymnodinium	0.9	1.28	179	0.48
	3.80	0.044	Green-blue algae	2–3	7.00	35	0.56
Rhincalanus cornutus Dana	3.17	0.067	Bacteria	1.4–2	1.10	42	0.07
	3.27	0.074	Gymnodinium	0.84	0.88	6.5	0.08
	3.00	0.028	Biddulphia	0.8	1.44	480	2.47
Haloptilus spiniceps Giesbrecht	2.20	0.006	Bacteria	1.4–2	1.15	10	0.34
Euchirella sp.	3.80	0.390	Biddulphia	0.6	1.60	119	0.05
	3.37	0.256	Green-blue algae	2–4	13.9	46	0.25
Pleuromamma abdominalis Lubbock	3.38	0.068	Biddulphia	0.5	1.0	4,255	9.4
	4.00	0.124	Green-blue algae	2–3	13.9	68	1.5
	4.00	0.057	Bacteria	0.3	1.04	100	0.18
	4.47	0.156	Animals	4.6	2,048.0	20	51.08
Pleuromamma gracilis Claus	3.00	0.059	Biddulphia	0.5	1.6	3,876	10.40
Scolecithrix danae Lubbock	2.13	0.035	Biddulphia	0.4	1.58	1,264	5.63
	2.50	0.034	Green-blue algae	2–3	13.94	331	13.56
Candacia aethiopica Dana	1.85	0.020	Biddulphia	0.5	1.60	50	0.41
	1.80	0.018	Animals	4.13	274.0	35.33	57.24
	1.80	0.018	Dead animals	1.1	133.0	45	33.22
Temora stylifera Dana	1.65	0.016	Green-blue algae	2–3	13.94	24	0.21
	1.76	0.021	Amphidinium	1.5	1.57	579	3.13

TABLE 1 (continued)

Animal	Size of specimen, mm	Weight of specimen, mg C	Kind of food	Concentration of food, g/m^3 wet biomass	C_r of food, mg C$\cdot 10^{-6}$/pulse	Ultimate radioactivity of animals' bodies, pulses/specimen$\cdot$day	C_{assim}/C, %
Cyclopoida	4.70	0.026	Bacteria	1.4–2	1.18	34	0.16
Copilia sp.	3.18	0.014	*Biddulphia*	0.8	1.44	16	0.15
	6.00	0.054	Animals	15.4	496.00	6	5.78
Oncaea sp.	1.19	0.005	Animals	5.7	231.25	6	24.73
	1.15	0.005	Dead animals	2.5	520.00	5	46.58
Sapphirina sp.	1.55	0.009	*Gymnodinium*	0.5	0.94	3	0.03
Corycaeus sp.	1.38	0.005	Animals	8.7	207.00	7	28.40
	2.50	0.030	Dead animals	6.3	500.00	7	11.6
Harpacticoida *Macrosetella* sp.	1.26	0.024	*Amphidinium*	1.0	1.025	38	0.16
Ostracoda	2.18	0.061	Bacteria	0.5	1.44	534	0.95
	1.30	1.300	*Biddulphia*	1.4–2	1.18	38	0.003
Decapoda	18.00	4.600	Green-blue algae	2–3	13.94	118	0.04
Appendicularia *Oikopleura* sp.	2.28	0.0009	Bacteria	1.4–2	1.09	256	34.17
Salpae	60.00	2.092	Bacteria	1.5	1.13	14,980	0.80
	60.00	2.092	*Amphidinium*	1.0	1.05	3,614	0.18
Larvae of flying-fish	4.1	0.068	Bacteria	1.4–2	1.25	8.3	0.01
	4.1	0.068	*Amphidinium*	1.0	1.46	6	0.01
	12.0	23.262	Animals	16.02	526.67	1,349	7.15

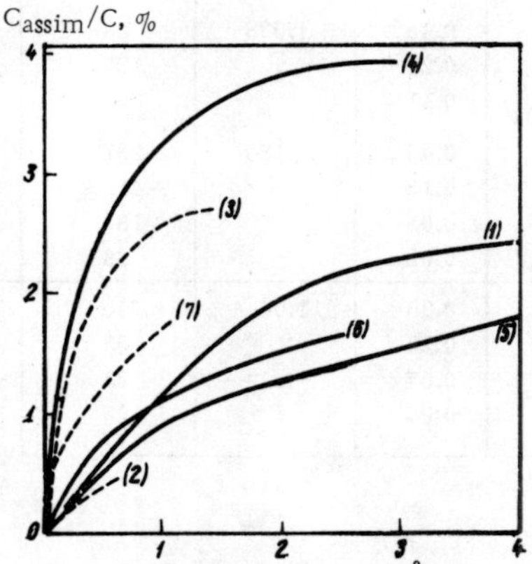

FIGURE 1. Index of assimilation in tropical copepods as a function of food concentration:

1 — *Undinula* + *Amphidinium*; 2 — *Undinula* + bacteria; 3 — mixture of Calanoida and bacteria; 4 — *Temora* + *Amphidinium*; 5 — *Temora* + blue-green algae; 6 — *Eucalanus* + *Biddulphia*; 7 — *Eucalanus* + bacteria.

TABLE 2. Dependence of feeding intensity (index of assimilation) of Copepoda in the tropical Pacific upon food concentration

Species	Weight of specimen, mg/C	Kind of food	Concentration of food, g/m³	c_r of food, mg/C·10^{-6}/pulse	Ultimate radioactivity of animals' bodies, pulses/specimen·day	C_{assim}/C, %
Undinula darwini Brady	0.0495	Amphidinium	4.0	1.504	804	2.44
	0.0524		2.0		698	2.00
	0.0495		1.0		640	1.26
	0.0618		0.5		193	0.47
	0.0495		0.1		63	0.19
	0.0440		0.05		14	0.05
	0.0343	Bacteria	0.60	1.085	141	0.45
	0.0261		0.30		115	0.48
	0.0300		0.10		43	0.16
	0.0300		0.06		22	0.08
	0.0225		0.03		6	0.04
	0.0193	Bacteria	0.9	1.42	413	2.53
	0.0178		0.3		298	1.98
	0.0193		0.1		82	0.50
	0.0193		0.04		12	0.07
Temora stylifera Dana	0.0207	Amphidinium	2.00	1.572	360	2.73
			1.50		626	4.75
			1.00		253	1.92
			0.50		339	2.57
			0.10		97	0.74
	0.1747	Blue-green algae Oscillatoria thiebautii	4.00	7.0	45	1.80
			2.00		33	1.32
			1.00		20	0.80
			0.50		18	0.72
Eucalanus attenuatus Dana	0.0436	Biddulphia	2.00	1.42	480	1.56
	0.0419		1.00		146	0.49
	0.0387		0.8		272	1.0
	0.0403		0.4		198	0.70
	0.0419	Bacteria	0.45	1.078	53	0.14
	0.04537		0.23		187	0.44
	0.06772		0.11		55	0.09
	0.0405	Bacteria	0.45	1.185	281	0.82
	0.0436		0.15		419	1.14
	0.0341		0.05		185	0.64
	0.0341		0.02		66	0.02
Pleuromamma abdominalis Dana	0.0570	Bacteria	0.30	1.04	100	0.18
	0.0654		0.20		53	0.08
	0.0812		0.07		23	0.03
	0.0812		0.04		26	0.03

The dependence of the ration upon concentration of food was evaluated from experimental data on the determination of food balance elements in coarse filter feeders Pleuromamma and Eucalanus feeding on large diatoms and small peridinians. Assuming that the amount of food consumed by the crustaceans per 24 hours at the maximal concentration of 2 g/m^3 approaches the maximum possible ration of this food, the dependence of the actual ration upon the concentration of food can be expressed by the equation $r = \frac{R}{1 - e^{k_0 - k}}$, where r is the actual ration, cal/specimen; k; the corresponding concentration of food, cal/m^3; R, the maximum ration, cal/specimen; k_0, the minimum concentration, cal/m^3, at which consumption of the food is begun (Vinberg and Anisimov, 1966; Parsons, LeBrasseur, and Fulton, 1967). The equation for Pleuromamma has the form $r \frac{0.07}{1 - e^{1.355 - \frac{K}{0.004}}}$; that for Eucalanus, $r = \frac{0.28}{1 - e^{150.7 - \frac{K}{0.013}}}$.

Elements of the food balance equation

The elements of the balance equation were investigated for representatives of different ecological groups in the tropical plankton. Classification of the plankton community into ecological groups and investigation of their typical representatives considerably facilitate the overall task of studying the trophic structure of a plankton community. This obviates the need to investigate the huge diversity of planktonic species, as their feeding and trophic relationships can be estimated from corresponding data for other species of the same ecological group.

The following ecological groups were selected for the balance experiments:

1) nonmigrant surface layer inhabitants, consumers of animal and plant food, mostly predators (Pontellidae);

2) nonmigrant or weakly migrant, mainly plant-eating inhabitants of the upper 50—100-m layer (Temora);

3) weakly migrant forms inhabiting the upper 50—200-m layer, consumers of animal food (Euchaeta);

4) nonmigrant species subsisting on plants and mixed food, inhabitants of the upper 100—200-m layer (Eucalanus);

5) vigorous migrants, consumers of plant and animal food, occurring at depths of 50—500-m (Pleuromamma).

The results of determining balance equation components are given in Table 3 and the results of calculations for diurnal balance indexes in Table 4.

The choice of maximum concentrations of forage organisms in the balance experiments was based on field quantitative data on the vertical distribution of chlorophyll, suspended organic matter, and small zooplankton in the investigated region. The maximum concentrations determined in the sea were 0.3—0.5 g/m^3 (140—234 cal/m^3) for bacteria, 0.5—2.0 g/m^3 (327—755 cal/m^3) for algae, and 0.1—0.2 g/m^3 (90—190 cal/m^3) for small animals.

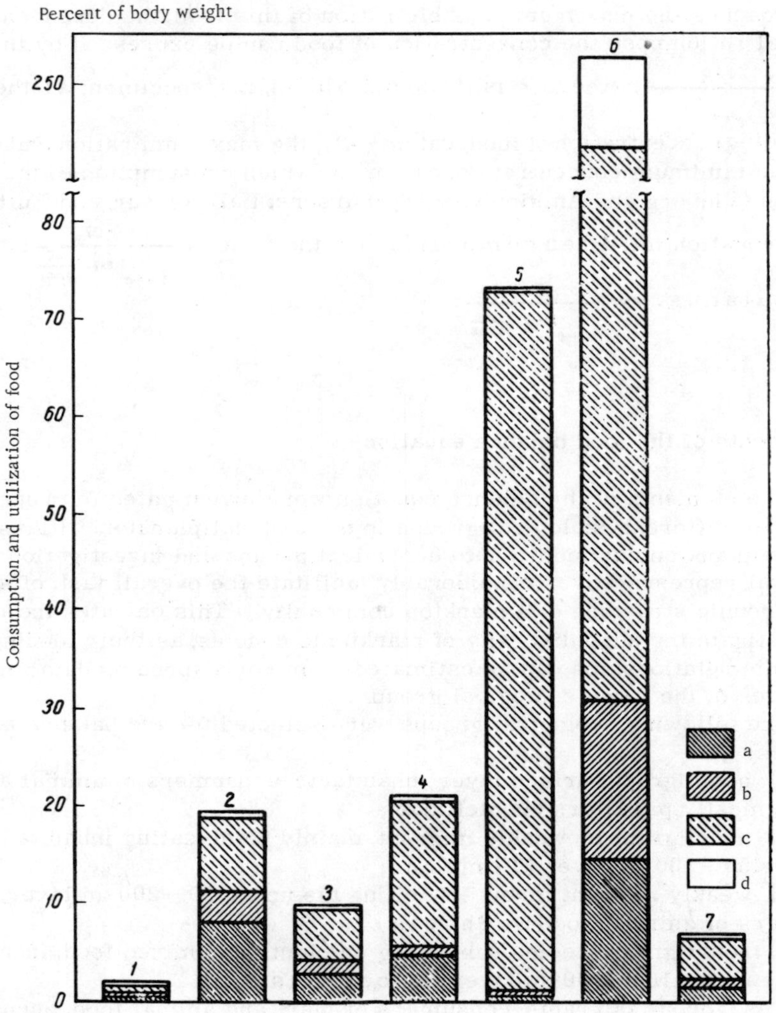

FIGURE 2. Ratio of the initial balance equation components in representatives of the different ecological groups:

1 — Pontellidae + algae; 2 — Temora + algae; 3 — Eucalanus + bacteria; 4 — Eucalanus + algae; 5 — Euchaeta + algae; 6 — Euchaeta + animals; 7 — Pleuromamma + algae; a — incorporation of matter in the body; b — losses of tagged food in respiration; c — feces; d — liquid organic excreta.

TABLE 3. Initial balance equation components (incorporation in the body R_2; liberated CO_2 – r_c ; feces – r_f ; excreted organic matter – r_d) for tropical Copepoda in a short-term experiment (temperature 27 – 28°)

Animal	Kind and concentration of food, g/m³	C_r of food, mg C·10⁻⁶/pulse	Exposure time	No. of crustaceans in experiment	R_2	r_c	r_f	r_d	R_2	r_c	r_f	r_d	$R_1 = r_c + r_f + r_d + R_2$
					pulses/specimen				mg C·10⁻⁶				
Pontellidae	Asterionella, 0.7	3.8	16.10–16.30	39	1.846	0.222	0.98125	0	8.8	1.05	4.66	0	14.51
	Biddulphia, 0.66	1.27	19.50–20.20	50	0.850	1.364	5.565	0.174	1.15	1.84	7.51	0.23	10.73
	Biddulphia, 0.66	1.27	19.55–20.25	50	0.809	0.867	6.955	0.174	1.09	1.17	9.4	0.24	11.90
	Gymnodinium, 0.66	0.96	20.15–20.45	50	0.344	0.942	0.329	0.099	0.32	0.96	0.32	0.1	1.73
	Asterionella, 0.7	3.8	16.40–17.00	31	0.830	0.371	1.145	0.124	3.94	1.76	5.34	0.57	11.61
Temora stylifera Dana	Amphidinium, 0.5	0.95	15.45–16.15	24	17.540	6.795	17.307	1.282	17.63	6.83	17.4	1.29	43.15
Eucalanus attenuatus Dana	Bacteria, 1.0	1.2	12.45–13.15	17	12.94	6.694	26.607	0.992	15.53	8.03	31.93	1.19	56.68
	Gymnodinium, 0.35	0.96	14.00–14.30	10	19.59	3.842	12.365	0.744	20.0	3.92	12.61	0.75	37.28
	Amphidinium, 0.8	1.75	20.25–20.55	7	19.200	1.50	164.400	0.550	33.6	2.62	287.7	0.96	324.88
	Amphidinium, 1.0	1.75	10.45–11.20	10	20.689	5.12	21.160	2.250	36.21	8.96	37.03	3.94	86.14
	Amphidinium, 2.0	1.75	20.30–21.00	6	60.480	26.88	234.00	11.250	105.84	47.04	409.50	19.69	582.07
Pleuromamma abdominalis Lubbock	Biddulphia, 0.1	1.27	1.30–2.00	10	2.4699	2.107	1.930	0.496	3.33	2.85	2.60	0.68	9.46
	Biddulphia, 0.13	1.27	1.05–1.35	10	20.99	5.206	12.864	0.620	28.34	7.03	17.36	0.84	53.57
	Biddulphia, 0.5	0.42	0.40–1.10	12	103.597	41.88	236.610	20.00	43.82	17.72	100.09	8.46	170.10
	Biddulphia, 0.5	0.42	0.45–1.15	10	75.97	80.25	232.209	40.12	32.13	34.10	98.22	16.93	180.38
	Biddulphia, 1.0	0.42	2.05–2.35	11	87.573	38.875	20.242	18.75	37.04	16.45	38.17	7.93	99.6
	Biddulphia, 1.0	0.42	2.08–2.38	11	124.26	27.50	513.940	16.25	52.56	11.63	217.40	6.87	288.46
	Biddulphia, 2.0	0.42	0.50–1.20	9	74.245	37.50	91.342	1.38	31.40	15.86	38.64	0.58	86.48
	Biddulphia, 2.0	0.42	0.55–1.25	10	145.899	49.50	507.336	35.375	61.71	20.94	214.60	14.96	312.20
	Amphidinium, 1.0	1.75	1.45–2.15	9	9.467	1.625	12.105	0.69	16.57	2.84	21.19	1.20	41.80
	Amphidinium, 1.0	1.75	1.50–2.20	10	5.562	2.125	11.005	0.88	9.73	3.71	19.26	1.54	34.24
Euchaeta marina Prestandrea	Biddulphia, 1.0	0.4	3.00–3.30	10	8.284	12.12	719.735	1.12	3.50	5.13	304.40	0.47	313.50
	Calanoida, 0.08	93.0	1.30–4.30	9	7.255	3.88	2.147	1,000	7.43	3.97	2.20	1.02	14.62
	Calanoida, 0.14	200.0	7.20–10.20	13	1.157	3.75	2.289	0.287	240.0	790.0	480.0	60.0	1570.0
	Calanoida, 0.16	130.0	15.00–18.00	9	2.288	4.25	70.066	2.00	314.0	582.0	9599.0	274.0	10770.0
	Calanoida, 0.16	130.0	15.00–18.00	2	41.182	36.25	411.821	23.125	5642.0	4970.0	56420.0	3167.0	70199.0

TABLE 4. Diurnal balance of the consumption and utilization of energy for tropical representatives of different ecological groups

Animal	Size, mm; weight, cal	Kind and concentration of food, g/m^3	Number of animals in experiment	Assimilated food U		Unassimilated food F		Ration R		Assimilability $\frac{U}{R}$%
				cal/specimen	% body weight	cal/specimen	% body weight	cal/specimen	% body weight	
Pontellidae	2.78 0.309	*Asterionella*, 0.7	35	0.0052	1.7	0.003557	1.2	0.008804	2.8	58
	2.78 0.309	*Biddulphia*, 0.66	50	0.00118	0.35	0.003904	1.3	0.005078	1.6	24
	2.84 0.328	*Gymnodinium*, 0.66	50	0.000590	0.2	0.000150	0.05	0.000740	0.2	80
Temora stylifera Dana	1.65 0.150	*Amphidinium*, 0.5	24	0.016475	11.0	0.012600	8.4	0.029075	19.4	57
Eucalanus attenuatus Dana	3.8 0.412	Bacteria, 1.0	17	0.015876	3.8	0.022317	5.4	0.038193	9.3	42
	3.65 0.365	*Gymnodinium*, 0.35	10	0.016120	4.4	0.009005	2.5	0.025125	6.9	64
	3.78 0.398	*Amphidinium*, 0.9	8	0.021324	5.5	0.062433	15.2	0.083756	20.8	32
	3.9 0.440	*Amphidinium*, 2.0	6	0.051514	11.7	0.144627	32.9	0.196141	44.6	26
Pleuromamma abdominalis Lubbock	3.08 90.473	*Amphidinium*, 1.0	10	0.003380	0.7	0.004446	0.94	0.007826	1.6	42

The daily rations of all the species feeding on plant food at its optimum concentrations varied from 0.2 to 74.1% body weight. The lowest ration values occurred in the surface layer inhabitants Pontellidae (not over 3% body weight) and in Pleuromamma, inhabiting the 50—500-m layer (up to 7% body weight). It rose to 20—40% in inhabitants of the upper 50-m layer (Temora) and the 100—200-m layer (Eucalanus) subsisting on plant food, and to 74% body weight in Euchaeta living at depths 50—200 m.

The ration of Eucalanus feeding on bacteria was about 10%. The ration values in the case of animal food were determined only for Euchaeta; the latter's daily ration was 42—53% of the body weight at the optimum concentration of animal food.

Assimilability ranged from 1 to 80% at the optimum concentration of all kinds of food, averaging 30—50% for different species.

The ratio of the different components of the balance equation elements is depicted in Figure 2. The incorporation of tagged matter in the body at the optimum concentrations of plant and bacterial foods was insignificant in all the species investigated, varying from fractions of one percent to ~ 8% body weight. The largest value, 7.9%, was obtained for Temora, inhabiting the upper 50-m layer. The losses of tagged plant food and bacterial food in respiration were still smaller, not exceeding 2%. On the whole, the assimilated matter (energy) at the optimum tagged food concentration amounted to 0.2—11% of the body weight and 1.4—80% of the ration in all the groups.

The amount of unassimilated plant and bacterial food varied considerably more widely, from 0.05 to 73% of the body weight. The largest amount of unassimilated plant food was determined in Euchaeta. The bulk of unassimilated food in all the species consisted of solid excreta (formed feces). Liquid excreta (dissolved organic matter) amounted to 0.01—0.7% of the body weight of crustaceans subsisting on plant and bacterial food, 3—10% of the entire unassimilated food, and 0.1—5.0% of the ration.

The ratio of balance equation components was different in animal and plant food. Incorporation of matter in the body and losses of tagged food in respiration proceeded at a uniform and fairly high rate in Euchaeta, amounting to 15—16% of the body weight, or 6—30% of the ration. Hence, the entire assimilated matter averaged 30% of the body weight and 35% of the ration. The amount of unassimilated animal food may be either very large (22.2% of body weight, 88% of ration) or nearly one-half of the amount of assimilated food, i. e. 12% body weight, 34% of the ration. Formed feces constitute the bulk of unassimilated food in the case of both animal and plant rations. Liquid excreta amounts to 3—9% body weight, 4—28% of the entire unassimilated food or 3—8% of the ration.

The effect of the concentration of plant food on the ratio of balance equation elements for Pleuromamma is shown in Figure 3. When the concentration rose to 0.5 g/m^3, the absolute values of all the food balance elements increased to their maxima, while assimilability declined; the balance elements and assimilability remained unchanged as the concentration of algae increased further to 2 g/m^3. The plant food concentration 0.5 g/m^3 is therefore optimal for Pleuromamma.

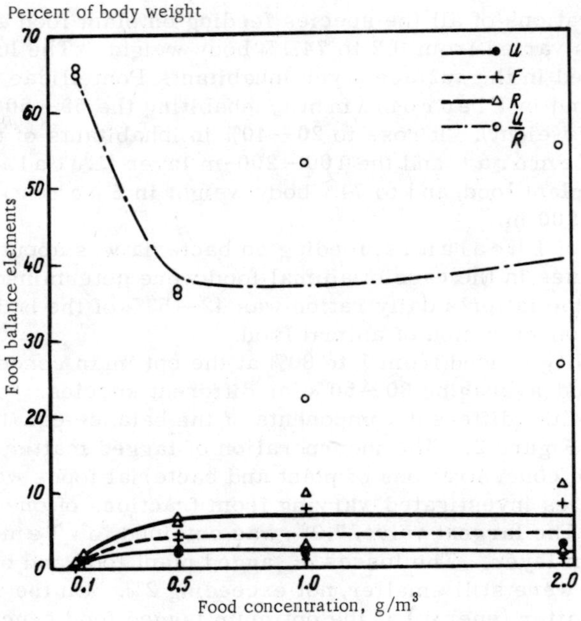

FIGURE 3. Dependence of food balance elements for Pleuromamma abdominalis upon food concentration:

U — assimilated matter; *F* — unassimilated matter; *R* — ration;

$\frac{U}{R}$ — assimilability.

DISCUSSION AND CONCLUSIONS

The results showed that the investigated animals of the tropical Pacific are capable of consuming all the groups of plankters, including bacteria, small and large algae, animals, and detritus. However, the degree of consumption and assimilation of all these foods differs, depending primarily on adaptation of the mouth parts and the entire mode of feeding to a particular type of food and the available composition and abundance of food in the water.

For unambiguous determination of feeding preferences in the food spectrum of each species, the researcher must eliminate the influence of different concentrations of food. Therefore, the indexes of assimilation of all the investigated animals for different foods were compared at the maximum food concentration occurring in nature (see Table 1). Values of the C_{assim}/C index reaching 5 — 10% and more denote satisfactory utilization and preference of the foods in question (Petipa, Sorokin, and Lanskaya, 1970). Such values in mature animals ⩾1.15 mm in size were observed in the case of animal food and large algae. Bacteria and small algae are also consumed by all the animals but are less important, C_{assim}/C = up 2 — 3%. Apparently, they are well utilized only by fine filter feeders such as Oikopleura, C_{assim}/C = 34% (Table 1) and Penilia, C_{assim}/C = 7 — 10% (Pavlova and Sorokin, 1970), and also by juvenile, nonpredatory small copepods.

Hence, animal food and large algae are preferred and better utilized by nearly all the large tropical copepods.

The maximum concentrations of forage organisms in the tropical and temperate waters occur in comparatively thin layers (Sorokin, 1971; Vinogradov, Gitel'zon and Sorokin, 1971). The total mass of formed food in such layers amounts to approximately 1 — 2 g/m^3, or 800 — 1,000 cal/m^3. These values may be two or three times higher in temperate waters.

Comparison of the dependence of feeding intensity upon food concentration (see Table 2 and Figure 1) showed that the feeding intensity of copepods became established at a more or less constant level at a food concentration of 0.5 — 4 g/m^3. Such optimum concentrations of forage items are maximal for the open ocean, although they may be one order of magnitude larger in some inland seas. The rates of consumption and assimilation of food drop sharply as the concentration falls below the optimum. Hence, the degree of consumption and assimilation of food bears a distinct relationship to its concentration in the environment.

Comparison of the food balance elements in representatives of different ecological groups of tropical plankton showed that the assimilability of a great variety of food is below 50% at the optimal food concentration and biomass (> 0.5 — 1.0 g/m^3). Assimilability above 70% was observed only at food concentrations that were considerably below the optimum, up to 0.1 — 0.3 g/m^3 (see Table 4). A similar pattern is exhibited by animals in temperate waters (Pavlova and Sorokin, 1970).

The results of calculating the theoretical food requirements of tropical plankters based on oxygen demand under laboratory conditions on board ship (Shushkina and Vilenkin, 1971) are presented in Table 5. Comparison of these data with the actual quantity of assimilated food (see Table 4) shows that none of the investigated animals satisfied its energy requirements with algal and bacterial foods even under laboratory conditions. The discrepancy should be even greater in nature, where their mobility is considerably higher (migrations!). Obviously, in order for the animals' energy requirements to be satisfied, their ration must include an obligatory animal component. This conclusion is completely in agreement with the results of experiments for determination of the food spectrum.

Opening of the intestines of certain typical phytophagous filter feeders (Calanidae and Eucalanidae) showed that these crustaceans actually subsist largely on animal food in the sea. The possibility that filter feeding copepods consume large animal food was also demonstrated by investigation of the functional properties of their oral limbs and by direct observation of live crustaceans (Petipa, 1967; Arashkevich, 1969; and others).

Hence, the relatively large (> 1.2 mm) tropical copepods may be regarded as consumers of mixed food (usually preferring large forms) irrespective of their morphological adaptation to any mode of feeding or kind of food. Apparently, the majority of copepods cannot satisfy their energy requirements in the absence of animal food. Such a broad food spectrum of tropical organisms, including larger forms and animals, is evidently due to the larger specific variety of tropical plankton (Hulbert, 1923, cited by Mullin, 1965), its quantitative paucity, and the small numerical abundance of each species. In the temperate zones, the same feeding pattern is occasionally displayed by copepods in summer, when the overall plankton abundance is low but its systematic variety is high (Zaika and Andryushchenko, 1969) or in the absence of the dominant single kind of food among the few available kinds (Petipa, 1959), and also in winter when phytoplankton is scarce (Digby, 1954; Marshall and Orr, 1955; and others). Usually, however, in contrast to tropical regions, the food requirements of animals in the temperate regions, with their large plankton biomass and its limited specific variety, can be satisfied by one kind of food, plant or bacterial. For instance Calanus hyperboreus and C. finmarchicus in northern seas subsist almost exclusively on phytoplankton (Marshall and Orr, 1955; Conover, 1962, and others). In the Black Sea Calanus helgolandicus lives and breeds the greater part of the year on account of certain groups of phytoplankton (Petipa, 1964), while Penilia avirostris subsists exclusively on small algae or bacteria (Pavlova, 1959; Pavlova and Sorokin, 1970).

TABLE 5. Respiratory requirements of tropical Copepoda — representatives of different ecological groups, at 27—28°

Animal	Size, mm; weight, cal.	Metabolic expenditures, cal/specimen	Respiratory requirements, % body weight
Pontellidae	2.78 0.309	0.05515	17.9
	2.78 0.309	0.05513	17.8
	2.84 0.328	0.05754	17.5
Temora stylifera Dana	1.65 0.150	0.03301	22.0
Eucalanus attenuatus Dana	3.8 0.412	0.06756	16.4
	3.65 0.365	0.06207	17.0
	3.78 0.398	0.06074	15.3
	3.9 0.440	0.07088	16.1
Pleuromamma abdominalis Lubbock	3.08 0.473	0.07460	15.8
	2.7 0.332	0.05746	17.3
	3.22 0.548	0.08270	15.1
	3.32 0.599	0.08824	14.7
	3.38 0.632	0.09168	14.5
Euchaeta marina Prestanbrea	2.5 0.281	0.5155	18.4
	2.55 0.332	0.05672	17.1
	4.18 1.638	0.17113	10.45

In this connection it is of interest to note the conclusions drawn by Timonin (1969), who identified several states of plankton communities differing in their trophic structure in hydrologically heterogeneous regions of the tropical region in the Indian Ocean. He demonstrated an increased zooplankton biomass, increased dominance of a few species, and an increased relative abundance of phytophagous copepods in zones with intensive upwelling, and on the other hand, a small plankton biomass, increased specific variety (no dominant species), and predominance of predatory forms in regions with stable water masses (as in our case); finally, the maximum abundance of euryphages, phytophages, and predators occurs in the zone of weak divergence and horizontal transport of waters, this ecosystem occupying an intermediate position between the above two.

Summarizing the above, it is possible to reach the following conclusion.

In the tropical regions of the ocean, with their relatively stable water masses, small biomass and large specific variety of the plankton, planktonic animals have a broad food spectrum and prefer relatively large forms, irrespective of their morphological adaptation to any specific kind of food and the mode of its capture. Animal food is obligatory for most zooplanktonic organisms. In its absence, their energy requirements are not satisfied even at high assimilation rates for plants and bacteria. The food spectrum of planktonic animals is usually narrower in the temperate regions, with their high plankton biomass, low specific variety, and dominance of a single or a few kinds of food. Their energy requirements can be satisfied by plant and/or bacterial food alone.

Therefore, the choice and consumption of specific kinds of food by planktonic animals in both the tropical and the temperate regions of the ocean depend significantly on the ratio between the concentration, variety, and partly the size of food items and consumers. Zooplankton consistently tends to be more omnivorous and predatory the higher the specific variety of the phytoplankton and zooplankton and the lower the total biomass.

Bibliography

Publications in Russian

Arashkevich, E. G. Kharakter pitaniya kopepod severozapodnoi chasti Tikhogo okeana (Feeding of Copepods in the Northwest Pacific). — Okeanologiya, Vol. 9, No. 5. 1969.

Delalo, E. P. O sutochnom ritme v pitanii *Pseudocalanus elongatus* (Boeck.) (On the Diel Feeding Rhythm of *Pseudocalanus elongatus* (Boeck.). — Trudy Sevastopol'skoi biologicheskoi stantsii, Vol. 15. 1964.

Geinrikh, A. K. O pitanii morskikh kopepod v tropicheskoi oblasti (On the Feeding of Marine Copepods in the Tropical Region). — Doklady Akademii Nauk SSSR, Vol. 119, No. 5. 1958.

Kabanova, Yu. G. Organicheskii fosfor kak istochnik pitaniya fitoplanktona (Organic Phosphorus as a Source of Food for Phytoplankton). Author's Summary of Candidate Thesis. Moskva. 1958.

Pavlova, E. V. Tsikl razvitiya i nekotorye dannye po rostu Penilia avirostris Dana v Sevastopol'skoi bukhte (The Life Cycle and Some Growth Data on Penilia avirostris in Sevastopol Bay). — Trudy Sevastopol'skoi biologicheskoi stantsii, Vol. 11. 1959.

Pavlova, E. V. and Yu. I. Sorokin. Bakterial'noe pitanie planktonnogo rachka Penilia avirostris iz Chernogo morya (Bacterial Feeding of the Planktonic Crustacean Penilia avirostris in the Black Sea). — In Sbornik: "Biologiya morya," No. 19. Produktsiya i pishchevye svyazi v soobshchestvakh planktonnykh organizmov. Kiev, "Naukova Dumka." 1970.

Petipa, T. S. Pitanie Acartia clausi Giesbr. and A. latisetosa Kritcz. v Chernom more (Feeding of Acartia clausi Giesbr. and A. latisetosa Kritcz. in the Black Sea). — Trudy Sevastopol'skoi biologicheskoi stantsii, Vol. 12. 1959.

Petipa, T. S. Sutochnyi ritm v pitanii i sutochnye ratsiony Calanus helgolandicus (Claus) v Chernom more (Diel Feeding Rhythm and Daily Rations of Calanus helgolandicus (Claus) in the Black Sea). — Ibid., Vol. 15, 1964.

Petipa, T. S. Sootnoshenie mezhdu prirostom, energeticheskim obmenom i ratsionami u Acartia clausi Giesbrecht (Relationship between Growth, Energy Metabolism, and Rations in Acartia clausi Giesbrecht). — In Sbornik: "Fiziologiya morskikh zhivotnykh." Moskva, "Nauka." 1966.

Petipa, T. S. O sposobakh dvizheniya i zakhvata pishchi u Calanus Helgolandicus (Claus) (On the Mode of Locomotion and Food Capture in Calanus helgolandicus (Claus)). — In Sbornik: "Biologiya i raspredelenie planktona yuzhnykh morei. Moskva, "Nauka." 1967.

Petipa, T. S., Yu. I. Sorokin, and L. A. Lanskaya. Issledovaniya po pitaniyu Acartia clausi Giesbr. s pomoshch'yu radiouglerodnogo metoda (Radiocarbon Studies of the Feeding of Acartia clausi Giesbr.). — In Sbornik: "Biologiya morya," No. 19. Produktsiya i pishchevye svyazi v soobshchestvakh planktonnykh organizmov. Kiev, "Naukova Dumka." 1970.

Sorokin, Yu. I. O primenenii radioaktivnogo ugleroda dlya izucheniya pitaniya i pishchevykh svyazei vodnykh zhivotnykh vnutrennikh vodoemov (Using Radioactive Carbon in Studies of the Feeding and Food Links of Aquatic Animals in Inland Waters). — Trudy Instituta biologii vnutrennikh vod Akademii Nauk SSSR, Vol. 12 (15). 1966.

Sorokin, Yu. I. Kolichestvennaya otsenka roli baterioplanktona v biologicheskoi produktivnosti tropicheskikh vod Tikhogo okeana (Quantitative Evaluation of the Role of Bacterioplankton in the Biological Productivity of Tropical Pacific Waters). (In this collection.)

Sorokin, Yu. I., T. S. Petipa, and E. V. Pavlova. Izuchenie pitaniya massovykh form planktona tropicheskoi oblasti Tikhogo okeana radiouglerodnym metodom (Radiocarbon Studies of the Feeding of Mass Plankton Species in the Tropical Pacific). — Okeanologiya, Vol. 10 No. 2. 1970.

Shushkina, E. A. and B. Ya. Vilenkin. Dykhanie planktonnykh rakoobraznykh tropicheskoi chasti Tikhogo okeana (Respiration of Planktonic Crustaceans in the Tropical Pacific). (In this collection.)

Timonin, A. G. Struktura pelagicheskikh soobshchestv. Kolichestvennoe sootnoshenie razlichnykh troficheskikh gruppirovok planktona v frontal'nykh zonakh tropicheskoi oblasti okeana (Structure of Pelagic Communities. Quantitative Ratios of Different Trophic Groups of Plankton in Frontal Zones of the Tropical Region of the Ocean). — Okeanologiya, Vol. 9, No. 5. 1969.

Troitskii, A. S. and Yu. I. Sorokin. K metodike rascheta biomassy bakterii v vodoemakh (On Techniques for Calculating Bacterial Biomass in Water Bodies). — Trudy Instituta biologii vnutrennikh vod Akademii Nauk SSSR, Vol. 19. 1967.

Vinberg, G. G. and S. I. Anisimov. Matematicheskaya model' vodnoi ekosistemy (Mathematical Model of an Aquatic Ecosystem). — In Sbornik: "Fotosinteziruyushchie sistemy vysokoi produktivnosti." Moskva, "Nauka." 1966.

Vinogradov, M. E., I. I. Gitel'zon, and Yu. I. Sorokin. O prostranstvennoi strukture soobshchestv evfoticheskoi zony tropicheskikh vod okeana (The Spatial Structure of Communities in the Euphotic Zone of the Ocean Tropics). (In this collection.)

Zaika, V. E. and A. A. Andryushchenko. "Taksonomicheskoe raznoobrazie" fito- i zooplanktona Chernogo morya ("Systematic Variety" of Phytoplankton and Zooplankton in the Black Sea). — Gidrobiologicheskii Zhurnal, Vol. 5. No. 3. 1969.

Publications in Other Languages

Conover, R. J. Metabolism and Growth in *Calanus hyperboreus* in Relation to its Life Cycle. — Rapp. Cons. Internat. Explor. Mer., Vol. 153. 1962.

Digby, P. S. The Biology of the Marine Planktonic Copepods of Scoresby Sound, East Greenland. — J. Anim. Ecol., Vol. 23. 1954.

Marshall, S. M. and A. P. Orr. The Biology of a Marine Copepod, *Calanus finnmarchicus* (Gunnerus). Edinburg, London, Oliver and Boyd. 1955.

Mullin, M. M. On the Feeding Behavior of Planktonic Marine Copepods and the Separation of their Ecological Niches. Research on Marine Food Chain, 1963—1964. University of California, Institute of Marine Resources. Report for the period June 1963 — December 1964, No. 65—6. 1965.

Mullin, M. M. Selective Feeding by Calanoid Copepods from the Indian Ocean. In: H. Barnes (Ed.). Some Contemporary Studies in Marine Science. London, Allen and Unwin. 1966.

Parsons, S. R., R. J. LeBrasseur and J. D. Fulton. Some Observations on the Dependence of Zooplankton Grazing on the Cell Size and Concentration of Phytoplankton Blooms. — J. Oceanogr. Soc. Japan, Vol. 23, No. 1. 1967.

BACTERIOPLANKTON AS FOOD FOR PELAGIC MARINE ORGANISMS

E.V. Pavlova, T.S. Petipa, and Yu.I. Sorokin
Institute of Biology of the Southern Seas of the Academy of Sciences of the Ukrainian SSR and Institute of Biology of Inland Waters of the Academy of Sciences of the USSR

The importance of bacteria as food for aquatic animals is not a new study topic (Clarke and Gellis, 1935; Fuller and Clarke, 1936; Beklemishev, 1954; Marshall and Orr, 1955; Lebedeva, 1959; Jørgensen, 1962; Wood, 1967). Most researchers have investigated representatives of the family Calanidae, mainly the genus Calanus, and have shown that the Calanidae consume bacteria to a varying extent but that the bacteria do not fully satisfy their food requirements. The possibility that purely bacterial food could be sufficient has been experimentally demonstrated for the fine filter feeder Penilia avirostris (Pavlova and Sorokin, 1970). However, the number of pure filter feeders among the marine zooplankton is very limited, as the majority of zooplankters are "graspers" — mixed feeders and predators. Hence, the significance of bacterioplankton in the food chains of the pelagic is still not clear.

The 44th cruise of the "Vityaz" included research into the role of bacterioplankton in the diet of pelagic marine organisms. Studies were conducted from December 1968 through February 1969 by the radiocarbon method (Sorokin, 1966).

The experimental techniques for marine animals and the subsequent calculations have been described by Sorokin, Petipa, and Pavlova, 1970; Petipa, Pavlova, and Sorokin, 1971; Sorokin, 1971.

RESULTS

Since bacteria are present in seawater as single cells and in aggregates (Jones and Jannasch, 1959; Jannasch and Jones, 1959 Sorokin, 1971), experiments were conducted to clarify the degree to which dispersed bacterial cells and their aggregates are assimilated. Adult crustaceans, Undinula darwini, and a mixture of small Calanoida were given bacterioplankton consisting of individual or aggregated bacteria. Both groups assimilated the aggregated bacterioplankton 10 times more intensively than the single cells (Figure 1). All subsequent experiments were performed with tagged natural bacterioplankton, i.e. bacterioplankton including aggregates.

TABLE 1. Initial data of experiments for determination of the assimilation index of bacterioplankton-fed tropical zooplankters with different modes of feeding

Animal	Length of body, mm	Weight, mgC	Food concentration, g/m³	C_r of food, mg C/pulse · 10^{-6}	Ultimate radioactivity of animal's body, pulses/day	C_{assim}/C
Filter feeders						
Appendicularia	2.50	0.0011	1.4	1.10	110	11.00
Same	2.50	0.0011	1.4	1.10	504	50.30
"	1.83	0.0004	1.4	1.06	156	41.20
Salpa maxina	60.00	2,0900	1.5	1.13	14,980	0.80
Mixture of Calanoida	2.14	0.0280	1.4–2.0	0.91	173	0.57
Same	1.75	0.0150	1.4–2.0	0.91	290	1.75
"	2.28	0.0340	0.6	1.085	141	0.45
"	1.85	0.0190	0.9	1.085	413	2.53
Eucalanus attenuatus	3.95	0.0490	1.4–2.0	0.90	519	0.95
Same	3.95	0.0490	1.4–2.0	0.90	350	0.95
"	3.90	0.0405	0.45	1.185	281	0.82
Rhincalanus cornutus	3.09	0.0620	1.4–2.0	1.10	26	0.05
Same	3.25	0.0720	1.4–2.0	1.10	58	0.09
Mixed feeders						
Pontellidae	2.50	0.0300	1.4–2.0	1.13	240	1.82
Same	2.50	0.0300	1.4–2.0	1.13	271	1.02
"	2.50	0.0300	1.4–2.0	1.13	372	1.40
Acartia sp.	1.28	0.0036	1.4–2.0	0.91	127	3.22
" "	1.28	0.0036	1.4–2.0	0.91	90	2.88
Pleuromamma abdominalis	4.00	0.0570	0.3	1.04	100	0.18
"Graspers"						
Eucaeta marina	3.20	0.0660	1.4–2.0	0.90	85	0.11
Euchaeta volfendeni	2.45	0.0280	1.4–2.0	0.90	27	0.09
Euchaeta, copepodites	2.05	0.0287	1.4–2.0	0.90	13	0.05
Haloptilus sp.	2.85	0.0118	1.4–2.0	1.10	1	0.01
Same	1.88	0.0034	1.4–2.0	1.18	16	0.56
"	1.87	0.0033	1.4–2.0	1.18	13	0.45
Copilia sp.	4.70	0.0260	1.4–2.0	1.18	34	0.16
Ostracoda	1.30	1.3000	1.4–2.0	1.18	38	0.003
Flyingfish larvae	4.10	0.0680	1.4	1.25	8.3	0.010

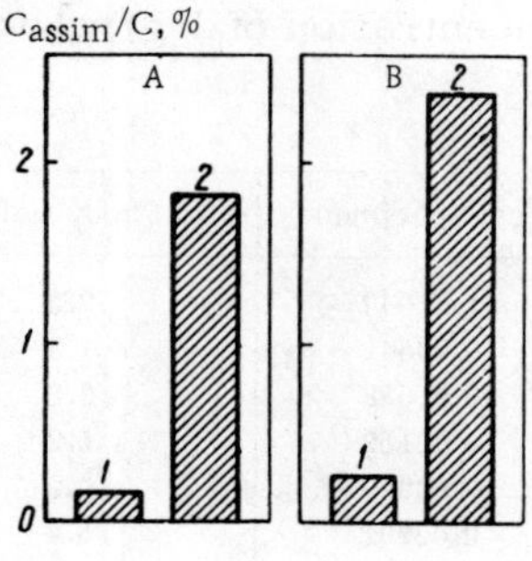

FIGURE 1. "Assimilation index" C_{assim}/C in Calanoida fed single bacterial cells (1) and natural bacterioplankton (2):

A — Undinula darwini; B — mixture of Calanoida.

The comparative rates of consumption and assimilation of bacterioplankton were determined for 14 species differing in mode of feeding (Table 1). The highest assimilation index was found for Appendicularia, 11.0 – 50.3%. Contrary to expectation, Salpae assimilated the bacterial food poorly, $C_{assim}/C = 0.8\%$. Apparently the large salps did not take well to the experimental conditions, as was shown by their very low activity in the vessels. The C_{assim}/C values were 0.05 – 2.53% for various filter-feeding Copepoda feeding on bacteria. The mean C_{assim}/C value for the entire group of plankton filter feeders was 8.57%. The assimilation index varied from 0.18 to 3.22% (average 1.75%) for the group of Crustacea with mixed feeding. The comparative rates of consumption and assimilation of bacterial food were still lower in the "graspers," $C_{assim}/C = 0.003 - 0.56\%$, averaging 0.16%. Flyingfish larvae consumed only very small amounts of bacteria, the assimilation index being 0.01%. Figure 2 shows the relative rate of consumption of different food items including bacteria in five species of tropical zooplankton and larvae of one species of flyingfish. The concentrations of bacterial and algal food were approximately the same in these experiments, 2 g/m^3, only the concentration of blue-green algae was 2 – 3 g/m^3. As a rule, the animal food was supplied to excess in the experiments. In the case of filter feeders (Calanidae, *Eucalanus*) the C_{assim}/C values for consumption of bacterioplankton were approximately of the same order of magnitude as those for algal food. In the case of grasper copepods (*Euchaeta*, *Copilia*) fed with animal food the C_{assim}/C values were one or two orders of magnitude larger than in the experiments involving bacterioplankton as food.

Figure 3 presents the findings on the dependence of the rate of feeding of different zooplankters on bacterioplankton upon the food concentration. The experimental bacterial concentration was chosen in accordance with the natural concentrations in the sea. The average biomass of bacteria in the water was about 30 mg/m^3, but it reached 300 mg/m^3 in the layers of maximum concentration (Sorokin, 1971). With the rise in the concentration of bacterioplankton within this range, i. e. from 0.03 to 0.3 g/m^3, in experiments with a mixture of small Calanoida (mean body weight 0.019 mgC), the assimilation index rose from 0.07 to 2.0%. For the same experimental food concentrations, the C_{assim}/C values increased from 0.06 to 0.5% for a mixture of large Calanoida (mean weight 0.03 mgC) and from 0.06 to 1.1% for *Eucalanus*; the C_{assim}/C values remained uniformly low throughout the range of bacterioplankton concentrations in experiments with *Pleuromamma*. The bacterioplankton concentration 0.3 g/m^3 is the lower boundary of the optimum, and its subsequent rise to 1.0 g/m^3 was not accompanied by any significant increase of the assimilation index in all the species investigated.

Assimilability and daily ration were determined in females of *Eucalanus attenuatus* at a bacterial concentration of 1 g/m^3, with the following results:

Balance equation elements	cal/specimen	% of body weight
Incorporated in the body	1.01047	2.5
Losses of tagged food in respiration	0.00541	1.3
Total assimilated matter	0.01588	3.8
Excreted as feces	0.02152	5.2
Excreted as dissolved organic substance	0.00079	0.2
Total unassimilated matter	0.02231	5.4
Daily ration	0.03819	9.2

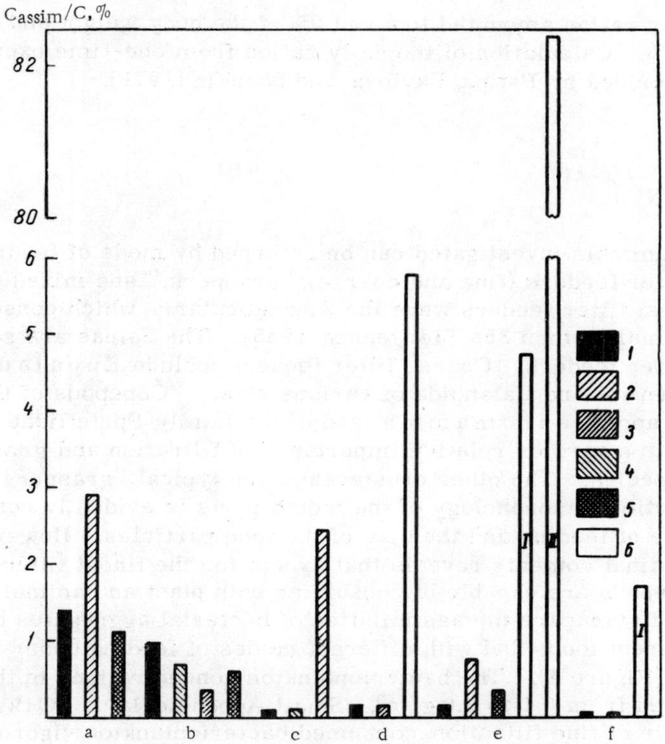

FIGURE 2. Comparison of the assimilation index C_{assim}/C for tropical zooplankters feeding on plant, animal and bacterial food:

1 — bacterioplankton; 2 — *Biddulphia catenata*; 3 — *Amphidinium klebsii*; 4 — *Gymnodinium lanskoi*; 5 — *Oscillatoria* sp.; 6 — animal food: I. Calanoida mixture, II — *Undinula darwini*; a — Calanoida mixture; b — *Eucalanus attenuatus*; c — *Rhincalanus cornutus*; d — *Copilia* sp.; e — *Euchaeta marina*; f — fish larvae.

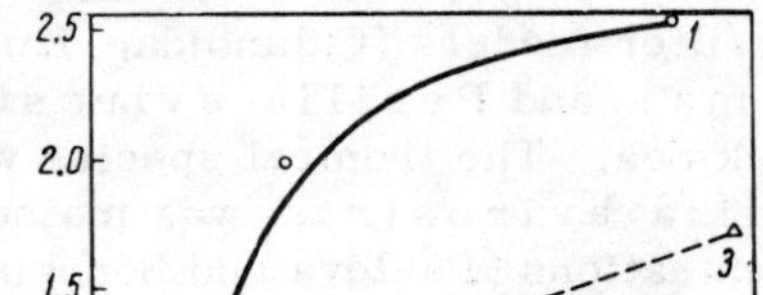

FIGURE 3. The assimilation index C_{assim}/C as a factor of the bacterioplankton concentration:

1 — Calanoida mixture, mean weight 0.019 mgC;
2 — Calanoida mixture, mean weight 0.03 mgC;
3 — *Eucalanus attenuatus*; 4 — *Pleuromamma abdominalis*.

The daily ration amounted to about 9% of the body weight and assimilability was 42%. Calculation of the daily ration from one-time experimental data is described by Petipa, Pavlova, and Sorokin (1971).

DISCUSSION

All the animals investigated can be referred by mode of feeding to three groups: filter feeders (fine and coarse), "graspers," and mixed feeders.

The finest filter feeders were the Appendicularia, which consume only particles smaller than 35μ (Jørgensen, 1955). The Salpae are somewhat coarser filter feeders. Coarse filter feeders include Eucalanus, Rhincalanus and Calanoida of various size. Copepods of the genera Acartia and Pleuromamma and of the family Pontellidae are mixed feeders, with a various relative importance of filtration and grasping in different species. The other crustaceans are typical "graspers."

The functional morphology of the mouth parts is evidently controlled by the mode of feeding and the size of the food particles. However, analysis of the intestinal contents reveals that except for the finest filter feeders, all the copepods are capable of consuming both plant and animal food. This enabled us to compare the assimilation of bacterial aggregates by animals eating different foods and with different modes of feeding under natural conditions (Figure 4). The bacterioplankton concentrations in the experiments varied from 0.1 to 2.0 g/m^3. Small Appendicularia (Oikopleura sp.), capable of fine filtration, consumed bacterioplankton vigorously, with an assimilation index of 34%. Obviously bacterioplankton may serve as the staple diet of Appendicularia in nature. Coarse filter feeders and mixed feeders fed bacterioplankton assimilated it to a much lower degree than fine filter feeders (assimilation index not above 3.0%). Bacteria can apparently provide only part of the food of animals in this group. For species which are chiefly consumers of animal food the assimilation indexes were low, 0.003 to 0.3%, when bacterial food was given. Hence, bacterioplankton does not play a significant role in the feeding of typical "graspers." The bacterioplankton assimilation indexes of fine filter feeders (mainly phytophages), coarse filter feeders + mixed feeders and graspers (mostly predators) are in the approximate ratios 100:10:1.

Figure 5 shows the assimilation indexes of bacterial food for three groups of tropical zooplankton — coarse filter feeders (Calanoida, Eucalanus), mixed feeders (Pleuromamma), and Penilia avirostris, a typical fine filter feeder from the Black Sea. The tropical species were fed natural bacterioplankton, while Penilia avirostris was maintained on disperse bacterial cells in four concentrations (Pavlova and Sorokin, 1970). When the bacterioplankton concentration is raised to 1.5 g/m^3, Penilia consumes and assimilates the bacterial food at a rate ten times higher than that of coarse filter feeders and mixed-feeding crustaceans. This ratio of the consumption rate for bacterioplankton may be assumed to be the same in nature and consequently bacterioplankton of special importance for organisms that are capable of filtering small food particles.

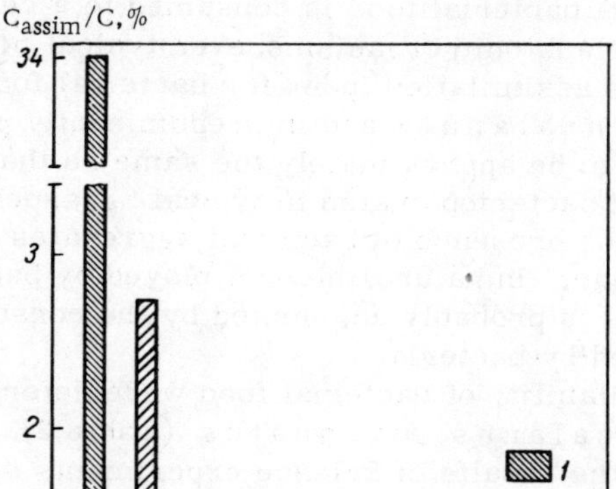

FIGURE 4. Comparison of the degree of assimilation of bacterioplankton of an equal concentration in various tropical organisms:

1 — phytophages; 2 — mixed feeders; 3 — predators; a — Appendicularia; b — Acartia; c — Calanoida; d — Pontellidae; e — Eucalanus; f — Rhincalanus; g — Haloptilus; h — Copilia; i — Ostracoda; j — fish larvae.

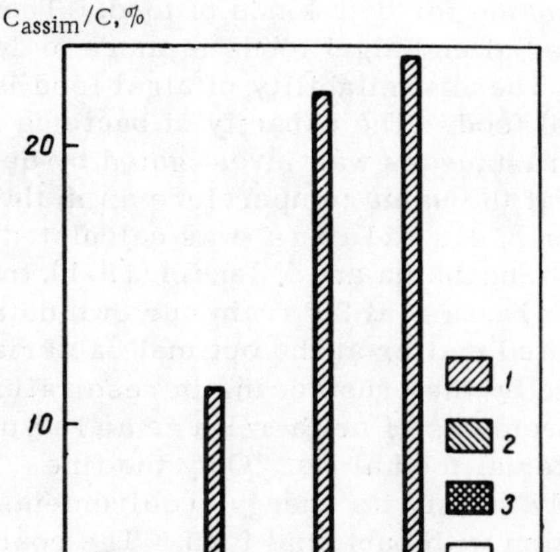

FIGURE 5. Assimilation indexes for bacterial food of planktonic crustaceans at different concentrations of bacteria:

I — 0.03—0.04 g/m^3; II — 0.1—0.2 g/m^3; III — 0.3—0.5 g/m^3; IV — 0.9—1.5 g/m^3; 1 — Penilia avirostris; 2 — Calanoida mixture; 3 — Eucalanus attenuatus; 4 — Pleuromamma abdominalis.

On the other hand, bacterial food is consumed to a certain extent by nearly all tropical planktonic organisms, even typical predatory "graspers." In certain cases, the assimilation index for bacterial food in mixed feeders (Eucalanus, Rhincalanus) and in predominantly predatory forms (Copilia) proves to be approximately the same as that for plant food. The accessibility of bacterioplankton to typical "graspers" is apparently due exclusively to the presence of bacterial aggregates attaining the size of large cells of algae. In nature, the part played by bacterial food in the diet of such animals is probably augmented by the consumption of detritus particles surrounded by bacteria.

Values of assimilability of bacterial food were determined in balance experiments on Eucalanus attenuatus (Table 2). For comparison Table 2 also gives the results of balance experiments on two planktonic crustaceans from the Black Sea at the same concentration (1.0 g/m^3) of single bacterial cells and algae. Each experimental consumer was fed algae of a suitable size: Eucalanus was fed on Amphidinium klebsii, Paracalanus on Exuviaella cordata and Penilia on Pedinella sp. All three consumer species are filter feeders, but they feed on different types and sizes of food items. The fine filter feeder Penilia consumes plant food not coarser than 8μ; Paracalanus is a phytophage feeding on algae larger than 20μ, while Eucalanus is an euryphage which consumes animal food in addition to algae. When fed bacteria, Penilia consumed a large daily ration (118% of its own weight), with a comparatively low assimilation rate, 27%. Given algae, the same crustacean consumed rather less daily (67% of its weight), with a considerably higher assimilation rate, 75%; however, the total amount of assimilated food was similar on both the bacterial and the algal diet — 17.7 and 28% of its own weight. Paracalanus daily consumed 24 and 5.0% of its body weight, respectively, when fed bacteria and algae; the assimilation rate was somewhat higher for algae than for bacteria, 77% as against 55%. The daily ration of bacteria for Eucalanus was still smaller, 9.2% of its weight. The proportion of assimilated food (2.5 and 4.0%) and the rate of assimilation (42 and 31%) were approximately the same for both kinds of food. These results suggest that the value of bacterial and algal foods is more or less the same, at an equal concentration. The assimilability of algal food is somewhat higher than that of bacterial food. The capacity of bacteria for satisfying the food requirements of crustaceans was investigated by determining total losses in respiration at the same temperature as in the feeding experiments. The respiration of Eucalanus was calculated at 27% according to the data reported by Shushkina and Vilenkin (1971), the respiration of Penilia and Paracalanus at 20° from our own data. Comparison of the energy of assimilated matter at the optimal bacterial concentration and the energy expended by the crustaceans in respiration showed (Figure 6) that the food requirements of neither Paracalanus nor Eucalanus can be met by bacterial food alone. Only the fine filter feeder Penilia can completely satisfy its energy requirements for respiration, growth and reproduction with bacterial food. The coarse filter feeders subsisting on mixed food are apparently unable to subsist on bacterial food alone even in the presence of an excessive concentration (up to 1.5 g/m^3).

TABLE 2. Daily ration and assimilation in sexually mature filter-feeding crustaceans fed on bacteria and algae (concentration, 1–2 g/m^3)

Species	Kind of food	Assimilated				Not assimilated				Ration		Rate of assimilation, %
		losses in respiration		incorporated in the body		feces		dissolved organic matter				
		cal/specimen	% of body weight	cal/specimen	% of body weight	cal/specimen	% of body weight	cal/specimen	% of body weight	cal/specimen	% of body weight	
Penilia avirostris	Bacteria	0.00260	14.5	0.00320	17.7	0.01550	85.8	—	—	0.0213	118.0	27
	Algae	0.00380	21.0	0.00500	28.0	0.00260	14.3	0.00080	4.7	0.0120	67.0	75
Paracalanus parvus	Bacterioplankton	0.00050	4.6	0.00100	8.7	0.00120	11.0	—	—	0.0027	24.3	55
	Algae*	0.00014	1.2	0.00031	2.7	0.00016	1.1	—	—	0.0006	5.0	77
Eucalanus attenuatus	Bacterioplankton	0.00500	1.3	0.01000	2.5	0.02100	5.2	0.00080	0.2	0.0380	9.2	42
	Algae	0.03500	0.9	0.01800	4.0	0.06100	14.8	0.00150	0.4	0.0840	20.7	31

* The crustaceans fed poorly in these experiments.

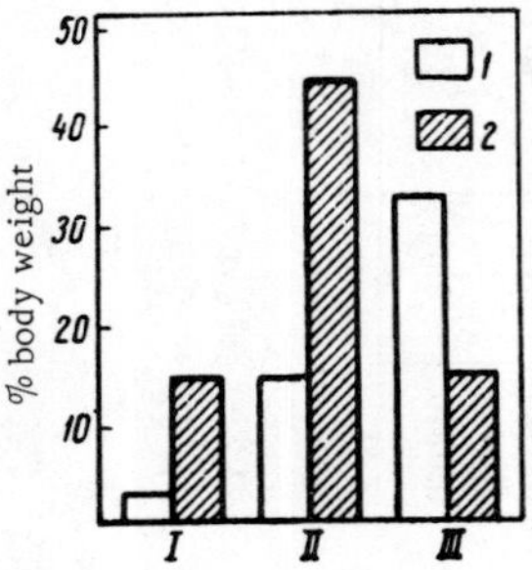

FIGURE 6. Comparison of the daily amount of assimilated bacterial food (concentration $1.0-1.5\ g/m^3$) in balance experiments (1) with food requirements calculated from daily oxygen consumption (2) in Eucalanus attenuatus (I), Paracalanus parvus (II) and Penilia avirostris (III), % body weight

The findings described suggest the following.

1. Tropical filter-feeding animal plankters assimilate natural bacterioplankton consisting of 30–40% bacterial aggregates at a rate 10 times higher than the assimilation rate for single bacterial cells at the same concentration.

2. The capacity of organisms with different modes of feeding for consuming and assimilating natural bacterioplankton was investigated in 14 mass species. The highest assimilation index was found in the fine filter feeders Appendicularia, averaging 34.0%. The index for the entire group of filter feeders was found to be 8.57%. The mean C_{assim}/C value is 1.75% for mixed feeders. The smallest amounts of bacterial food are consumed by the "graspers," $C_{assim}/C = 0.16\%$.

3. The values of the assimilation index for bacterioplankton in the case of phytophagous fine filter feeders, mixed feeders and predators were in the ratio 100:10:1.

4. An increase of the bacterial food concentration from average values ($0.03\ g/m^3$) to the maximum values encountered in the sea ($0.3\ g/m^3$) raises the index of assimilation by different plankters by a factor of 5–20, but a further increase of the concentration up to $1.0\ g/m^3$ leads only to an insignificant increase of the C_{assim}/C values.

5. The bacterial and algal foods are of equal value, at an equal concentration, for filter-feeding crustaceans. The assimilability of algal food is somewhat higher than that of bacterial food.

6. Planktonic crustaceans which are mixed feeders cannot satisfy their food requirements (energy losses in respiration) by consuming only natural bacterioplankton at the optimum concentration of $0.1 - 2.0\ g/m^3$. Bacterial food alone is not sufficient for the normal functioning of these animals. Bacteria can satisfy the food requirements of only fine filter feeders.

Bibliography

Publications in Russian

Beklemishev, K. V. Pitanie nekotorykh massovykh planktonnykh kopepod v dal'nevostochnykh moryakh (Feeding of Some Mass Planktonic Copepods in Far Eastern Seas). — Zoologicheskii Zhurnal, Vol. 33, No. 6. 1954.

Lebedeva, M. N. Bakterial'nye niti, vynesennye iz serovodorodnykh glubin Chernogo morya, kak vozmozhnyi ob"ekt pitaniya zooplankterov-fil'tratorov na primere Calanus helgolandicus (Bacterial Filaments Carried From the Hydrogen Sulfide-Contaminated Depths of the Black Sea as a Possible Forage Item for Filter-feeding Zooplankters as Exemplified by Calanus helgolandicus). — Trudy Sevastopol'skoi biologicheskoi stantsii, Vol. 11. 1959.

Pavlova, E. V. and Yu. I. Sorokin. Bakterial'noe pitanie planktonnogo rachka *Penilia avirostris* Dana iz Chernogo morya (Bacterial Feeding of the Planktonic Crustacean *Penilia avirostris* Dana from the Black Sea). — In Sbornik: "Biologiya morya," No. 19. Produktsiya i pishchevye svyazi v soobshchestvakh planktonnykh organizmov. Kiev, "Naukova Dumka." 1970.

Petipa, T. S., E. V. Pavlova, and Yu. I. Sorokin. Izuchenie pitaniya massovykh form planktona tropicheskoi oblasti Tikhogo okeana radiouglerodnym metodom (Radiocarbon Studies of the Feeding of Mass Plankton Forms in the Tropical Zone of the Pacific). (In this collection.)

Sorokin, Yu. I. O primenenii radioaktivnogo ugleroda dlya izucheniya pitaniya i pishchevykh svyazei vodnykh zhivotnykh (Radioactive Carbon as an Aid to Studying the Feeding and Food Relations of Aquatic Animals). — Trudy Instituta biologii vnutrennikh vod Akademii Nauk SSSR, Vol. 12 (15). 1966.

Sorokin, Yu. I. Kolichestvennaya otsenka roli bakterioplanktona v biologicheskoi produktivnosti tropicheskikh vod Tikhogo okeana (Quantitative Evaluation of the Role of Bacterioplankton in the Biological Productivity of Tropical Pacific Waters.) (In this collection.)

Sorokin, Yu. I., T. S. Petipa, and E. V. Pavlova. Kolichestvennoe issledovanie pishchevoi roli morskogo bakterioplanktona (Quantitative Studies of the Trophic Role of Marine Bacterioplankton). — Okeanologiya, Vol. 10, No. 2. 1970.

Shushkina, E. A. and B. Ya. Vilenkin. Dykhanie planktonnykh rakoobraznykh tropicheskoi chasti Tikhogo okeana (Respiration of Planktonic Crustaceans in the Tropical Pacific). (In this collection.)

Publications in Other Languages

Clarke, G. L., and S. S. Gellis. The Nutrition of Copepods in Relation to the Food-Cycle of the Seas. — Biol. Bull., Vol. 68; No. 2. 1935.

Fuller, G. L. and G. L. Clarke. Further Experiments on the Feeding of *Calanus finmarchicus*. — Biol. Bull., Vol. 70, No. 2. 1936.

Jannasch, H. W. and G. E. Jones. Bacterial Populations in Sea Water as Determined by Different Methods of Enumeration. — Limnol. and Oceanogr., Vol. 4, No. 2. 1959.

Jones, G. E. and H. W. Jannasch. Aggregates of Bacteria in Sea Water as Determined by Treatment with Surface Active Agents. — Limnol. and Oceanogr., Vol. 4. No. 2. 1959.

Jørgensen, C. B. Quantitative Aspects of Filter Feeding in Invertebrates. — Biol. Reviews, Cambridge, Vol. 30, No. 4. 1955.

Jørgensen, C. B. The Food of Filter-Feeding Organisms. — Rapp. cons. internat. Explor. Mer., Vol. 153. 1962.

Marshall, S. M. and A. P. Orr. The Biology of a Marine Copepod, *Calanus finmarchicus* (Gunnerus). Edinburgh, London. 1955.

Wood, E. I. F. Microbiology of Oceans and Estuaries. Amsterdam, London. 1967.

CONSUMPTION OF PHYTOPLANKTON AND BACTERIA BY TROPICAL EUPHAUSIIDS

L.A. Ponomareva, E.A. Tsikhon-Lukanina, and Yu.I. Sorokin

Shirshov Institute of Oceanology of the Academy of Sciences of the USSR and Institute of Biology of Inland Waters of the Academy of Sciences of the USSR

The ecology of tropical euphausiids, a group of substantial importance in the food chains of the sea, has been little studied, and as a rule knowledge of their biological characteristics is fragmentary or lacking. This applies to feeding patterns in particular. Some researchers (Paulsen, 1909; Lebour, 1924; Barkley, 1940) thought that the euphasiids feed mainly on phytoplankton, especially diatoms, while others (Hickling, 1925) claimed that euphausiids are detriphages. According to Einarsson (1945), Mauchline (1967) and Ponomareva (1963), the Euphausiacea are mixed feeders, although some species feed exclusively on phytoplankton or zooplankton.

The composition of food and modes of feeding are well known for antarctic, Far Eastern and some North Atlantic species; in contrast, there is only one work on tropical species (Ponomareva, Naumov, and Zernova, 1962), and that only of a tentative nature.

Most authors who have studied the feeding of euphasiids have dealt mostly with the morphology of the mouth parts and the quantitative composition of the food. Lasker (1966) was the first to carry out a quantitative investigation of the feeding of Euphausia pacifica. He fed the euphausiids with animal food, although the representatives of the genus Euphausia are for the most part filter feeders.

METHODS

The work was carried out in January — February 1969 in the Coral and Solomon seas during the 44th cruise of the "Vityaz," involving three euphausiids: Euphausia diomedeae, E. brevis and Thysanopoda tricuspidata. Feeding was quantitatively assessed by the radiocarbon method. The experimental animals were adults caught at depths down to 30 — 50 m, this because the survivial rate of euphausiids taken from deeper layers is poor under laboratory conditions (Ponomareva, 1967). The crustaceans were caught with a plankton net and kept in aquariums for 4 — 12 hours, after which they were transferred to vessels with capacities of 1 — 2 liters containing seawater that had been previously passed through a membrane filter. One or two animals were used for each experiment.

They were fed green algae Chlamydomonas sp. peridinians Amphidinium klebsii, diatoms Biddulphia catenata and bacteria, both accumulated on suspended particles and single cells. All the food had previously been tagged with C^{14}. The algae were separated from the water on membrane filters and treated with 0.005% hydrochloric acid for removal of residual radioactive carbonate. The washed tagged phytoplankton was used to prepare a mixture that was fed to the crustaceans. The concentratior of food in the experiments was around 0.5 g/m^3. The water temperature was 27–28°. The experimental procedures have been described by Sorokin (1966, 1971) and Petipa, Pavlova, and Sorokin (1971).

We carried out 82 experiments for determination of the rations and 7 for determination of the assimilability of algae and bacteria. The variability of the empirical results was evaluated by means of standard deviations (σ) and the coefficient of variation (CV).

RESULTS

Determination of the daily rations of crustaceans in short-term experiments calls for studies of their diurnal feed rhythm. One of the authors had performed previous studies of the feeding of the same euphausiids using material from the Indian Ocean. It was shown that a diurnal rhythm is virtually absent in E. diomedeae aged one year, whereas two-year-olds feed most intensively at night and in the morning. We carried out observations of the intensity of feeding of E. diomedeae on the peridinian Amphidinium klebsii at different times during the 24-hr period, in the morning, during the day and at night. A slight increase of feeding intensity was noted in the night and morning hours, thus:

Time, hr	Experiment 1	Experiment 2	Average
5.00 – 7.00	118 $mgC \cdot 10^{-6}$ hours/specimen	122	120
14.00 – 16.00	106 " "	101	103
24.00 – 02.00	95 " "	130	113

In scaling the amount of food consumed by the euphausiids to the diurnal period, we assumed there to be no diurnal feeding rhythm in the animals.

Tables 1 and 2 give the relative mean diurnal rations of Euphausia diomedeae, E. brevis and Thysanopoda tricuspidata. The algal and bacterial rations of these euphausiid species are very small, 0.06 – 1.80%.

E. diomedeae of 10–12 mm feeds best on bacteria aggregated on particulate matter and blue-green algae, whereas green algae, diatoms and peridinians are consumed by young E. diomedeae in much smaller amounts. The relative mean diurnal rations are lower in the case of older E. diomedeae of 14–18 mm. The larger crustaceans also prefer diatoms and bacteria aggregated on particles.

TABLE 1. Relative mean diurnal rations in *Euphausia diomedeae*

Kind of food	Number of experiments	C_r of food, mgC/pulse · 10^{-6}	Size of animal, mm	Weight of one specimen, mgC	Daily amount of food consumed by one specimen		Ration, %	
					pulses/min	mgC · 10^{-3}	average	coefficient of variation
Green algae	6	1.25	11	0.72	2,538 ± 284	3.38 ± 0.37	0.46 ± 0.05	11
Blue-green algae	7	7.5	12	0.77	1,240 ± 124	9.30 ± 0.94	1.20 ± 0.11	10
Diatoms	10	1.48	11	0.70	2,860 ± 687	4.82 ± 1.16	0.62 ± 0.25	24
			15	1.32	3,819 ± 382	5.00 ± 0.50	0.34 ± 0.036	10
Peridinians	2	1.72	11	0.70	1,872	3.22	0.46	—
			14	1.30	854	0.81	0.06	—
Bacteria on particles	8	1.20	10	0.62	7,112 ± 1,560	12.72 ± 2.80	1.80 ± 0.40	22
			18	3.10	10,149 ± 1,015	12.17 ± 1.22	0.40 ± 0.04	10
Single bacterial cells	7	1.20	16	1.93	976 ± 244	1.17 ± 0.30	0.08 ± 0.02	25

TABLE 2. Relative mean diurnal rations of *Thysanopoda tricuspidata* and *Euphausia brevis*

Species	Kind of food	Number of experiments	C_r of food, mgC/pulse · 10^{-6}	Size of animal, mm	Weight of one specimen, mg C	Daily amount of food consumed by one specimen		Ration, %	
						pulses/min	mgC · 10^{-3}	average	coefficient of variation
Thysanopoda tricuspidata	Green algae	8	1.25	15	2.0	4,900 ± 177	6.113 ± 2.2	0.22 ± 0.08	36
	Blue-green algae	10	7.5	22	3.5	1,700 ± 595	12.52 ± 4.4	0.40 ± 0.14	35
	Diatoms	5	1.48	20	2.9	3,308 ± 1,650	5.65 ± 1.8	0.19 ± 0.10	50
	Peridinians	9	1.72	22	3.4	9,573 ± 3,830	16.55 ± 6.6	0.50 ± 0.20	40
	Bacteria on particles	4	1.20	20	2.7	949 ± 315	1.14 ± 0.4	0.06 ± 0.02	33
Euphausia brevis	Peridinians	6	1.72	12	0.7	2,013 ± 202	3.47 ± 0.4	0.55 ± 0.06	10

Specimens 15 — 22 mm in size were used in experiments with Thysanopoda tricuspidata. This species consumed large amounts of peridinians and blue-green algae while green algae and diatoms were eaten less vigorously. In contrast to the case with E. diomedeae, less bacterioplankton was consumed by Th. tricuspidata than all other kinds of food.

E. brevis rejected blue-green and green algae, while the amounts of peridinians consumed by this species were the same as those consumed by the two other euphausiids.

We determined the assimilation of green algae, peridinians and bacterioplankton in the case of Euphausia diomedeae (Table 3). Over the experimental period (8 — 10 hours), a small fraction of assimilated carbon, 2 — 17%, is liberated as CO_2 by respiration. However, the bulk of consumed food, 50 — 90%, is incorporated in the body. Most of the unassimilated food, 8 — 19%, is excreted with the solid fraction (feces) and only 0.4 — 8% as liquid excreta. The assimilation rates (*A*/*R*) of the food consumed by this species is high, 74 — 91%. It is highest for green algae, 91% and slightly lower for peridinians, 82% and bacterioplankton, 74%. The coefficients of variation of the assimilation rate are small, 2 — 4%.

According to Lasker (1966), Euphausia pacifica assimilated animal food (nauplii of Artemia salina) to the average extent of 84% (varying from 66 to 95%).

TABLE 3. Assimilation of algae and bacteria by Euphausia diomedeae aged one year (13 — 14 mm)

Kind of food	Number of experiments	pulses/min					mgC · 10^{-6}			Assimilation, %	
		Consumed food, *R*	unassimilated food: liquid excreta	unassimilated food: feces	assimilated food: in the body	assimilated food: lost in respiration	consumed food, *R*	unassimilated food	assimilate food, *A*	average	coefficient of variation
Green algae	3	838	3	65	750	20	1144	94	1050	91 ± 4	4
		100	0.4	8	90	2	—	—	—		
Peridinians	2	367	6	55	289	17	349	58	291	82 ± 3	4
		100	2	15	78	5	—	—	—		
Bacteria on particles	2	519	40	100	292	87	620	164	456	74 ± 2	2
		100	8	19	56	17	—	—	—		

Hence, the rate of food consumption by the tropical euphausiid Euphausia diomedeae under experimental conditions is practically the same at different hours during the diurnal period. The relative mean daily rations of the investigated euphausiid species feeding on phytoplankton and bacteria are very low. E. diomedeae consumes bacteria aggregated on particles in larger amounts than single bacterial cells. The rates of assimilation of phytoplankton and bacteria by E. diomedeae is fairly high, algae being assimilated better than bacteria.

Bibliography

Publications in Russian

Petipa, T. S., E. V. Pavlova, and Yu. I. Sorokin. Izuchenie pitaniya massovykh form planktona tropicheskoi oblasti Tikhogo okeana radiouglerodnym metodom (Radiocarbon Studies of the Feeding of Mass Plankton Forms in the Tropical Zone of the Pacific). (In this collection.)

Ponomareva, L. A. Evfauziidy severo-zapadnoi chasti Tikhogo okeana, ikh rasprostranenie i ekologiya massovykh vidov (Distribution of Euphausiids of the Northwestern Pacific and the Ecology of Mass Species). Moskva, Izdatel'stvo Akademii Nauk SSSR. 1963.

Ponomareva, L. A. Otchet o rabotakh biologicheskogo otryada v 40 reise e/s "Vityaz'" (Report of the Work of the Biological Division during the 40th Cruise of the R/V "Vityaz"). — Biblioteka Instituta Okeanologii Akademii Nauk SSSR. 1967.

Ponomareva, L. A., A. G. Naumova, and V. V. Zernova. Pitanie nekotorykh vidov evfauziid v Indiiskom okeane (Feeding of Some Euphausiid Species in the Indian Ocean). — Trudy Instituta okeanologii Akademii Nauk SSSR, Vol. 58. 1962.

Sorokin, Yu. I. O primenenii radioaktivnogo ugleroda dlya izucheniya pitaniya i pishchevykh svyazei vodnykh zhivotnykh (Radioactive Carbon as an Aid in Studying the Feeding and Food Relations of Aquatic Animals). — Trudy Instituta biologii vnutrennikh vod Akademii Nauk SSSR, Vol. 12(15). 1966.

Sorokin, Yu. I. Kolichestvennaya otsenka roli bakterioplanktona v biologicheskioi produktivnosti tropicheskikh vod Tikhogo okeana (Quantitative Evaluation of the Role of Bacterioplankton in the Biological Productivity of Tropical Pacific Waters). (In this collection.)

Publications in Other Languages

Barkley, E. Nahrung und Filterapparat des Walkrebses Euphausia superba Dana. — Ztschr. Fischer. u. Hilfswissenschaften, Beiheft 1, Walforsch., 1. 1940.

Einarsson, H. Euphausiacea. Northern Atlantic Species. — Dana Report, Vol. 27. 1945.

Hickling, C. Notes on Euphausiids. — J. Mar. Biol. Assoc. U. K., Vol. 13, No. 3. 1925.

Lasker, R. Feeding, Growth, Respiration and Carbon Utilization of an Euphausiid Crustacean. — J. Fish. Res. Bd. Canada, Vol. 23, No. 9. 1966.

Lebour, M. The Euphausiidae in the Neighbourhood of Plymouth and their Importance as Herring Food. — J. Mar. Biol. Assoc. U. K. Vol. 13, No. 2. 1924.

Mauchline, J. Feeding Appendages of the Euphausiacea (Crustacea). — J. Zool., Vol. 153, No. 1. 1967.

N e m o t o, T. Feeding Pattern of Euphausiids and Differentiations in their Body Characters. — Inform. Bull. Planktonology, Japan. Comm. No. to Dr. J. Matsue's 60th birthday. 1967.

P a u l s e n, O. Plankton Investigations in the Waters around Iceland and in the North Atlantic in 1904. — Meddr. Komm. Danm. Fisk. — og Havunders., Ser. Plankton, Vol. 1, No. 8. 1909.

EVALUATION OF THE PRODUCTION OF TROPICAL ZOOPLANKTON

E.A. Shushkina
Shirshov Institute of Oceanology of the Academy of Sciences of the USSR

The need to study production processes of oceanic zooplankton at different trophic levels arose in connection with the general trend of research in marine and freshwater hydrobiology.

Research along these lines, performed mainly for fresh waters, resulted in several methods for determination of the production of aquatic animals (Vinberg, 1968).

The chief methods for calculating the production of animal populations with continuous recruitment may be considered to be the graphic method as modified by Vinberg (Pechen' and Shushkina, 1963) and Greze (Greze and Baldina, 1964), the physiological method (Vinberg, 1966) and the radiocarbon method (Chmyr, 1967). The most reliable is the graphic method, based on analysis of the growth curves and age composition of populations. Although the method is fully applicable to freshwater animals, it cannot be applied yet to tropical zooplankters because attempts to raise open ocean zooplankters, especially from tropical regions, in aquariums or breeding chambers, necessary to obtain the growth curves, have so far been unsuccessful. We therefore had to resort to the other methods, radiocarbon and physiological, for evaluation of the rate of production of tropical zooplankton during the 44th cruise of the "Vityaz" in the southwest Pacific.

METHODS

Radiocarbon determinations of zooplankton production were performed in situ at stations 6029 (polygon No. 1), 6052 (polygon No. 3), 6064 (polygon No. 4) and 6065 (polygon No. 5) (Vinogradov, 1971).

A plexiglass water sample of 100-liter capacity was used as the breeding chamber. An ampule with a volume of about 15 ml fixed in the upper part of the sampler was filled with a mixture of tagged algae before the experiment as food for phytophagous animals caught in the sampler. A rubber plug served as the bottom of the ampule. Before the sampler was cast overboard, the ampule with the tagged algae was stoppered on top with a similar rubber plug pierced loosely by a glass rod, the tip of which protruded above the upper edge of the sampler's walls. When the sampler was closed at a predetermined depth its lid struck the protruding tip of the rod which slid into the upper stopper and pushed the bottom stopper from the ampule, so that the latter's content flowed into the sampler. It was established by preliminary experiments that the algae so introduced were distributed uniformly throughout the sampler. The sampler containing the water sample and tagged algae was exposed at depths of 5 to 50 m for 9–12 hours. The experiments were carried out at night in order to avoid photosynthesis by phytoplankton in the sampler, since this would alter the tag of the food (Shushkina and Sorokin, 1969). The water temperature at 0–50 m ranged between 28 and 30°.

The experimental setup and conditions are shown in Table 1.

TABLE 1. Experimental setup and conditions for radiocarbon measurements of microzooplankton production

Polygon No.	Experiment No.	Sampling depth, m	Duration of experiment, hours	Tagged experimental algae	Biomass, mg/m³: filter feeders, B_1	Biomass, mg/m³: predators, B_2	Biomass, mg/m³: total, B_{tot}	$B_2 : B_{tot}$, %	Mean length (l), mm: phytophages, l_1	Mean length (l), mm: predators, l_2	Mean wet weight (W_1), µg: phytophages, W_1	Mean wet weight (W_1), µg: predators, W_2
1	1	5	9	Gymnodinium	27.5	35.6	63.1	56	0.26	0.47	2.0	7.5
3	2	5	11	Chlamydomonas	2.7	4.0	6.7	60	0.60	0.66	5.5	15.0
	3	25	9		6.5	15.8	22.3	71	0.55	2.60	13.9	103.5
	4	10	12		6.4	3.5	9.9	36	0.53	0.66	6.4	31.8
	5	50	16		8.5	9.0	17.5	51	0.45	1.14	9.0	32.4
4	6	50	9	Amphidinium	36.6	8.1	44.7	17	0.48	0.23	23.0	22.8
	7	50	13		25.8	19.6	45.4	43	0.50	1.20	19.6	51.9
	8	50	13		50.5	16.9	67.4	25	0.77	0.97	45.7	36.6
5	9	10	11	Amphidinium	62.4	68.8	131.2	52	0.47	0.59	7.8	40.8
Mean								46	0.5	1.0	14.8	38.0

The radioactively tagged foods for experimental animals were C^{14}-tagged cultures of the algae Gymnodinium, Chlamydomonas and Amphidinium. The algae were concentrated prior to the experiments by filtration through a No. 5 membrane filter and washed with slightly acidulated water, followed by determination of radioactivity and biomass. The quantity of tagged algae introduced into the sampler was adjusted so that their biomass in the course of the experiments should not be more than double the natural phytoplankton concentration in the region of the investigations, and that the radioactivity of the water in the sampler should not be less than 500 pulses/min·l.

In the morning following exposure the sampler was hoisted aboard and a water sample of 7–8 liters was taken for determination of the concentration of organic suspended matter (k) and its radioactivity (r). The water remaining in the sampler, ~90 liters, was filtered through a funnel made of No. 76 bolting cloth (mesh 0.12×0.12 mm). The zooplankton remaining on the cloth was washed with acidulated water to remove dissolved radioactive carbonate and fixed with formalin. The sample was then examined under a binocular microscope for counting the number of animals by genera and for determination of their size (l). All the animals brought up by the sampler were divided into two trophic groups: phytophages with biomass B_1 and predators with biomass B_2. A certain number of animals from each group was transferred to a glass plate (1/3 of an ordinary slide), lightly washed with distilled water for removal of salts, comminuted and uniformly distributed over the glass. The preparations were dried and the respective radioactivities of the phytophages (r_1) and predators (r_2) were measured on a PP-8 counter. The results of these measurements are listed in Tables 1 and 2.

TABLE 2. Results of measurements of specific radioactivities of phytophagous microzooplankters and organic particulate matter in experimental radiocarbon determinations of production

Polygon No.	Experiment No.	Duration of experiment, hours	Organic particulate matter: mass (k), cal/l*	radioactivity (r) pulses/l	r/k, pulses/cal	Phytophages: r_1, pulses/g	biomass B_1, µg in the preparation	r_1/B_1: pulses/mg	pulses/cal
1	1	9	2.61	1,500	575	28	345	81	162
3	2	11	1.50	220	147	5	242	21	41
	3	9	2.85	10,410	3,650	181	588	308	616
	4	12	1.42	13,680	9,480	575	314	1,813	3,060
	5	16	1.56	11,370	7,275	1,517	729	2,080	4,160
4	6	9	1.21	4,080	3,967	951	3,604	264	528
	7	13	3.12	7,640	2,450	1,208	2,215	545	1,090
	8	13	1.00	2,660	2,660	1,088	2,500	435	870
5	9	11	4.10	11,360	2,770	86	485	177	354
Mean	—	11	2.15	—	3,666	—	—	—	1,276

* One gram wet weight of zooplankton corresponds to 500 cal.

It was assumed that in the course of exposure the phytophagous animals with known biomass B_1 caught in the sampler, consuming the C^{14}-tagged food of known concentration k and radioactivity r, in turn acquired radioactivity r_1. The quantity k includes, in addition to the introduced biomass of tagged algae, the phytoplankton, bacterioplankton and detritus already present in the sample. Having determined the quantities k, r, B_1 and r_1 as described above, it is possible to calculate the diurnal accretion of C^{14} in the bodies of phytophages by the following formula (Sorokin, 1966, our notation):

$$\frac{C_{\text{assim}}}{C} = \frac{r_1}{B_1} \cdot \frac{k}{r} \cdot \frac{1}{t}. \qquad (1)$$

The quantity C_{assim}/C apparently does not characterize the production intensity of the animals and cannot be regarded as close to the diurnal *P/B* coefficient as was assumed by Sorokin, who proposed it (1966).

If the animals under investigation do not grow during the experimental period, i. e. their production is nil, or if they even lose weight under adverse experimental conditions, i. e. their production is "negative," they may still consume some quantity of the tagged food, thereby acquiring radioactivity. However, in this case one cannot speak of any production or *P/B* coefficients, although positive and even fairly high values may be obtained from measurements of the animals' radioactivity in conjunction with equation (1).

Therefore, in order to calculate the production and *P/B* coefficient from the measured C_{assim}/C value, formula (1) was modified by introducing the correction for metabolism derived from Vinberg's equation (1964) with certain assumptions, thus:

$$\frac{C_{assim}}{C} = \frac{r_1}{W_1} \cdot \frac{k}{r} \cdot \frac{1}{1-K_2}[1 - e^{-\left(\frac{T}{W_1}\right)\cdot t}], \tag{2}$$

where

$$K_2 = \frac{P}{P+T}. \tag{3}$$

In equation (3), *P* and *T* respectively stand for production and metabolism stated in identical units. Then the desired *P/B* coefficient for the phytophages caught in the sampler can be determined by the formula

$$\frac{P}{B} = \frac{T_1}{W_1} \cdot \frac{C_{assim}/C - \varkappa}{\varkappa}, \tag{4}$$

where $\varkappa = 1 - e^{-\left(\frac{T_1}{W_1}\right)}$, while $\frac{T_1}{W_1}$ is the respiration intensity stated in fractions of body weight W_1.

Therefore, in order to calculate the desired *P/B* coefficient of the zooplankton under investigation, the C_{assim}/C value determined by the radiocarbon method must be corrected for respiration according to equation (4).

Here and further the term "production" or "rate of production" is designated *P* and stated in terms of weight (g, mg) or energy (cal) units referred to unit time (in this case, the 24-hr period). The term "production intensity" signifies the rate of production per unit biomass and is designated *P/W* in the case of single animals and *P/B* in the case of groups of animals or a population. Production intensity *P/B* is stated in terms of percentages or fractions usually for a 24-hr period. Analagous notation is applied to feeding and respiration, *T* and *R* respectively denoting the rates of respiration and feeding (ration), while $\frac{T}{W}$ and $\frac{R}{W}$ denote the respiration (metabolism) and feeding intensities.

The other method used for evaluation of the production intensity of tropical zooplankton was the physiological method (Vinberg, 1966), which does not involve direct measurements of production. Production intensity is calculated from known relationships of physiological indexes, some of which were determinated experimentally.

One such index is the animals' rate of respiration Q, the relationship of which to the individual weight is usually expressed by an equation of the type

$$Q = aW^{\alpha}, \tag{5}$$

where Q is the rate of oxygen consumption by a single animal (ml O_2/hr); W, weight of animal (g wet weight); a, proportionality factor equal to the respiratory rate of an animal weighing 1 g; α is a constant.

If Q and W are stated in the same units, for instance calories, formula (4) becomes

$$T = a_1 W^{\alpha}. \tag{5a}$$

The sought production rate P is related to the metabolic rate T through the coefficient of utilization of assimilated food by growth K_2 (formula (3)). Then from (3) and (5a) it is possible to derive the following equation for determination of the magnitude of production:

$$P = a_1 W^{\alpha} \cdot \frac{K_2}{1-K_2}, \tag{6}$$

and hence

$$P/W = a_1 W^{\alpha-1} \cdot \frac{K_2}{1-K_2}. \tag{7}$$

Thus, it is seen from (7) that calculation of production using the physiological method requires knowledge of three quantities, respiratory rate of investigated animals (T), their weight (W) and coefficient of utilization of assimilated food for growth (K_2). The first two quantities were determined during the 44th cruise of the "Vityaz."

The dependence of respiration rates T upon body weight W, obtained in collaboration with B. Ya. Vilenkin for eleven species of tropical copepods, can be described by the formula

$$T = 1.01\, W^{0.71}, \tag{5b}$$

where $[T]$ = mcal/specimen · day and $[W]$ = mcal/specimen. The body weight W was calculated from the relationships $W(l)$ derived in collaboration with A. P. Ostapenya for the same copepods. The dependence formulas for weight (g, mg) or mass of body stated in mcal (cal · 10^{-3}) proved different for copepods with different forms of the body (Ostapenya and Shushkina, 1971). They were used for determination of the body weight of investigated animals from measured length l in calculations of regression $T(W)$. The third quantity necessary for production calculation by the physiological method — K_2 — was assumed to be constant and equal to 0.25 — 0.3 for the entire population of planktonic copepods. This assumption, based on the author's and published data, should be checked and defined more accurately.

The radiocarbon method for measuring the production of populations, though apparently promising, has not yet been finally elaborated, while the physiological method is based on as yet untested assumptions. Therefore the results obtained by these two methods can yield only an approximate evaluation of the intensity of production of tropical zooplankton.

RESULTS

The modified radiocarbon method for zooplankton production measurements made it possible to estimate the production rate of filter-feeding animals subsisting on small algae, bacterioplankton and detritus (Shushkina and Sorokin, 1969). Since the 100-liter water sampler caught more or less only small zooplankton (large animas were very seldom caught), the values of ***P/B*** coefficients obtained characterized the production intensity of small zooplankton comprising animals of average size 0.5 — 1.0 mm. Paracalanus, Clausocalanus, Macrosetella, and Microsetella were the most common components of this microzooplankton. All were classified in the group of phytophages. All species of the genera Acartia, Eucaeta and Candacia and all Cyclopoida and Chaetognatha were classified as predators. The biomass of predators averaged 45% of the total microzooplankton biomass. This is somewhat better than the ratio of predators to phytophages in fresh waters, where the biomass of predators averages some 3% of the total biomass of zooplankton (Shushkina, 1966). In our observations the average size of tropical phytophages was approximately one-half that of predators. The average weight of predators was the higher the greater the average weight of the phytophages in nearly all the polygons and in all the experiments (see Table 1).

The total microzooplankton biomass in the 1st, 4th and especially the 5th polygon was higher than in the 3rd polygon, where it amounted to about 14 mg wet weight per m^3. With such a low biomass, the number of animals caught by the 100-liter sampler was so small as to make measurements of their radioactivity difficult and inaccurate. The radioactivities and specific radioactivities (r_1/B_1) of the phytophages in Table 2 were therefore measured in only one replication. In experiment No. 2, with the lowest microzooplankton biomass (about 7 mg wet weight per m^3), the radioactivity of the preparation of phytophages r_1 exceeded the background by only 5 pulses (see Tables 1 and 2). A 100-liter sampler cannot, therefore, be used for measurements of the production of phytophages by the radiocarbon method in the described modification unless the microzooplankton biomass is at least 30 — 50 mg/m^3.

The results of measurements of the specific radioactivity of nonpredatory microzooplankton (see Table 2) and organic suspended matter were used to calculate the diurnal incorporation of C^{14} in the crustaceans, C_{assim}/C using equation (1).

The phytophages were assumed to feed on organic suspended matter, the experimental concentration k and radioactivity r of which are given in Table 2. This is one of the dubious assumptions that had to be adopted for the radiocarbon method, and it may have led to very high C_{assim}/C values in practically all the experiments, since the quantity r in equation (1) stands for the mass of the entire organic suspended matter (calories), whereas r signifies the radioactivity of only the introduced tagged algae. If the experimental animals are assumed to consume only the tagged algae, the latter's biomass being k, then the quantity k/r and consequently also C_{assim}/C will be much smaller, approaching the C_{assim}/C values determined by Petipa and Pavlova in the same cruise (Petipa, Pavlova, and Sorokin, 1971). However, one can hardly agree with such an assumption.

The correct procedure seems to be to measure the quantity k as the total concentration of introduced tagged algae and the phytoplankton and bacterioplankton already in the sampler. Furthermore, the constancy of k must be monitored throughout the experiment and throughout the volume of water in the sampler. But all these measurements are very labor-consuming and were impossible to carry out. Moreover, the experimental animals do not always actively consume the tagged algae from the laboratory cultures offered to them. The use of tagged algae from laboratory cultures as radioactive food for animals under investigation may well have to be given up eventually and it will be necessary to try and attach the C^{14} tag to natural phytoplankton. This, however, will involve methodological difficulties related to the need to concentrate the tagged phytoplankton and separate it from the zooplankton and detritus. Once all these difficulties are eliminated, the radiocarbon method will be more reliable.

In view of the above, the P/B coefficients determined by the radiocarbon method must be regarded merely as provisional values, although for reasons not yet clarified, they turned out to be close to the results obtained by the physiological method, which seems to us more reliable (Table 3).

Therefore the diurnal P/B coefficients were calculated by the physiological method, using the ratio of metabolic rate to weight $T = 1.01W^{0.71}$ and at $K_2 = 0.25 - 0.3$ for predators and phytophages of the tropical micro- and mesozooplankton (Table 3).

TABLE 3. Results of measurements of production intensity of tropical microzooplankton according to the radiocarbon and physiological methods ($\frac{T}{W}$, C_{assim}/C and $\frac{P}{B}$ stated in terms of daily percentages)

Experiment No.	Phytophages							Predators			
	Weight W, mcal	$\frac{T}{W}$	by C^{14}			Physiological, P/B,		Weight W, mcal	$\frac{T}{W}$	Physiological, P/B,	
			C_{assim}/C	P/B	K_2	$K_2 = 0.25$	$K_2 = 0.3$			$K_2 = 0.25$	$K_2 = 0.3$
1	1.0	100	75	18	15	33	43	3.8	69	23	30
2	2.3	80	61	10	11	26	34	7.5	56	18	24
3	7.0	53	45	2	3	19	25	51.8	32	11	14
4	3.2	72	76	20	22	24	31	15.9	45	15	20
5	4.5	66	91	59	47	22	28	16.2	45	15	19
6	12.5	48	42	4	1	16	21	11.4	50	17	22
7	9.8	52	82	54	51	17	22	25.9	39	13	17
8	22.9	41	60	33	45	14	18	18.3	44	14	19
9	3.9	68	28	—	—	23	29	20.4	42	14	18
Mean	7.4	65	67	25	24	22	28	19.0	47	16	20

As already pointed out, animals up to 1 mm long caught poorly with plankton nets but comparatively satisfactorily with the 100-liter sampler were classified as microzooplankton.

Animals longer than 1 mm, escaping the sampler with comparative ease but caught with plankton nets of No. 38 bolting cloth, were classified as mesozooplankton.

The group of mesozooplanktonic predators included juvenile and adult specimens of Candacia aethiopica, adult Centropages gracilis, all Eucaeta, Pontellina and Chaetognatha. Animals such as Calanus papuer, Lucicutia flavicornis, Pyrocypris sp., Euphausia sp., Pleuromamma sp., Undinula vulgaris, Temora discaudata, Scolecithrix danae, Eucalanus sp. and Neocalanus sp. were classified as mesoplanktonic phytophages.

The ***P/B*** coefficients for both microzooplanktonic and mesozooplanktonic predators proved to be lower than those for nonpredatory zooplankton, since according to preliminary data the mean weight of predators was 2.5 times that of phytophages in the investigated plankton community (Table 4).

The important part played by microzooplankton in the life activities of the entire plankton community is often underestimated because of the considerable undercatching of oceanic microzooplankton with common plankton nets of No. 38 bolting cloth. By tentative estimates, production of nonpredatory microzooplankton is close to that of nonpredatory mesozooplankton, although the latter's biomass per m^3 is considerably larger.

TABLE 4. Results of determinations of size ($\bar{l}$), weight (W), metabolic rate (T) and intensity (T/W) and P/B coefficients for different trophic groups of microzooplankton and mesozooplankton (B_2, biomass of predators)

Size group of zooplankton	Trophic level	$\bar{l}$, mm	$\bar{W}$, μg	$\frac{W_2}{W_1}$	$\frac{B_2}{B_{tot}}$, %	$T(W)$ mcal/daily	P/B, % daily	P/B, by C^{14} %	K_2,%	P/B by the physiological method
Microzooplankton	Phytophages	0.5	15	2.6	54	$1.0W^{0.71}$	65	25	24	22–28
	Predators	1.0	38		46		47	—	—	15–20
Mesozooplankton	Phytophages	—	140	2.3	42	$1.0W^{0.71}$	22	—		6–10
	Predators	—	320		58		16	—		5–8

The dependence of the diurnal production values ***P*** determined by the physiological and radiocarbon methods for microzooplankton and mesozooplankton upon the mean body weight is shown in Figure 1. Having determined the mean weight ***W*** of animals of a particular size group, species, genus, it is possible to make an approximate estimate of the diurnal production of each individual and then of an entire group and population from the above relationship ***P(W)***.

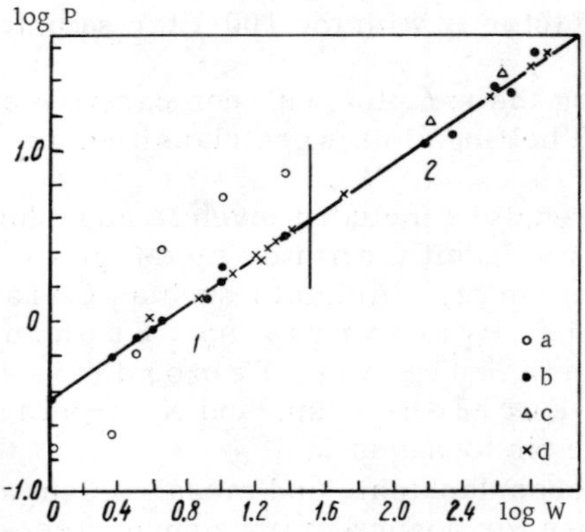

FIGURE 1. Production of tropical zooplankton determined by the radiocarbon and physiological methods in relation to the animals' mean weight W:

Region 1 — for nonpredatory microzooplankton in the sampler; Region 2 — for mesozooplankton; a — radiocarbon method; b - d — physiological method: b — phytophages, c — euryphages, d — predators.

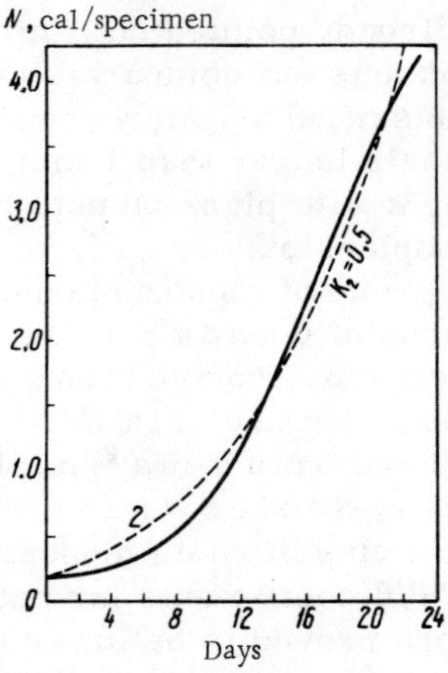

FIGURE 2. Growth of mysids (Neomysis mirabilis) from the Sea of Japan (Peter the Great Bay):

1 — empirical curve; 2 — curve calculated by the physiological method for $K_2 = 0.5$ and $T = 0.14\,W^{0.82}$.

As already pointed out, neither the radiocarbon nor the physiological method constitutes direct methods for measuring the production of aquatic animals. Moreover, the radiocarbon method must be better elaborated while the physiological method involves certain assumptions that must be verified. Therefore it was necessary to compare the results obtained by the physiological method with direct measurements of the animals' production based on analysis of growth curves and the size composition of the investigated population. This was done in 1969 for mysids (Neomysis mirabilis) from the Sea of Japan (Peter the Great Bay).

Mysids raised in breeding chambers in the sea at a depth of 1.5 m were subjected to periodic measurements of size, body weight and oxygen consumption in relation to age from the moment of hatching almost to sexual maturity. By using the relationship $T = 0.14\,W^{0.82}$ established in these experiments and the coefficient $K_2 = 0.5$, which is usual for intensively growing animals, the growth curve of the mysids (shown with dashes in Figure 2) was calculated by the physiological method. The solid line in Figure 2 depicts the growth curve for the same mysids constructed from direct measurements of body weight in the course of growth. As was to be expected, there was a small discrepancy between the empirical and calculated growth curves, especially in the first and last experimental periods. However, these discrepancies were not serious, and when these two curves were used to calculate the daily P/B coefficient for the entire growing mysid population, the results were quite similar, 13% with the direct, graphic method, 17% with the physiological method. A similar agreement of the results obtained with these two methods occurred in the case of the freshwater mollusk Sphaerium corneum with S-type growth (Arabina and Rubinova, 1968).

Comparison of the results of evaluating production intensity by the physiological and direct methods applied to such systematically and ecologically dissimilar animals as mysids from the Sea of Japan — mixed feeders — and freshwater mollusks *Sphaerium corneum* — filter feeders — suggests that the physiological method permits an approximate evaluation of the intensities of production in populations of aquatic animals with continuous recruitment, provided the weight of the animals and their intensity of respiration are known. The coefficient of utilization of assimilated food for growth (K_2) for the entire population may be taken as $K_2 = 0.25 - 0.3$, on the basis of the following considerations. Analysis of the author's and published data (Shushkina, Anisimov, and Klekowski, 1968; Shushkina, 1968; Welch, 1968) indicates that it is possible to adopt a K_2 value for growing animals close to 0.5. This coefficient is obviously lower for an entire population consisting of growing individuals, adult females and males. However, $K_2 = 0.5$ may be assumed for all individuals in a population by taking into account the increment of the biomass of adult animals due to formation of sexual products. This assumption is apparently more justified for marine and freshwater copepods forming populations with continuous recruitment, i. e. with a fairly continuous production of eggs and young. Therefore the K_2 value may be taken to be close to 0.5 for individuals of all ages in a population for copepods and possibly for other aquatic animals. It should be noted, however, that in practice, K_2 calculations are based on respiratory rates determined for sluggish animals in the respirometer, i. e. basal metabolism (Sushchenya, 1969). Obviously, the metabolism of animals actively moving in their natural state is somewhat higher, but probably not more than 2 — 3 times higher (Sushchenya, 1969). Then assuming the basal metabolism to be $1/2 - 1/3$ of the active metabolism, the K_2 value calculated by the formula $K_2 = \frac{P}{P+T}$ is found to be rather $0.25 - 0.3$ instead of 0.5.

The possibility of estimating the intensity of production of oceanic copepods from their body length is attractive. An attempt at deriving such a relationship for tropical copepods was undertaken on the basis of relationships between body weight W (stated in mcal) and body length (l, mm) established in collaboration with A. P. Ostapenya (Table 5). By using these relationships and also the dependence of metabolic rate upon individual weight $T = 1.01\, W^{0.71}$ and the value $K_2 = 0.25$, equation (7) yielded a formula for P/B as a function of body length l. These relationships were derived for three groups of tropical calanoids with different form of the body.

TABLE 5. Body weight W and diurnal P/B coefficient as functions of body length l for certain tropical calanoids ($K_2 = 0.25$; $T = 1.01W^{0.71}$), $[l]$ = mm, $[W]$ = mcal/specimen, $[P/B]$ = daily fraction

Geneva	Ratio of body length to width	$W(l)$	$P/B(l)$
Scolecithrix, Temora, Euchirella, Candacia	2.5–2.8	$32.8\, l^{3.76}$	$0.12\, l^{-1.1}$
Neocalanus, Undinula, Euchaeta, Pleuromamma, Pontelina	3.5–3.7	$49.1\, l^{2.59}$	$0.11\, l^{-0.75}$
Eucalanus, Euaugaptilus	4–5	$23.1\, l^{2.41}$	$0.14\, l^{-0.7}$

Although such an attempt probably appears to be premature and to some extent absurd, it is nevertheless quite possible to presume that these or at least similar relationships between production intensity and one or several comparatively readily determinable parameters characterizing an investigated population or biocenosis will be taken up before long.

It is worthwhile to note that according to tentative estimates, the values of the *P/B* coefficients for zooplanktonic organisms in ponds, lakes, seas and oceans do not differ as sharply as was expected, although the phytoplankton biomass in ponds and the tropical oceanic regions differ by an average factor of 50,000 and the zooplankton biomass by a factor of 2,000.

Bibliography

Publications in Russian

Arabina, I. P. and L. A. Rubinova. Skorost' rosta i intensivnost' obmena u nekotorykh predstavitelei presnovodnogo bentosa (Growth and Metabolic Rate of Some Representatives of Freshwater Benthos). — Doklady Akademii Nauk BSSR, No. 2. 1968.

Chmyr, V. D. Radiouglerodnyi metod opredeleniya produktsii zooplanktona (Radiocarbon Determination of Zooplankton Production). — Doklady Akademii Nauk, Vol. 173, No. 1. 1967.

Greze, V. N. and E. P. Baldina. Dinamika populyatsii i godovaya produktsiya Acartia clausi Giesbr. i Centrophages kröyeri Giesbr. v neriticheskoi zone Chernogo morya (Population Dynamics and Annual Production of Acartia clausi Giesbr. and Centrophages kröyeri Giesbr. in the Neritic Zone of the Black Sea). — Trudy Sevastopol'skoi biologicheskoi stantsii, Vol. 17. 1964.

Ostapenya, A. P. and E. A. Shushkina. Kaloriinost' setnogo planktona i energeticheskie ekvivalenty massy tela nekotorykh tropicheskikh planktonnykh rakoobraznykh (Caloricity of Net Plankton and Energy Equivalents of the Body Mass of Some Tropical Planktonic Crustaceans). (In this collection.)

Pechen', G. A. and E. A. Shushkina. Produktsiya rachkovogo zooplanktona v ozerakh raznogo tipa (Production of Crustacean Zooplankton in Lakes of Different Types). — Tezisy X konferentsii po vnutrennim vodoemam Pribaltiki. Minsk, Izdatel'stvo Belorusskogo gosudarstvennogo universiteta. 1963.

Petipa, T. S., E. V. Pavlova, and Yu. I. Sorokin. Izuchenie pitaniya massovykh form planktona tropicheskoi oblasti Tikhogo okeana radiouglerodnym metodom (Radiocarbon Studies of the Feeding of Mass Plankton Forms in the Tropical Zone of the Pacific). (In this collection.)

Shushkina, E. A. Sootnoshenie produktsii i biomassy zooplanktona ozer (Production to Biomass Ratio in Lake Zooplankton). — Gidrobiologicheskii Zhurnal, Vol. 2, No. 1. 1966.

Shushkina, E. A. Raschet produktsii kopepod na osnove zakonomernostei obmena i koeffitsienta izpol'zovaniya usvoennoi pishchi na rost (Calculation of the Production of Copepods Based on Metabolic Regularities

and the Coefficient of Utilization of Assimilated Food for Growth). — Okeanologiya, Vol. 8, No. 1. 1968.
Shushkina, E. A. and Yu. I. Sorokin. K metodike opredeleniya produktsii zooplanktona radiouglerodnym metodom (Contribution to the Radiocarbon Method for the Determination of Zooplankton Production). — Okeanologiya, Vol. 9, No. 4. 1969.
Sorokin, Yu. I. O primeneni radioaktivnogo ugleroda dlya izucheniya pitaniya i pishchevykh svyazei vodnykh zhivotnykh (Radioactive Carbon as an Aid in Studying the Feeding and Food Relations of Aquatic Animals). — Trudy Instituta biologii vnutrennikh vod Akademii Nauk SSSR, Vol. 12(15). 1966.
Sushchenya, L. M. Kolichestvennye zakonomernosti metabolizma i transformatsii veshchestva i energii rakoobraznymi (Quantitative Patterns of Metabolism and Transformation of Matter and Energy by Crustaceans). Doctoral Thesis. Moskva. 1969.
Vinberg, G. G. Puti kolichestvennogo izucheniya potrebleniya i usvoeniya pishchi vodnymi zhivotnymi (Methods for a Quantitative Study of the Consumption and Assimilation of Food by Aquatic Animals). — Zhurnal Obshchei Biologii, Vol. 25, No. 4. 1964.
Vinberg, G. G. Skorost' rosta i intensivnost' obmena u zhivotnykh (Growth and Metabolic Rate in Animals). — Uspekhi sovremennoi biologii, Vol. 61, No. 2. 1966.
Vinberg, G. G. (editor). Metody opredeleniya produktsii vodnykh zhivotnykh (Methods for Determining the Production of Aquatic Animals). Minsk, Izdatel'stvo "Vysheishaya Shkola." 1968.
Vinogradov, M. E. Issledovaniya funktsionirovaniya okeanicheskikh biologicheskikh sistem (Studies of the Life Activity of Oceanic Biological Systems). (In this collection.)

Publications in Other Languages

Shushkina, E. A., S. I. Anisimov, and R. Z. Klekowski. Calculation of Production Efficiency of Plankton Copepods. — Polsk. Arch. Hydrobiol., Vol. 15(28), No. 3. 1968.
Welch, H. E. Relationships between Assimilation Efficiencies and Growth Efficiencies for Aquatic Consumers. — Ecology, Vol. 49, No. 4. 1968.

RESPIRATION OF PLANKTONIC CRUSTACEANS IN THE TROPICAL PACIFIC

E.A. Shushkina and B.Ya. Vilenkin
Shirshov Institute of Oceanology of the Academy of Sciences of the USSR

The intensity of respiration of tropical planktonic crustaceans was measured in all polygons during the 44th cruise of the "Vityaz" in order to obtain energy metabolism characteristics for mass species of the epipelagic.*

The animals were obtained from a plankton sample dredged at low speed from a depth of 200 m and distributed in vessels of 3 — 5-liter capacity. They were immediately classified by species and left for a few hours in water that had been passed through No. 6 filter paper. The crustaceans were then transferred to calibrated glass flasks with ground glass stoppers of capacity 30 — 50 ml filled also with filtered water. The number of animals per respirometer depended upon the latter's volume, size of the animals, and the intended duration of the experiment. From 2 — 20 crustaceans of the same species and size were as a rule placed in each flask. The control was provided by the same water in the same flasks and under identical conditions but without animals. All the operations with the animals on shipboard were conducted at 28 — 30°, this corresponding to the temperature of the upper 50-m layer.

After the end of exposure (which usually lasted 5 — 6 hours), the concentrations of dissolved oxygen in the control and experimental respirometers were determined by two methods, according to Winkler and polarographically.

For polarographic determinations an oxygen transducer with a platinum-lead pair of electrodes was used. The electrodes were inserted in a plexiglass tube with nipples at the ends which served for attachment of cambric tubes, one of which was shut with a Mohr clamp while the other was inserted in a glass tube; the tip of this was covered with No. 49 bolting cloth and was lowered into the vessel with the water to be analyzed. After 10 — 15 ml water had been passed through the transducer, the clamp was closed, the measuring circuit was switched on and the glavonometer reading taken. The strength of current in the circuit is proportional to the oxygen concentration in the water. The entire apparatus was mounted on a stand and kept constantly filled with water. The electrodes were short-circuited between determinations. One determination took about one minute.

* For a description of the region the reader is referred to Vinogradov's paper in this collection.

This polarographic apparatus yielded stable reproducible results at a constant temperature for a working period of 7 — 10 days, after which it had to be washed with 0.1 N HCl, immersed in seawater for 24 hours and recalibrated; the shunt resistance was usually established anew.

The theoretical sensitivity of the apparatus is higher than that of Winkler's method and in our case it was improved at the expense of a small complication of the measuring circuit. However, for very accurate measurements, the transducer must be rigorously thermostated. We did not resort to this because the sensitivity of the apparatus was quite adequate for our purpose.

Winkler's method was also used in parallel with the polarographic determination. In this case fixation of oxygen in the respirometers was accompanied by the fixation also of the animals in the flasks. After the precipitate had been dissolved and titrated together with the crustaceans, the experimental animals were measured for their body length, which remained practically unchanged by the fixatives in the first hours after titration.

We carried out 13 experiments on 11 common species of copepods. The measurements in each experiment were performed in 2—10 replications for each of 5 — 8 species. In all, about 300 oxygen concentration measurements were carried out.

In the first three experiments the oxygen concentration was measured polarographically and according to Winkler; the results obtained with both experiments were very similar and therefore oxygen concentration measurements in the respirometers were limited to the polarographic method in all subsequent experiments. With this method, the animals were fixed in formalin after the experiment and their body length (l, mm) was measured under a binocular microscope. This was followed by determinations of the mean values ($\overline{W}$) for animals in each experimental flask by resorting to the relationships between the energy equivalent of body mass W and body length l established by Ostapenya and Shushkina (1971) for these species of crustaceans. The energy body mass equivalent was measured in millicalories (mcal) and designated for convenience as "body weight stated in calories," or simply "body weight" (W).

The experimental crustaceans formed three fairly distinct groups with respect to their body form according to the ratio of length to width. The relationship $W=f(l)$ was determined for all three groups (Table 1).

TABLE 1. Energy body mass equivalent (W, mcal) of copepods as a function of their length (l, mm) for different forms of body

Group	Genus of Copepoda	Ratio of body length to width	$W(l)$
I	Scolecithrix, Temora, Euchirella, Candacia	2.5—2.8	$32.8l^{3.76}$
II	Neocalanus, Undinula, Euchaeta, Pleuromamma, Pontellina	3.5—3.7	$49.1l^{2.59}$
III	Eucalanus, Euaugaptilus	4 — 5	$23.1l^{2.41}$

The relationships $W(l)$ given in Table 1 were applied to Undinula vulgaris, Euchaeta marina, Pontellina sp., Pleuromamma xiphias, Scolecithrix danae, and Eucalanus attenuatus. For the other five copepods it was more convenient to calculate and use specific regressions that were somewhat different from the general group ones, thus:

Gandacia aethiopica	$W = 33.6\, l^{3.50}$
Temora discaudata	$W = 23.8\, l^{4.12}$
Pleuromamma gracilis Undinula darwini Neocalanus sp.	$W = 34.4\, l^{3.05}$

The oxygen consumption in mg O_2 was converted to calories with the oxycaloric coefficient $K_c = 3.40$ cal/mg O_2. The thus calculated T values (mcal/day) are depicted in Figure 1 as functions of W (each point represents a separate measurement of the rate T of the oxygen consumption by a crustacean weighing W.

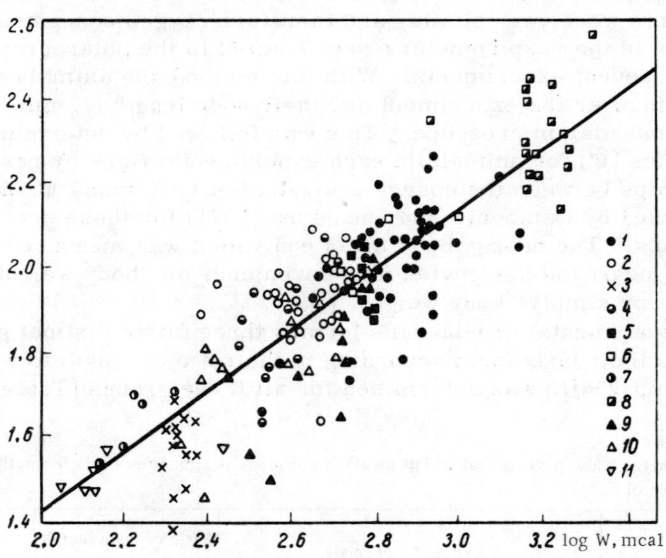

FIGURE 1. Respiratory rate T (mcal/day) as a function of body weight W (mcal//specimen) for 11 copepod species in the tropical zooplankton:

1 — Eucaeta marina; 2 — Scolecithrix danae; 3 — Undinula darwini; 4 — Undinula vulgaris; 5 — Pontellina sp.; 6 — Neocalanus sp.; 7 — Pleuromamma gracilis; 8 — Pleuromamma xiphias; 9 — Eucalanus attenuatus; 10 — Candacia aethiopica; 11 — Temora discaudata.

The regression $T(W)$ calculated by the method of least squares can be described by the equation

$$T = 1.01W^{0.71} \quad \text{at} \quad 30°; \qquad (1)$$

where $[T]$ = mcal/day and $[W]$ = mcal/specimen.

It was of interest to examine the results for any differences in the respiratory rates of animals belonging to the different trophic levels. The intensity of respiration had been reported to be dependent upon the crustaceans' type of feeding (Conover, 1960; Pavlova, 1967). In order to test this, all the experimental animals were divided into three groups: phytophages, euryphages, and predators (Table 2). The data on respiratory rate T and body weight W were then averaged by the species for each group and plotted on the graph $T=f(W)$ (Figure 2). As seen from Figure 2 and Table 2, our experiments did not reveal any appreciable differences in metabolic level between animals with different types of feeding. Apparently, the relationship $T(W)$ can be described by the same formula (1) for all the copepod species investigated.

TABLE 2. Respiratory rate of animals with different modes of feeding

Type of feeding	Species	Dimensions, mm		Number of measurements	$\overline{T}$, mcal/ /day	$\overline{W}$, mcal/specimen	$\overline{T}/\overline{W}$, % daily
		extreme	average				
Phytophages	Temora discaudata	1.5–1.8	1.6	6	33	149	22
	Undinula darwini	1.8–1.9	1.8	13	37	213	17
	Undinula vulgaris	2.0–2.4	2.2	12	71	369	19
	Eucalanus attenuatus	2.8–3.9	3.5	13	63	470	13
	Neocalanus sp.	2.5–3.1	2.7	4	113	642	18
Mean			2.4			369	18
Euryphages	Pleuromamma gracilis	1.6–1.8	1.7	5	43	167	26
	Scolecithrix danae	1.7–2.1	2.0	27	85	430	20
	Pleuromamma xiphias	3.5–5.2	4.3	17	208	1,527	14
Mean			2.7			708	20
Predators	Candacia aethiopica	1.7–2.3	2.0	14	62	353	18
	Pontellina sp.	2.6–2.7	2.6	8	93	595	16
	Euchaeta marina	2.4–3.7	2.9	44	110	743	15
Mean			2.5			558	16

In order to compare the relationship derived between metabolic rate and body weight with published material (Pavlova, 1967; Sushchenya, 1969), we had to switch from the dimensionalities $[T]$ = mcal/day and $[W]$ = mcal/ /specimen to those commonly used in measurements of respiration of animals, i. e. $[Q]$ = ml O_2/hour and $[W]$ = g wet weight.

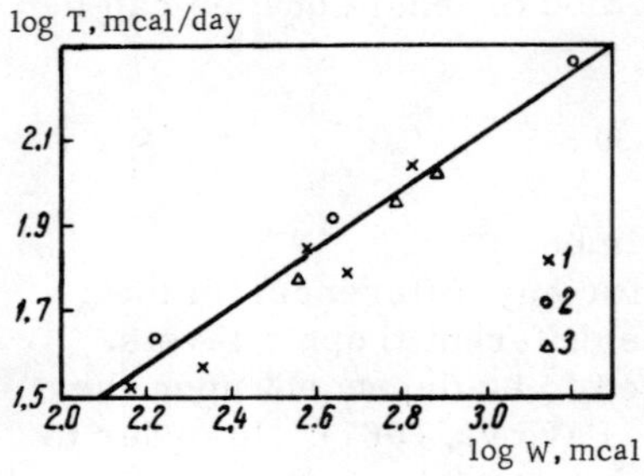

FIGURE 2. Respiratory rate T as a function of body weight W for planktonic animals of different trophic levels:

1 — filter feeders; 2 — euryphages; 3 — predators.

By equation (1) (general form: $T=aW^b$), at $W = 1$ mcal, $T = 1.01$ mcal/day. Then, assuming 1 ml O_2 to correspond to 4.86 cal (1 cal = 0.206 ml O_2) and 1 cal to correspond to 2 mg wet weight with average caloricity of crustaceans 500 cal/g wet weight, it is found that

$$Q=0.097\ W^{0.71} \text{ at } 30°, \qquad (1\text{A})$$

where $[Q]$ = ml O_2/hr · specimen and $[W]$ = g wet weight.

In Figure 3 equation (1A) is represented by the dashed straight line running somewhat above the regression line $Q(W)$ for the entire class of crustaceans at 20° (Sushchenya, 1969). The metabolic level of copepods from the tropical Atlantic at 30° runs somewhat lower Pavlova, 1967). The results of investigations of the respiratory rate of copepods from the tropical regions of the Atlantic and Pacific oceans proved to be practically identical.

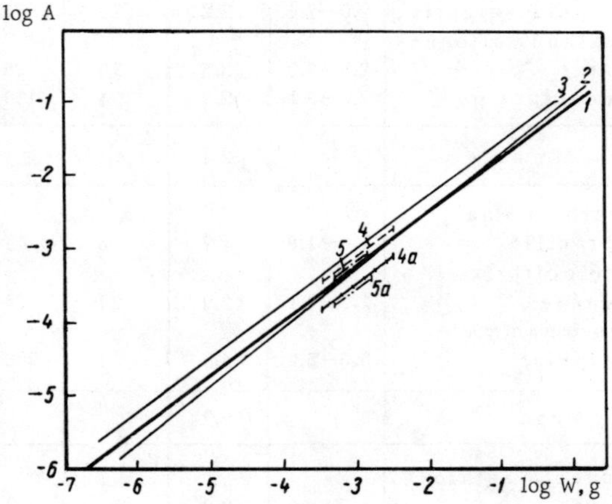

FIGURE 3. Respiratory rate Q (ml O_2/hour) as a function of body weight W (g wet weight):

1 — $Q = 0.125W^{0.159}$ at 20° (Sushchenya, 1969, for the class Crustacea); 2 — $Q = 0.165W^{0.61}$ at 20° (Vinberg, 1956, the same); 3 — $Q = 0.200W^{0.777}$ at 20° (Sushchenya, 1969, for Copepoda); 4 — $Q = 0.178W^{0.81}$ at 30° (Pavlova, 1967, for tropical Copepoda); 4a — $Q = 0.082W^{0.81}$ at 20° (same); 5 — $Q = 0.100\ W^{0.71}$ at 30° (our data for tropical Copepoda in the Pacific); 5a — $Q = 0.045W^{0.71}$ at 20° (same).

In our observations, the intensity of respiration of tropical copepods at 30° proved to be similar to that of crustaceans in the temperate latitudes at 20°. If the respiratory rates of tropical copepods determined by ourselves and Pavlova(1967) are scaled to 20°, as was done by Sushchenya (1969), the metabolic level of the species we investigated proved to be somewhat below the mean metabolic level for the entire class Crustacea at 20°. Our results on the respiratory rate of tropical crustaceans were used to determine the intensity of production of tropical zooplankton using the radiocarbon and the physiological methods (Shushkina, 1971).

Bibliography

Publications in Russian

Ostapenya, A. P. and E. A. Shushkina. Kaloriinost' setnogo planktona i energeticheskie ekvivalenty massy tela nekotorykh tropicheskikh planktonnykh rakoobraznykh (Caloricity of Net Plankton and Energy Equivalents of the Body Mass of Some Tropical Planktonic Crustacea). (In this collection.)

Pavlova, E. V. Uroven' obmena nekotorykh vidov kopepod tropicheskoi Atlantiki (Metabolic Level of Some Copepods in the Tropical Atlantic). — In Sbornik: "Biologiya i raspredelenie planktona yuzhnykh morei." Moskva, "Nauka." 1967.

Shushkina, E. A. Otsenka intensivnosti produtsirovaniya tropicheskogo zooplanktona (Evaluation of the Production of Tropical Zooplankton). (In this collection.)

Sushchenya, L. M. Kolichestvennye zakonomernosti metabolizma i transformatsii veshchestva i energii rakoobraznymi (Quantitative Patterns of Metabolism and Transformation of Matter and Energy by Crustaceans). Doctoral Thesis. Moskva. 1969.

Vinberg, G. G. Intensivnost' obmena i pishchevye potrebnosti ryb (Metabolic Intensity and Food Requirements of Fish). Minsk, Izdatel'stvo Belorusskogo gosudarstvennogo universiteta. 1956.

Vinogradov, M. R. Issledovaniya funktsionirovaniya okeanicheskikh biologicheskikh sistem (Studies of the Life Activity of Oceanic Biological Systems). (In this collection.)

Publications in Other Languages

Conover, R. J. The Feeding Behavior and Respiration of Some Marine Planktonic Crustacea. — Biol. Bull., Vol. 119, No. 3. 1960.

CALORICITY OF NET PLANKTON AND ENERGY EQUIVALENTS OF THE BODY MASS OF SOME TROPICAL PLANKTONIC CRUSTACEA

A. P. Ostapenya and E. A. Shushkina

Belorussian Lenin University and Shirshov Institute of Oceanology of the Academy of Sciences of the USSR

Following experimental determinations of the energy balance of mass species of tropical zooplankton conducted during the 44th cruise of the "Vityaz"*, measurements of the caloricity and energy equivalents of the body mass of planktonic crustaceans had to be performed.

Data on caloricity (energic value of unit mass, cal/mg) and energy body mass equivalents (energy value of a single specimen, cal/specimen) are widely used in analysis of production processes in aquatic ecosystems and to a certain degree characterize the food value of organisms.

The body mass of organisms can be stated not only in terms of wet or dry weight but also in quantities that are proportional to these, such as the contents of carbon or nitrogen, the amount of oxygen required for complete oxidation of organic matter or energy equivalent of body mass.

In using energic principles in studies of production processes, the expression of mass through its energy equivalent has appreciable advantages. The energy equivalent of mass is fully comparable with such parameters describing the energy balance of animals, as increment, ration, metabolic losses, etc., stated in energy units. This method of expressing mass automatically takes into account differences, sometimes very considerable ones, in the chemical composition of compared organisms.

In expedition work on small planktonic organisms weighing only tenths of a milligram, determination of energy mass equivalent can be performed with comparative ease and with an accuracy that is sufficient for many purposes, whereas determination of their mass by direct weighing on shipboard is practically impossible.

The plankton for analyses was collected with Juday plankton nets (No. 38 bolting cloth) and "BR" nets (No. 140 bolting cloth) and fixed in 4% formalin. The animals were then sorted according to species and measured under a binocular microscope. Altogether, we sorted and measured over 3,500 animals belonging to 14 mass species of tropical plankton and carried out 76 determinations of energy equivalents. The caloricity of net plankton was determined in 13 samples obtained from different horizons.

* For a description of region the reader is referred to Vinogradov's paper in this collection.

Some of the measurements (45 analyses) of the energy body mass equivalents of planktonic animals were carried out on board, while the other measurements, as well as determinations of the caloricity of total plankton and relative ash content, were executed on shore since these required accurate weighing of ignited samples.

The caloricity of net plankton and the energy body mass equivalent of planktonic animals were determined by the modified dichromate oxidation method recommended in manuals for determination of the production of aquatic animals (Vinberg, 1968); this modification is extensively used for determinations of the caloricity of body substance of freshwater organisms (Ostapenya, 1965; Ostapenya et al., 1967; Kryuchkova, 1969).

When marine organisms are digested by dichromate oxidation the results are liable to be considerably distorted by interaction of the oxidizing sulfuric acid-dichromate mixture with chlorides in the seawater. In order to avoid this, the samples must be painstakingly washed free of seawater. Since washing samples to be oxidized with distilled water is undesirable because it may cause plasmolysis in some planktonic organisms, the animals were washed with an ammonium acetate solution that is isotonic to seawater. The washing was carried out on watch glasses or in Petri dishes until the reaction with silver nitrate testified to the absence of chlorides.*

After the washing, the animals were dried on filter paper, transferred to trays made of cover glasses and dried for 1 — 1.5 hr at 80°. Total plankton samples were thoroughly crushed in a mortar on drying.

The trays with dry samples were ignited in refractory chemical beakers of 100 ml, with addition of 10 ml 0.1 N potassium dichromate solution in concentrated sulfuric acid, 100 mg silver sulfate being added to every beaker as oxidation catalyst. The beakers were covered with foils and placed in a boiling water bath for 2 — 3 hours (on board) or in a drying oven at 150° for 15 min (on shore). A special check confirmed satisfactory agreement between the results obtained with both methods for the heating of samples undergoing oxidation. After cooling of the analytical samples, each beaker received an additional 15 ml distilled water, as a result of which the mixture became heated. After another cooling, the samples were titrated with 0.02 N Mohr's salt in the presence of phenylanthranylic acid as indicator.

The analytical results were stated in milligrams of oxygen consumed in oxidation of organic matter, then converted to the number of calories accumulated in the oxidized sample. The oxicaloric coefficient of 3.38 cal/mg O_2 is now commonly used (Vinberg et al., 1934; Maciolec, 1962). Since the organic substance of the body of aquatic organisms is about 90% oxidized by the dichromate digestion method used (Ostapenya, 1965), we employed the oxicaloric coefficient of 3.75 cal/mg O_2, taking into account the correction for incomplete oxidation.

The ratio of mineral to organic fraction in the bodies of planktonic animals and in total plankton samples was determined by igniting the samples in a muffle furnace at 500°.

Data on the caloricity of the dry matter and ash content in total plankton samples are listed in Table 1.

* More careful analyses carried out on shore showed that seawater chlorides are better removed with a solution of ammonium carbonate or with tap water passed through a No. 1 membrane filter.

TABLE 1. Caloricity of dry matter and ash content in total plankton samples

Station No.	Collecting device	Horizon, m	Caloricity, cal/mg	Ash, % dry weight
6036	Juday net	200—0	5.55	6.03
6052	"	200—0	5.50	5.03
6055	"	200—0	4.83	8.03
6064	"BR" net	200—0	5.26	3.32
6064	"	50—0	5.07	3.60
6064	"	100—50	5.44	4.90
6064	"	200—100	5.32	9,40
6064	"	500—200	4.91	7.80
6064	"	1000—500	5.11	5.00
6064	"	2000—0	5.07	8.70
6067	Juday net	200—0	4.92	6.00
6068	"	200—0	4.82	10.13

The caloricity of the dry matter of plankton from the surface layer (0—200m) proved to be similar in different regions of the Coral Sea (stations 6036, 6052, 6055, 6064), varying from 4.83 to 5.55 cal/mg, with 5.28 cal/mg as the mean value.

The caloricity was somewhat lower, 4.92 and 4.82 cal/mg, in the open ocean (stations 6067 and 6068).

The caloricity of dry planktonic matter was 5.07 cal/mg in the 2,000—0-m layer at Station 6064, and the mean caloricity for samples collected from different horizons, from the surface to 1,000 m, averaged 5.17 cal/mg. Thus, the results do not suggest any significant or regular differences in the energic value of dry substance of planktonic organisms inhabiting different depths.

For determination of the caloricities of different size fractions of total plankton, samples collected at stations 6052 and 6067 in the 200—0-m layer were sorted into size groups using sieves with bottoms made of bolting cloth of different numbers (Table 2).

TABLE 2. Dry matter caloricities and ash contents in different size fractions of total plankton

Station No.	Horizon, m	Sieve No.	Mesh size, mm	Caloricity, cal/mg	Ash, %
6052	200—0	8	1.00	5.53	3.05
		10	0.80	5.56	4.87
		14	0.50	5.41	—
		21	0.40	5.66	5.04
		38	0.20	5.81	4.11
		50	0.15	5.26	14.81
6067	200—0	8	1.00	4.84	4.85
		10	0.80	5.17	6.09
		14	0.50	5.46	2.97
		21	0.40	5.68	3.20
		38	0.20	4.89	11.11
		50	0.15	5.06	16.69

The caloricity of dry substance of different size fractions varied from 4.84 to 5.81 cal/mg. The results obtained for the two stations suggest that there is no distinct reproducible dependence of caloricity on the size of the plankton fractions.

The caloricity of dry plankton substance in the Coral Sea is appreciably higher than in some other regions of the tropical zone of the ocean. According to Ostapenya et al. (1967), the caloricity of dry plankton substance averages 4.44 in the Gulf of Mexico, 3. 91 off South America in the equatorial zone of the Atlantic Ocean, 4.05 in the Gulf of Aden and 3.01 cal/mg in the Red Sea.

The literature has reported higher values of caloricity. Thus, the dry matter caloricity of total plankton in the equatorial Atlantic calculated by Vinogradova (1967) from the chemical composition by means of Rubner's energy equivalents and converted to physical caloricities (Ostapenya, 1968) was found to be 5.17 cal/mg, i. e. practically identical with the caloricity of the dry substance of the total plankton in the Coral Sea.

The high caloricity values of the dry plankton matter in the Coral Sea are largely due to the comparatively low mineral fraction in the dry Plankton matter, averaging 6.04% (with extreme values 3.32 and 10.13%) in the Coral Sea and the investigated part of the open ocean, as against the average total plankton ash contents of 13.1% in the Gulf of Mexico, 44.9% in the Red Sea and 22.7% in the Gulf of Aden (Ostapenya et al., 1967) (see Table 1). The relative mineral content in total plankton samples is heavily dependent upon the specific composition of the plankton and is liable to vary considerably even in comparatively closely lying regions. For instance, the mineral fraction of plankton collected off the shores of South America in the equatorial zone of the Atlantic Ocean is 20 – 40% (Ostapenya et al., 1967) as against only 6.2% in the central part of the equatorial Atlantic (Vinogradova, 1967), which is close to the ash content of plankton in the Coral Sea. Fluctuations of ash content in the bodies of individual crustaceans do not exceed those in the total plankton.

Species		Length, mm	Ash, % of dry weight
Neocalanus sp.		3.27	12.3
Euchaeta marina	average sample	3.32	11.0
Euchaeta marina	adult females	3.10	16.1
Euchaeta marina	IV – V copepodite stages	2.14	5.1
Undinula vulgaris	adult females	2.50	13.2
Undinula vulgaris	IV – V copepodite stages	2.02	5.5
Undinula darwini		1.90	3.3
Lucifer sp.		10.70	4.0
Candacia aethiopica		1.93	4.6

It should be noted that the relative ash contents differ markedly in animals of different age groups. Thus, in Euchaeta marina copepodites of stages IV – V contained only 5.1% ash, but this increased to 16.1% in mature females of this species. A similar pattern was observed in Undinula vulgaris.

The ash content increased appreciably in plankton samples collected on small-mesh sieves (see Table 2). The ash content was 14.8% in the plankton collected on a No. 50 sieve (station 6052), but averaged 4.3% in plankton collected on sieves with large meshes. The plankton fractions collected on No. 38 and 50 sieves (station 6067) had ash contents of 11.1 and 16.7%, respectively, while the ash content of the other fractions averaged 4.3%. Such a significant rise of ash content in samples collected on sieves with small meshes was due to the considerable admixture of phytoplankton, which has a much higher ash content than zooplankton.

The caloricity of plankton organic matter calculated from the caloricity of dry matter and ash content averaged 5.5 cal/mg for all the samples.

The caloricity of the organic matter of aquatic organisms inhabiting the temperate latitudes has now been firmly established at 5.6 – 5.8 cal/mg (Slobodkin and Richman, 1961; Paine, 1966). The average caloricity of organic matter of tropical plankton has been reported to be somewhat lower (Vinogradova, 1967; Ostapenya et al., 1967) on account of the lower content of lipids in the organic matter of planktonic organisms of low latitudes (Bogorov, 1960; Venhataraman, 1953; Subrahmanyan and Sen Gupta, 1962). Nevertheless, the organic matter of some tropical plankters may have a fairly high caloricity. For instance, while the average caloricity of organic matter of total plankton collected off Cuba was comparatively low (5.0 cal/mg), that of planktonic stages of polychaetes of the family Nereidae was 5.7 cal/mg, that of crab larvae in the megalops stage 5.8 cal/mg and shrimps of the genus *Lucifer* 6.0 cal/mg. Hence it follows that the caloricity of organic matter of total plankton is largely dependent upon the quantitative ratio of organisms belonging to different systematic groups which may vary appreciably in their energetic values.

The comparatively high value obtained by us for the caloricity of organic matter of plankton in the Coral Sea and the investigated region in the open ocean, 5.5 cal/mg, seems to indicate that the samples collected consisted mostly of organisms possessing a high energetic value of organic matter.

As was to be expected, the results of determining energic equivalents of the mass of certain planktonic copepods revealed appreciable differences between specimens of different size within the same species. The length and body weight of planktonic crustaceans are known to be related, and the relationship can be approximately described by the equation $W = al^b$, where W is the weight, l the length, and a and b are coefficients. The relationship between the energic mass equivalent and length of planktonic animals should evidently be described by an analogous equation.

Figure 1 shows the relationship between body length and energic mass equivalent for 8 species of planktonic Copepoda – *Undinula darwini*, *Undinula vulgaris*, *Euchaeta marina*, *Pleuromamma gracilis*, *Pleuromamma xiphias*, *Eucalanus subtenuis*, *Pontillina* sp., and *Neocalanus* sp. The distribution of points in this figure suggests that the relationship can be described with a linear equation. Calculated by the method of least squares, this equation has the form $Y = 1.69 + 2.59\,X$, where Y is the logarithm of energic equivalent of body mass and X is the logarithm of length. The data for animals of different species shown in Figure 1 fluctuate insignificantly and irregularly above the calculated straight line, denoting the absence of appreciable differences among the animals with respect to the caloricity of dry substance.

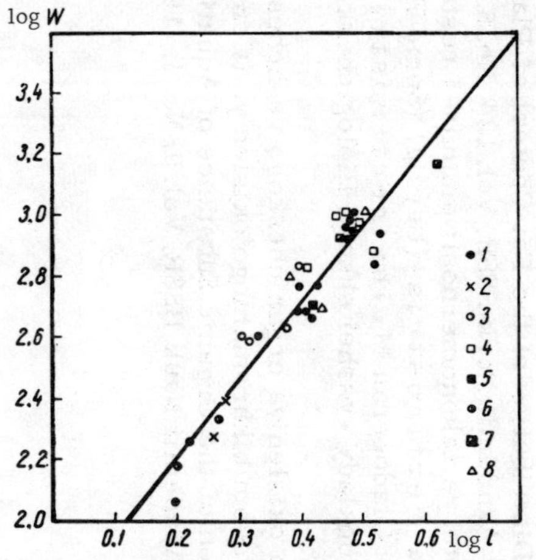

FIGURE 1. Energy mass equivalent W as a function of body length l for tropical copepods with length to width ratio 3.5–3.7. $W = 49.1\, l^{2.59}$:

1 — Euchaeta marina; 2 — Undinula darwini; 3 — Undinula vulgaris; 4 — Neocalanus sp.; 5 — Pontellina sp.; 6 — Pleuromamma gracilis; 7 — Pleuromamma xiphias; 8 — Eucalanus subtenuis.

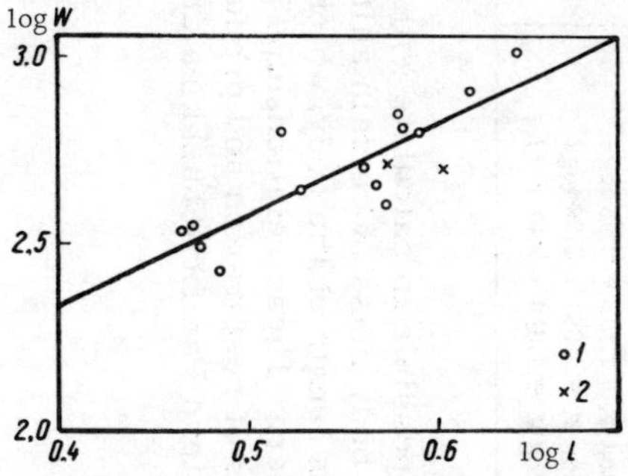

FIGURE 2. Energy mass equivalent W as a function of body length l for copepods with length to width ratio 4–5. $W = 23.1\, l^{2.41}$:

1 — Eucalanus elongatus and P. attenuatus; 2 — Euaugaptilus sp.

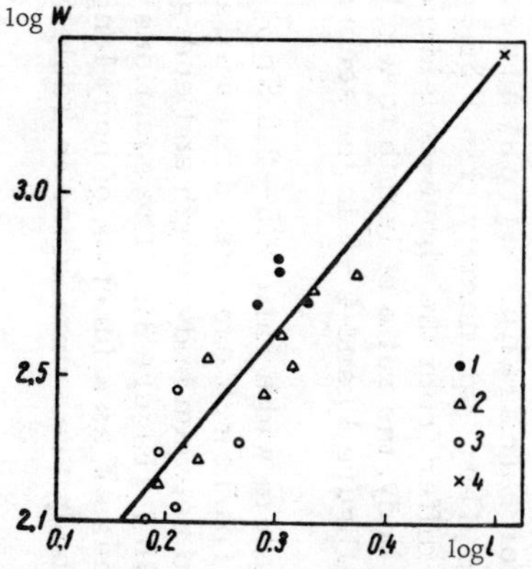

FIGURE 3. Energy mass equivalent W and body length l for copepods with length to width ratio 2.5–2.8. $W = 32.8\, l^{3.76}$:

1 — Scolecithrix danae; 2 — Candacia aethiopica; 3 — Temora discaudata; 4 — Euchirella sp.

A different equation describes the relationship between energic mass equivalent and body length for Eucalanus attenuatus, Eucalanus elongatus and Euaugaptilus sp., namely, $Y = 1.36 + 2.41X$ (Figure 2). These animals differ from the above-considered group of copepods in the form of the body, the ratio of length to width being 3.5–3.7 in the first group (Figure 1) and 4–5 in the second group (Figure 2).

For copepods with a length to width ratio 2.5–2.8 Scolecithrix danae, Candacia aethiopica, Temora discaudata, Euchirella sp.) the relationship between body length and energic mass equivalent is $Y = 1.52 + 3.76X$ (Figure 3). The equations describing the energy equivalent of body mass W as a function of body length l for animals with different forms of the body are summarized in Table 3.

TABLE 3. Relationships between the energy body equivalent (W) and length (l) for some species of tropical Copepoda with different forms of body

Length to width ratio	$W = f(l)$	
	$\log W = \log a + b \log l$	$W = aW^b$
2.5–2.8	$\log W = 1.516 + 3.76 \log l$	$W = 32.8\ l^{3.76}$
3.5–3.7	$\log W = 1.691 + 2.59 \log l$	$W = 49.1\ l^{2.59}$
4–5	$\log W = 1.364 + 2.41 \log l$	$W = 23.1\ l^{3.41}$

These equations make it possible to calculate with sufficient accuracy the energy equivalent of the body mass of certain animals in the tropical plankton community from the length of the body, which is a comparatively readily determined parameter. These relationships have been applied to determinations of intensities of respiration and production in tropical crustaceans of the southwestern Pacific (Shushkina and Vilenkin, 1971).

Bibliography

Publications in Russian

Bogorov, V. G. Geograficheskoe raspredelenie zhirnosti planktona v okeane (Geographical Distribution of the Fat Content of Plankton in the Ocean). — Doklady Akademii Nauk SSSR, Vol. 134, No. 6. 1960.

Kryuchkova, N. M. Nekotorye zakonomernosti pitaniya i rosta vetvistousogo rachka Moina rectirostris (Leydig) (Some Feeding and Growth Patterns of the Cladoceran Moina rectirostris (Leydig). — Nauchnye doklady vysshei shkoly. Biologicheskie nauki, No. 9. 1969.

Ostapenya, A. P. Polnota okisleniya organicheskogo veshchestva vodnykh bespozvonochnykh metodom bikhromatnogo okisleniya (Completeness of Dichromate Oxidation of the Organic Substance of Aquatic Invertebrates). — Doklady Akademii Nauk BSSR, Vol. 9, No. 4. 1965.

Ostapenya, A. P. 1968. Kaloriinost' total'nogo planktona Meksikanskogo zaliva i kubinskoi platformy (Caloricity of the Total Plankton of the Gulf of Mexico and Cuban Platform). — In Sbornik: "Issledovaniya Tsentral'no-Amerikanskikh morei," No. 2. Kiev, "Naukova Dumka." 1968.

Ostapenya, A. P., L. M. Sushchenya, and N. N. Khmeleva. Kaloriinost' planktona tropicheskoi zony okeana (Caloricity of Plankton in the Tropical Zone of the Ocean). — Okeanologiya, Vol. 7, No. 6. 1967.

Shushkina, E. A. and B. Ya. Vilenkin. Dykhanie planktonnykh rakoobraznykh tropicheskoi chasti Tikhogo okeana (Respiration of Planktonic Crustaceans in the Tropical Pacific). (In this collection.)

Vinberg, G. G. (ed.). Metody opredeleniya produktsii vodnykh zhivotnykh (Methods for Determining the Production of Aquatic Animals). Minsk, Izdatel'stvo "Vysheishaya Shkola." 1968.

Vinberg, G. G., V. S. Ivlev, T. P. Platova, and L. L. Rossolimo. K metodike opredeleniya organicheskogo veshchestva. Opyt kaloricheskoi otsenki kormovykh zapasov vodoema (On Ways to Determine Organic Matter. An Attempt at a Caloric Appraisal of Food Resources in a Water Body). — Trudy Limnologicheskoi stantsii v Kosino, Vol. 18. 1934.

Vinogradov, M. E. Issledovaniya funktsionirovaniya okeanicheskikh biologicheskikh sistem (Studies of the Life Activity of Oceanic Biological Systems). (In this collection.)

Vinogradova, Z. A. K biokhimii planktona ekvatorial'noi Atlantiki (On the Biochemistry of Plankton in the Equatorial Atlantic). — In Sbornik: "Biokhimiya morskikh organizmov." Kiev, "Naukova Dumka." 1967.

Publications in Other Languages

Maciolec, J. A. Limnological Organic Analyses by Quantitative Dichromate Oxidation. —Res. Rept. U. S. Fish a Wildlife Serv., Vol. 60. 1962.

Paine, R. T. Endotermy in Bomb Calorimetry. — Limnol. and Oceanogr., Vol. 7, No. 1. 1966.

Slobodkin, L. B. and S. Richman. Calories/gm in Species of Animals. — Nature, Vol. 191, No. 4785. 1961.

Subrahmanyan, R. and R. Sen Gupta. Studies on the Plankton of the East Coast of India. 1. Seasonal Variation in the Fat Content of the Plankton and its Relationship to Phytoplankters and Fisheries. — Proc. Indian Acad. Sci., Vol. 57, No. 1. 1962.

Venhataraman, R. Chan. Studies on the Mackerel Fat Variations: Correlation of Plankton with that of Fish. — Proc. Indian Acad. Sci., Vol. 37. 1953.

DIURNAL VARIATIONS IN THE VERTICAL DISTRIBUTION OF THE PLANKTON BIOMASS IN THE EQUATORIAL WEST PACIFIC

M. E. Vinogradov and Yu. A. Rudyakov

Shirshov Institute of Oceanology of the Academy of Sciences of the USSR

The life activity of pelagic biological systems is governed largely by the spatial distribution and migrations of plankton. The diurnal shifts of the plankton are especially interesting from this standpoint as they lead to a regular and often very intensive vertical redistribution.

Detailed studies of these migrations have been carried out in the temperate and cold-water zones of the ocean, but the tropical regions have been rather neglected in this respect. The bulk of information was obtained by comparing observations performed at different stations at different times during the 24-hr period or at diurnal stations with large intervals between series of catches, so that detailed studies of the pattern of migrations were impossible (King and Hida, 1954, 1957; Bsharah, 1957; Legand and Derosieres, 1959; Naumov and Ponomareva, 1964; Bogorov and Vinogradov, 1960; Geinrikh, 1961; Brinton, 1962; Vinogradov and Voronina, 1964; Vinogradov, 1968; and others).

One of the tasks of the 44th cruise of the "Vityaz" was to make a thorough investigation of the diurnal vertical migrations of plankton in the surface and subsurface layers. The following prerequisites for obtaining full-value information suitable for statistical processing were recognized: a) a number of observation points obtained at equal intervals during the diurnal period that should be sufficient for harmonic analysis and b) minimization of sampling observation error. It was also considered desirable to obtain information that would make it possible to organize a variance complex of the type of randomized blocks with two replications adapted to solving the problem of periodic regression.

It was decided that the expansion of the variation curves for plankton abundance in each layer into Fourier series should be limited to the first two harmonics, the existence of which is obvious and is readily, though not always unambiguously, explicable. The existence of the first harmonic, i. e. the presence of one maximum and one minimum during the diurnal period, proves that the analyzed layer comprises the depth range of ascent or descent of the migrating plankton mass. Naturally, the abundance variations should be passing through opposite phases in the layers of maximum ascent and descent. Existence of the second harmonic, i. e. two maxima and two minima during the diurnal period, may indicate either that the population ascends or descends twice during 24 hours or that the layer under consideration lies between the depths of maximum ascent and descent, the first brought about by the passage of organisms through the layer during their ascent and the second maximum during their descent.

It was decided to limit the measurements to 12 per diurnal period for each layer (in other words, the series of catches should be carried out every two hours), and also to fractionate the layers as much as possible since the pattern should become more distinct the thinner the layers are.

We tried to increase the accuracy and number of measurements in order to reduce the sampling observation error. The determination of plankton concentration was made more precise by more accurate lowering and closing of the net at predetermined depths and by measuring the path traveled by the net. The number of measurements was increased by increasing the number of replications, i. e. simultaneous investigation of each layer with two nets operated strictly synchronously and by maximizing the number of investigated cycles with observations conducted for two diurnal periods instead of one.

Since different species may migrate not only at different intensities but also nonsynchronously, fluctuations in the total amount of plankton should be less prominent than fluctuations in the numbers of different species. Nevertheless, they give an idea of the nature of plankton migrations and are of definite interest from this point of view.

MATERIAL AND METHODS

The investigations were conducted at two test area stations (polygons) lying at the northern periphery of the South Equatorial Current (station 6033: 04°10' S, 168°31' E) and in the divergence zone at the northern boundary of the Coral Sea (station 6052: 09°50' S, 156°43' E). The observations were conducted for one diurnal period at station 6033 and for two diurnal periods (with an interval of one diurnal period) at station 6052.

The material was collected with Juday 80/113 nets (No. 38 bolting cloth with meshes of 0.18 mm) operated with two winches simultaneously (No. 4 and No. 5) located 11 m apart. For each haul both nets were lowered simultaneously and proceeded at identical speed. The hoisting of both nets was also simultaneous and the difference in hoisting speeds was very small. When the ship was drifting, the necessary corrections were introduced by paying out the cable. In sending the "messenger," the operating speed of hoisting was taken into account and the net was closed at the predetermined depth with an error usually not exceeding 1 – 2 m.

Hauls were made down to 300 m for the following 7 layers: 0 – 25, 25 – 50, 50 – 75, 75 – 100, 100 – 150, 150 – 200, and 200 – 300 m. A series was carried out in 1 hr 05 min to 1 hr 10 min under favorable conditions.

The volume of plankton in the samples was measured in a Yasnov volumeter, except in the case of salps, siphonophores, jellyfish and macroplanktonic animals longer than 3 cm. The measurements were carried out in two replications. The error involved in determination of the sample volumes was up to 0.1 cm^3.

The volume of plankton in the samples taken at station 6033 was measured directly on board 10 – 20 days after the material was obtained, while the volume of plankton in the samples taken at station 6052 was measured after six months' storage.

RESULTS

Station 6033. The shifting of waters at this station is of a distinct three-layer nature. During the observation period, on December 22, 1969, the current flowed mostly in the northeastern direction at a velocity of up to 20 cm/sec in the layer extending from the surface down to 50 – 75 m, in the southern direction at the maximum velocity (at depths of about 150 m) of 24 cm/sec in the 75 – 200-m layer, and in the northwestern direction at a velocity that increased with increasing depth, reaching 18 cm/sec at depths of about 300 m in the layer below 200 m.

The vertical flows are mostly descending to 250 m (although the vertical velocity vector is directed upward at depths of 75 m) and ascending below 250 m. The vertical transport velocity is $10^{-2} - 10^{-5}$ cm/sec (Sapozhnikov and Galerkin, 1971).

Changes in the directions of the currents are not accompanied by any abrupt change in the thermohaline characteristics. The thermocline is distinct. The largest temperature gradient is 0.12 – 0.13 degrees/m. The upper boundary of the thermocline lies at 125 – 150 m. On the whole, the structure of the waters is characterized by a TS-curve of an intermediate type combining the structural features of the equatorial and tropical waters of the southern hemisphere. The weather was favorable; the wind did not exceed 3 – 4 m/sec on the average, occasionally dropping to 0 – 0.5 m/sec.

The work was started at 2300 hr on December 21, and the series were performed every 2 hours until 0100 hr on December 23. If the haul from one of the winches was unsuccessful, it was repeated on the same winch, and the next haul was started from both winches simultaneously. Such instances, however, were few. The net on winch No. 4 was lost only once (series 10) and the three upper layers were investigated from winch No. 5 additionally after the main series was completed.

Although the nets, hauling time and conditions were identical, the volumes of samples taken with the two nets in parallel were as a rule different. The differences were especially large for the 0 – 25-m layer in series 2, when the net operated from winch No. 5 was used on a concentration of the hyperiid Anchylomera blossevillei making a catch of 5,000 mm^3/m^3 plankton in the 0 – 25-m layer, whereas the net operated from winch No. 4 did not catch these crustaceans at all* and the volume of plankton was only 30 mm^3/m^3, that is, only 1/180 the volume in the other net. (For this reason, series 1 and 2 were excluded from further consideration.) The volumes of catches differed fairly significantly also in other cases, as is clear from the scatter of points in Figure 1. The net operated from winch No. 5 usually brought up larger catches, probably because the catching capacity of the net operated from winch No. 4 was reduced by the filtered water counter installed on that net. The net was lost during series 10 and was replaced with one without a counter. Among the 67 parallel catches taken before the accident, the net operated from winch No. 4 (with the counter) yielded smaller catches than the net operated from winch No. 5 (without the counter) in 43 cases. The differences in the catching capacities of the two nets should be regarded

* The crustaceans may have concentrated in the spot illuminated by the searchlight.

as significant, $\varkappa^2 = 5.40$, and the probability of the larger values less than 2.5%. Out of the 31 parallel catches taken after the accident, the net operated from winch No. 4 gave smaller catches in 16 cases. The difference in the catching capacities of the two nets was now no longer significant, $\varkappa^2 = 0.03$ and the probability of the larger value exceeding 75%. It is true that especially large differences in the catches, affecting the position of the 50% level for the entire plankton mass in the water column, occurred precisely in the series 11 — 13, when both nets were operated without a filtered water counter. However, the cause of these differences is not clear and we consider the scatter of values as random. In further analysis we shall utilize chiefly the mean results of parallel hauls.

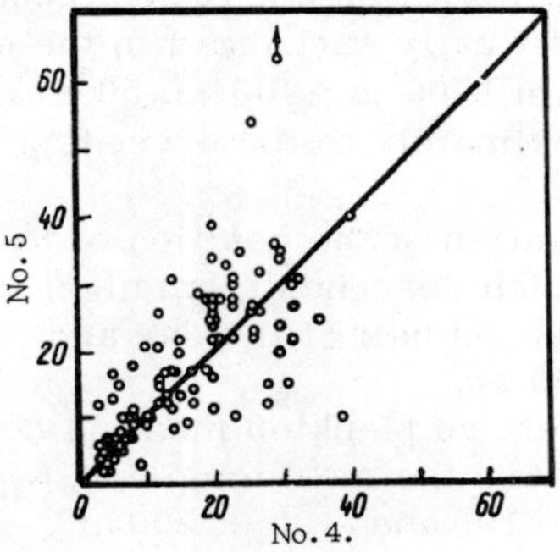

FIGURE 1. Comparison of plankton biomass in parallel catches taken with nets operated from winches No. 4 and No. 5 at station 6033, mm^3/m^3

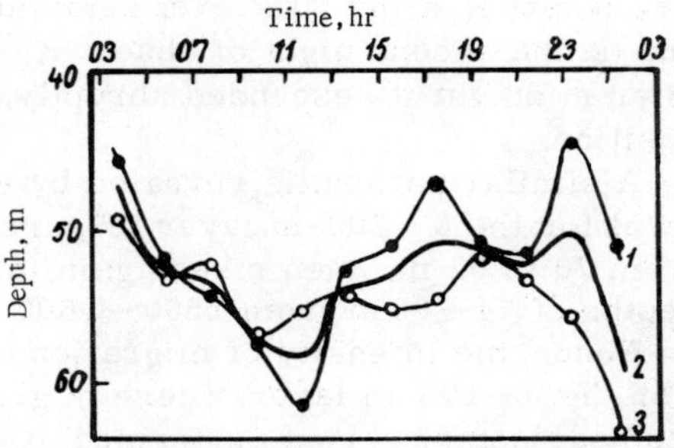

FIGURE 2. Variation of the depth of location of the 50% plankton abundance level in the 0—100-m layer at station 6033:

1 — winch No. 4; 2 — mean for the two winches; 3 — winch No. 5.

The vertical position of the plankton concentration maximum varies in the course of the diurnal period. In the first night (2300 — 0400 hr) the concentration maximum occurred in the 0 — 25-m layer; it then descended and was at 75 — 100 m in the first half of day (0700 — 1200 hr). It rose somewhat in the afternoon and was located in the 50 — 75-m layer between 1300 and 2200 hr, and returned to the 0 — 25-m layer toward midnight (2300 — 0000 hr). The total amount of plankton in the 0 — 100-m layer varied little over 24 hours and scarcely any regular differences were observed between daytime and night catches:

Series No.	3	4	5	6	7	8
Time, hr	0310—0413	0508—0606	0708—0822	0905—1000	1105—1211	1304—1401
Replaced plankton volume, mm^3/m^3	27.0	22.2	22.2	19.1	23.3	23.4
Series No.	9	10	11	12	13	14
Time, hr	1507—1604	1707—1825	1915—2014	2108—2206	2305—0003	0106—0214
Replaced plankton volume, mm^3/m^3	24.7	23.8	25.8	27.5	22.5	24.7

The total amount of plankton in the 0 — 100-m layer averaged 22.8 mm^3/m^3 during daylight and 25.0 mm^3/m^3 in the dark period. There is a possibility that the difference was partially due to more active avoidance of the nets by the plankton in the daytime. Hence, the layer did not become significantly enriched in plankton at night on account of a rise of animals from deeper waters. The main diurnal plankton migrations were confined to the upper 100-m layer. At most 10% of the animals found in this layer at night descended below its boundary in the daytime.

A clearer picture of the migration of plankton in the upper 100-m layer is furnished by the variation of the depth of the 50% level of the entire plankton mass in the layer (Figure 2). This depth varies only slightly, the 50% level being at 47 — 50 m at night and descending to 57 — 60 m toward noon; in other words, the amplitude of its diurnal migration does not exceed 100 m. The position of the 50% level remained practically unchanged in the evening and on the second night of observations from 1700 to 2400 hr (50 — 52.5 m); after midnight it descended abruptly, approximately to the preceding noon position.

A similar pattern is revealed by examination of the position of the 50% level for the 0 — 200-m layer (Figure 3), which descends from night to noon from 70 to 92 m, then rises again in the second half of the day and lies at depths of 74 — 83 m from 1500 — 1600 to 0200 hr.

Hence, the intensity of migration of the entire plankton mass is not high. For the 0 — 100-m layer, where migrations are the most intensive, $K_{100}\% = 15\%$ (comparison of series performed at 0000 — 0100 and 1100 — 1200 hr). For the entire 0 — 300-m layer, $K_{380}\% = 13\%$. It is of interest to note that a similar migration coefficient ($K_{500}\% = 10.3\%$) was determined by comparison of groups of nighttime and daytime plankton stations performed in the waters of the South Equatorial Current (Vinogradov and Voronina, 1964).

A more detailed pattern of the diurnal vertical migrations of the entire plankton mass can be obtained by analyzing the diurnal variations of plankton abundance in each of the layers investigated. Such an analysis was carried out by fitting the two first harmonic components to the experimental data (Bliss, 1955), testing their significance and constructing confidence intervals for the mean fluctuation level, semiamplitude, and time of attaining their maximum. The deviations from the normal distribution were corrected by performing all calculations on transformed data, using transformation of the form $\log(10x + 1)$ where x signifies the plankton volume in mm^3/m^3.

The diurnal variations of plankton abundance in the 0 — 25-m layer can be described by the equation

$$Y = 0.078 + 0.0192 \cos(0.5236\,t - 5.2290),$$

where 0410 hr (zonal time) is taken as the origin.

The sinusoid described by this equation is responsible for 34% of the plankton abundance variation in this layer, and attains 5% significance level (see table).

The mean level of fluctuations is 19.7 mm^3/m^3 with a 5% confidence interval 17.8 — 21.7 mm^3/m^3. The semiamplitude of fluctuations is 4.5 mm^3/m^3 with confidence interval 2.9 — 7.4 mm^3/m^3. The plankton abundance maximum was observed at 0010 hr at confidence interval 2134 — — 0224 hr. The second harmonic proved insignificant.

TABLE. Variance analysis of data on plankton abundance variation for the 0—25-m layer

Source of variation	Number of degrees of freedom	Sum of squares	Mean square
Nets	1	0.00030	0.00030
1st harmonic	2	0.00442	0.00221
2nd harmonic	2	0.00108	0.00054
Deviations	7	0.00360	0.00052
Nets × 1st harmonic	2	0.00080	0.00040
Nets × 2nd harmonic	2	0.00057	0.00029
Nets × deviations	7	0.00220	0.00032
Total	23	0.1297	

In the 25—50-m layer the mean plankton concentration was 22.8 mm^3/m^3 with confidence interval 20.0—25.8 mm^3/m^3. The first harmonic was weak and nonsignificant, while the second was responsible for 13% of the variation but did not reach the 5% significance level.

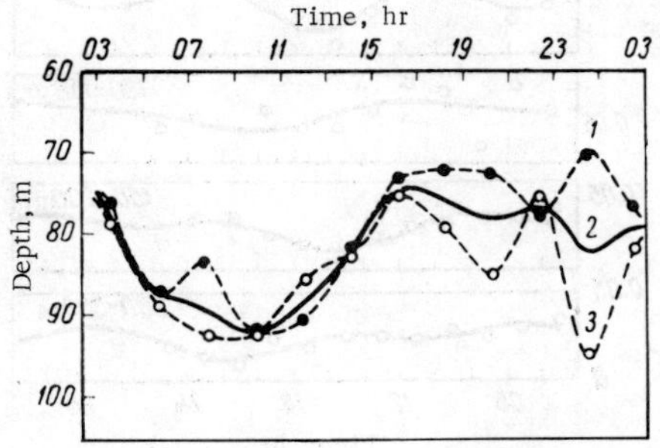

FIGURE 3. Variation of the depth of location of the 50% plankton abundance level in the 0—200-m layer at station 6033:

1 — winch No. 4; 2 — mean for the two winches; 3 — winch No. 5.

The plankton abundance variation in the 50—75-m layer is described by the equation

$$Y = 0.1004 + 0.0174 \cos(0.5236\, t - 3.7604),$$

where 0354 hr was the time datum. The sinusoid described by this equation was responsible for 29% of total variation and reached the 5% significance level. The mean fluctuation level was 26.0 mm^3/m^3 with confidence interval 23.0—28.7 mm^3/m^3. The semiamplitude was 4.0 mm^3/m^3 with confidence interval 1.0—7.0 mm^3/m^3. The plankton abundance maximum was observed at 1807 hr with confidence interval 1523 — 2034 hr. The second harmonic was nonsignificant.

The mean plankton concentration in the 75—100-m layer was 26.0 mm³/m³ with confidence interval 23.7—28.5 mm³/m³. The first harmonic proved to be nonsignificant; the second harmonic (with maxima at approximately 1100 and 2300 hr) was responsible for 14% of the variation but did not attain the 5% significance level. However, the plankton concentration at 1600 hr proved to be significantly lower than at 0200 and 1200 hr.

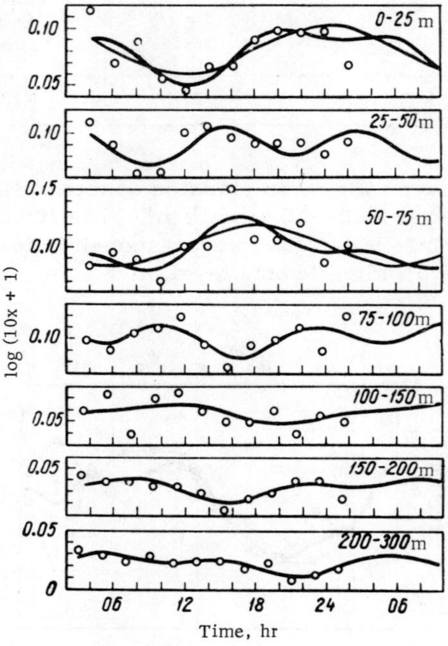

FIGURE 4. Diurnal variations of plankton abundance by layers at station 6033, mm³/m³.

The diurnal variations are extremely small in the 100—150, 150—200 and 200—300-m layers (Figure 4). The confidence intervals for mean plankton abundance here are as follows:

Layer, m	100—150	150—200	200—300
Mean plankton concentration, mm³/m³	13.7	8.1	6.3
5% confidence interval for the mean concentration	12.2—15.0	6.8—9.3	5.2 — 7.4

Thus, analysis of the diurnal variations of plankton concentration by layers also shows that these variations can be detected only in the upper 100-m layer. The fact that only the first harmonic with a midnight maximum is significant for the 0—25 m layer suggests that the majority of migrating species rise to the surface layer simultaneously. The marked manifestation of the second harmonic component in the 75—100-m layer apparently suggests that the plankton rises into the layer from depths below 100 m.

Station 6052 was executed at the northern boundary of the Coral Sea in the divergence zone of the Solomon Islands (Rotschi and Lemasson, 1967), which is characterized by upwelling in the 200 — 1,000-m layer. The 0 — 100-m layer is occupied by equatorial waters of low salinity underlain by the subsurface central southern Pacific water mass with its core at 150 m (Arsen'ev et al., 1971). The currents at the station are chiefly meridional. In the first diurnal period (January 17) the current in the upper 100 — 125-m layer flowed south-southwest, attaining a speed of 37 cm/sec at 75 m. Currents of the northwestern direction with speeds of 10 — 25 cm/sec were predominant at 150 — 250 m, while currents flowing northeast at 11 — 18 cm/sec were predominant below 250 m. On January 19, the direction of currents in the upper 125-m layer remained on the whole the same, but the maximum speed was measured at 50 m. The current flowed northwest below 150 m, with an initial speed of 17 m/sec at 200 m which reached 3 — 5 m/sec at 400 — 500 m (Sapozhnikov and Galerkin, 1971).

The pycnocline had the largest gradient at 100 — 125 m. The surface layer contained an abundance of blue-green algae (Oscillatoria thiebauti and Katagnymena sp.), as is characteristic of most of the Coral and Solomon seas in this season of the year.

The work was begun at 2300 hr on January 16. As with the preceding station, the series were performed every two hours throughout the 24 hours of January 17. After an interval of one diurnal period, the entire observation cycle was repeated, from 2300 hr on January 18 to 2400 on January 19. During the two diurnal observation periods, 26 series of hauls were made with each net. Because of the fairly strong currents and wind of force 2 — 3, the angle of slope of the cable was as a rule 20 — 30°, occasionally increasing to 35° and only occasionally decreasing to 10 — 15°.

As with station 6033, the volumes of parallel samples taken with the two nets were different (Figure 5), and although both nets were operated without counters and prepared for operation with special attention to identical mounting (length of guide ropes, tension of the cone, etc.), the catches yielded by one net were significantly richer than those by the other, the nets operated from winch No. 4 bringing in larger catches on both days of observations in spite of the replacement of nets. We do not know of the causes of these differences and regard them as random.

No regular diurnal variations of plankton biomass could be detected either in the first or in the second diurnal period. The concentration maximum was stably located in the 0 — 25-m layer, shifting to the 25 — 50-m layer only in a few series, sometimes during the daytime, and at other times at night, but the plankton abundance was only slightly greater than in the 0 — 25-m layer.

Similarly, no regularity could be detected in fluctuations of biomass in the different layers (Figure 6).

Hence, there were no distinct diurnal plankton migrations at station 6052. Even if such migrations did exist in some species, they could not be detected from the variation in the total plankton mass as determined by the volumetric method.

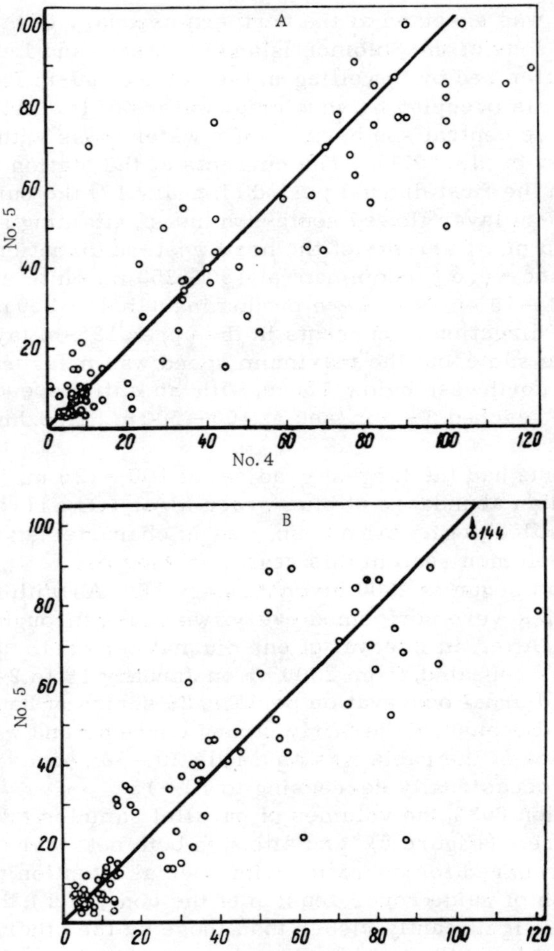

FIGURE 5. Comparison of plankton biomass in parallel catches taken with winches No. 4 and No. 5 at station 6052, mm^3/m^3:

A — January 17; B — January 19.

The reasons for this may be various. Firstly, it has been demonstrated (Vinogradov and Voronina, 1964) that the intensities of migration of all zooplanktonic species decrease sharply in divergence zones, where the plankton continuously (round the clock) concentrates in the surface layer and, as already mentioned, station 6052 was executed in the divergence zone. Secondly, the large quantity of blue-green algae continuously found in the surface layer might obliterate the slight diurnal fluctuations of zooplankton biomass that possibly occured. Finally, the station was executed in a region with fairly strong currents of variable direction that might cause a significant exchange of waters over the observation period. Under the variable conditions of a closed sea, this might entrain differences in the plankton biomass in different waters.

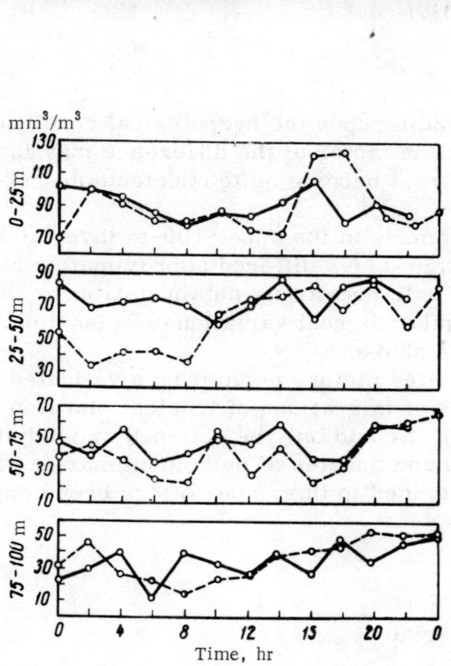

FIGURE 6. Diurnal variations of plankton abundance by layers at station 6052. Solid line — January 17, dashed line — January 19.

In fact, the total concentration of plankton in the upper 100-m layer varies appreciably over the diurnal period, the variation being aperiodic, as can be demonstrated in the instance of station 6052, where the replaced plankton volumes (mm³/m³) for the 0—100-m layer were as follows:

Series No.	1	2	3	4	5	6
Time, hr	2300—0000	0100—0200	0300—0400	0500—0600	0700—0800	0900—0100
January 17	64	59	65	51.5	56.5	56
January 19	48	56	48.5	42	38	57

Series No.	7	8	9	10	11	12	13
Time, hr	1100—1200	1300—1400	1500—1600	1700—1800	1900—2000	2100—2200	2300—0000
January 17	56	69	57	61.5	67.5	64.5	—
January 19	50	57	67	66	67	69	69.5

These variations in the total mass of plankton apparently obliterate the diurnal fluctuations of its biomass. The dissimilarity in the absolute values of plankton biomass and its diurnal variation in the different layers investigated during the first and second diurnal observation period also suggest that their fluctuations should be ascribed in the first place to an exchange of waters with different population densities.

Data on migration intensities of tropical zooplankton compiled by Vinogradov (1968) show that the nocturnal biomass is usually 1.2—1.8 times the daytime biomass in the upper 100-m layer in different regions of the tropical zone, depending upon the hydrological conditions and composition of the plankton, but the factor of the difference may increase to 4 in some regions and in contrast become quite undetectable in other (divergence zones).

The plankton biomass in the upper 100-m layer at station 6033 in the daytime and nighttime series differed approximately by a factor of 1.2, which was in accordance with previously obtained data for the central regions of the ocean. No regular diurnal variations of plankton mass were detected at station 6052 (see above).

Certainly, a detailed picture permitting a validated appraisal of the distinctive features of migrations of tropical plankton and the effect of different factors on it could only be obtained by studying the migrations of different groups and species of pelagic animals. The material collected at the stations described in this paper offers broad opportunities for such studies.

Bibliography

Publications in Russian

Arsen'ev, V. S., L. I. Galerkin, E. A. Plakhin, and V. V. Sapozhnikov. Osnovnye cherty gidrologii i gidrokhimii raiona issledovanii 44-go reisa "Vityazya" (Major Hydrological and Hydrochemical Features of the Region Investigated during the 44th Cruise of the "Vityaz"). (In this collection.)

Bogorov, V. G. and M. E. Vinogradov. Raspredelenie biomassy zooplanktona v tsentral'noi chasti Tikhogo okeana (Distribution of the Zooplankton Biomass in the Central Pacific). — Trudy Vsesoyuznogo gidrobiologicheskogo obshchestva, Vol. 10. 1960.

Geinrikh, A. K. O vertikal'nom raspredelenii i sutochnykh migratsiyakh kopepod v raione k yugo-vostoku ot Yaponii (Vertical Distribution and Diurnal Migrations of Copepoda in the Region Southeast of Japan). — Trudy Instituta Okeanologii Akademii Nauk SSSR, Vol. 51. 1961.

Naumov, A. G. and L. A. Ponomareva. Vertikal'noe raspredelenie i sutochnye migratsii osnovnykh predstavitelei zooplanktona v severnoi chasti Indiiskogo okeana (Vertical Distribution and Diurnal Migrations of the Chief Representatives of Zooplankton in the North Indian Ocean). — Trudy Instituta Okeanologii Akademii Nauk SSSR, Vol. 64. 1964.

Sapozhnikov, V. V. and L. I. Galerkin. K raschetu pervichnoi produktsiii tropicheskikh vod po vertikal'nomu vynosu biogennykh elementov (Calculation of Primary Production in Tropical Waters from the Upward Influx of Biogenic Elements. (In this collection.)

Vinogradov, M. E. Vertikal'noe raspredelenie okeanicheskogo zooplanktona (Vertical Distribution of Oceanic Zooplankton). Moskva, "Nauka." 1968.

Vinogradov, M. E. and N. M. Voronina. Raspredelenie planktona v vodakh ekvatorial'nykh techenii Tikhogo okeana. II. Vertikal'noe raspredelenie otdel'nykh vidov (Plankton Distribution in the Equatorial Pacific Currents. II. Vertical Distribution of Individual Species). — Trudy Instituta Okeanologii Akademii Nauk SSSR, Vol. 65. 1964.

Publications in Other Languages

Bliss, C. I. Periodic Regression in Biology and Climatology. — Bull. Connecticut Agricult. Exper. Station, Vol. 615. 1958.

Brinton, E. Variable Factors Affecting the Apparent Range and Estimated Concentration of Euphausiids in the North Pacific. — Pacific Sci., Vol. 16, No. 4. 1962.

Bsharah, L. Plankton of the Florida Current. V. Environmental Conditions, Standing Crop, Seasonal and Diurnal Changes at a Station Forty Miles East of Miami. — Bull. Marine Sci. Gulf and Carribean, Vol. 7, No. 3. 1957.

King, J. and T. S. Hida. Variations in Zooplankton Abundance in Hawaiian Waters. — U. S. Fish and Wildlife Serv. Spec. Sci. Rep. Fisheries, Vol. 118. 1954.

King, J. and T. S. Hida. Zooplankton Abundance in the Central Pacific, pt. II. — Fish Bull. U. S. Fish and Wildlife Serv., Vol. 57. 1957.

Legand, M. and R. Derosieres. La répartition du zooplankton des croisières "Astrolabe" et "Boussole", ses rélations avec divers éléments biologiques et océanographiques. — Inst. Franc. Océanie, Centre Océanogr., Rapp. Sci., No. 10. 1959.

Rotschi, H. and L. Lemasson. Oceanography of the Coral and Tasman Seas. — Oceanogr. Mar. Biol. Ann. Rev., Vol. 5. 1967.

VERTICAL DISTRIBUTION OF SOME PERIDINIANS IN THE EQUATORIAL PACIFIC

I. N. Sukhanova
Shirshov Institute of Oceanology of the Academy of Sciences of the USSR

This paper presents the data processed from collections at the diurnal station 6033 (4°10' S, 168°31' E) in the South Equatorial Current. The samples were obtained with a Juday net (diameter 80/113 cm, No. 38 bolting cloth) simultaneously from two winches 11 m apart. The series included catches from the layers 0—25, 25—50, 50—75, 75—100, 100—150, 150—200, and 200— 300 m and were repeated at two-hour intervals from 2343 hr on December 21 to 0214 hr on December 23. The phytoplankton was examined in parallel samples down to 200 m. The cells were counted in $^1/_{25}$ part of a sample. In all, 168 samples were processed.

The low density and limited specific composition of the net phytoplankton and the low seston biomass (Vinogradov and Rudyakov, 1971) characterize this as one of the oligotrophic regions of the World Ocean. Typical of such regions are sporadic finds of diatoms and a predominance of peridinians of the main tropical complex consisting of Pyrocystis pseudonoctiluca, Ceratium trichoceros, C. massiliense, and C. carriense (Sukhanova, 1962). Other species were represented by small numbers of specimens and then only in a few of the series.

The parallel catches differed sometimes very significantly (by factors of 2—3) with respect to the total amount of phytoplankton. The differences were even more marked with respect to the numbers of cells of different species. Thus, in the case of the three species Ceratium trichoceros, C. massiliense, and C. carriense there were fairly large numbers of cells in one of the samples, while the parallel sample did not contain any. Furthermore, the maximum numerical abundance of Ceratium trichoceros in the 0—25-m layer (series 10) and the 50—75-m layer (series 11, 12, and 14) and of C. massiliense in the 25—50-m layer (series 8) in catches brought up with one of the winches occurred in the series where no cells at all were found in the parallel catches (Figure 1).

These substantial differences in the cell counts in parallel catches are somewhat leveled if we compare the cell counts in the vertical series of catches encompassing the layer inhabited by the main part, or core, of the population. The positions of population cores* (between the 25 and 75% levels) were calculated for five species of peridinians in parallel catches of all 14 series, making it possible to evaluate the degree of spatial non-uniformity of populations, determine the depth inhabited by the bulk of a population, and reveal the asymmetry of the vertical distribution.

* The method of calculation was proposed by Moore et al. (1953).

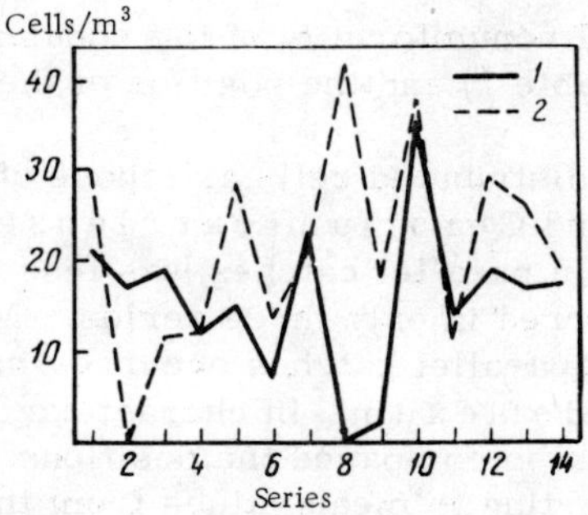

FIGURE 1. Cell counts in Ceratium massiliense in the 25–50-m layer in parallel catches:
1 – winch No.4; 2 – winch No.5.

TABLE 1. Ratio of cell counts (cell/m^3) in two parallel catches (the smaller value is taken as unity)

Species	Series													
	1	2	3	4	5	6	7	8	9	10	11	12	13	14
Pyrocytosis pseudonoctiluca	1.6	1.2	1.2	1.2	1.3	1.2	1.4	1.2	1.2	1.5	1.5	2.0	1.3	1.2
Ceratium deflexum	4.0	1.7	1.1	1.6	1.2	1.8	2.0	2.7	1.9	4.5	1.2	2.1	1.2	1.1
Ceratium carriense	1.3	2.1	1.2	1.8	1.1	1.2	1.2	1.1	2.4	1.2	1.1	1.5	1.6	2.1
Ceratium massiliense	2.5	2.6	1.3	1.8	1.6	1.4	1.5	2.6	2.6	1.2	1.4	1.4	1.2	1.6
Ceratium trichoceros	2.9	1.4	1.6	2.0	2.3	2.9	1.9	1.9	1.9	1.2	5.5	4.6	2.5	5.1

TABLE 2. Discrepancies in parallel catches with respect to the position of the upper (A) and lower (B) boundary and the spread of the core (C) (means from 14 series)

Species	A		B		C	
	m	%	m	%	m	%
Pyrocystis pseudonoctiluca	9.0	21	4.0	10	8.0	18
Ceratium carriense	9.0	22	6.5	15	10.5	25
Ceratium massiliense	10.0	24	7.4	18	14.0	34
Ceratium trichoceros	9.0	23	22.0	56	17.0	44
Ceratium deflexum	1.5	9	5.5	55	4.0	27

The degree of spatial nonuniformity of the populations can be estimated from the cell counts (Table 1) and the position of the core in parallel catches.

The most uniformly distributed cells are those of Pyrocystis pseudonoctiluca and Ceratium carriense. In the first species the ratio of cell counts in parallel catches was less than 2:1. In the second species ratios > 2 occurred in only three series. Marked fluctuations of numerical abundance in parallel catches occurred in the case of Ceratium trichoceros and C. deflexum. In characterizing the spatial nonuniformity of populations, we also compared the positions of boundaries and the spread of cores by resorting to mean values from the 14 series. The results were stated in absolute values and also in comparison to the mean spread of the core of the specific population in question (Table 2).

The above data show the highest uniformity of quantitative distribution and the highest constancy of inhabited layer in Pyrocystis pseudonoctiluca and Ceratium carriense.

Diurnal observations made it possible to analyze the variations in the numerical characteristics and vertical distribution of populations with respect to time. Means for two parallel catches were used. A daytime maximum of cell count was detected in four out of the five species investigated (Table 3, Figure 2).

TABLE 3. Schedule of the daytime maximum of numerical abundance of four species of peridinians in the 0–200-m layer

Species	Onset of the daytime maximum, hr	End of the daytime maximum, hr
Pyrocystis pseudonoctiluca	0700 – 0800	1500 – 1600
Ceratium carriense	0900 – 1000	1300 – 1400
Ceratium trichoceros	1100 – 1200	1500 – 1600
Ceratium deflexum	1300 – 1400	1900 – 2000

The onset of the maximum depends upon the depth of occurrence of the population. In the surface-dwelling Ceratium deflexum the increase in the cell count was recorded at later hours than in Pyrocystis pseudonoctiluca which inhabits the deep layers of the euphotic zone. The same trend was revealed by analysis of the layerwise vertical distribution of each of the four species. The population's maximum numerical abundance in the lower horizonz (75–100, 100–150 m) was observed in the early hours, whereas the increase of cell counts in the upper layers occurred later. We should probably look for a relationship between the time at which the population recover their numerical abundance after nighttime consumption by the zooplankton and the specific behavioral characteristics of the consumers.

The cell counts of Pyrocystis pseudonoctiluca, Ceratium carriense, C. massiliense, and C. trichoceros increased in the 50–75-m layer in the night series No. 13. The denser part of the population may well appear in a fairly thin layer in view of the stratified structure of the currents in this region (Arsen'ev et al., 1971).

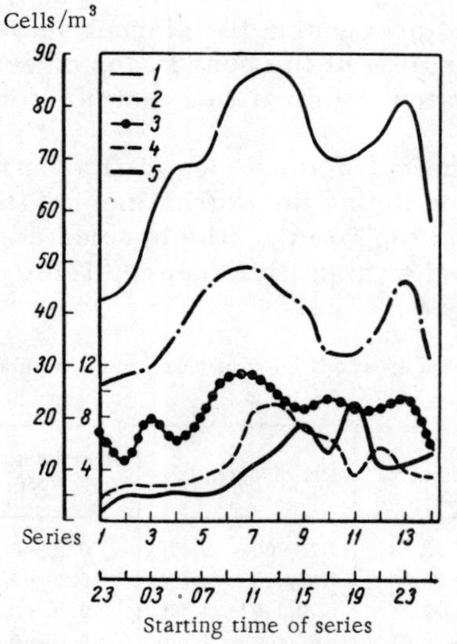

FIGURE 2. Diurnal variation of the numerical abundance of phytoplankton, cells/m³, in the 0—200-m layer:

1 — total abundance; 2 — Pyrocystis pseudonoctiluca; 3 — Ceratium carriense; 4 — C. trichoceros; 5 — C. deflexum. The small scale is for species 3, 4, 5.

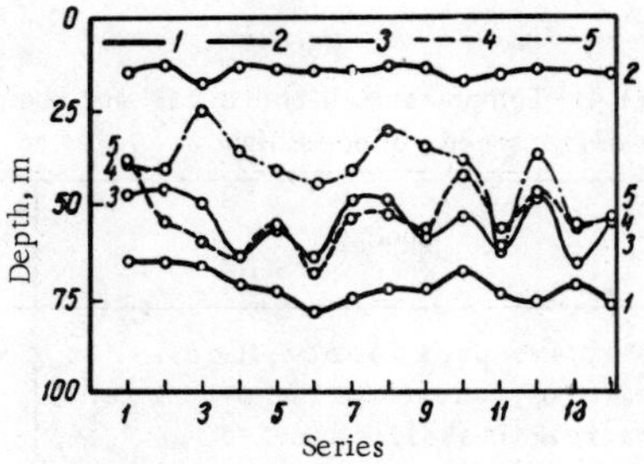

FIGURE 3. Position of the 50% level of the populations of five species:

1 — Pyrocystis pseudonoctiluca; 2 — Ceratium deflexum; 3 — C. carriense; 4 — C. massiliense; 5 — C. trichoceros.

The mass of each species occupied a specific layer over the diurnal period (Figure 3). The species could be arbitrarily divided into "surface" (Ceratium deflexum), "deepwater" (Pyrocystis pseudonoctiluca) and "midwater" (Ceratium trichoceros, C. massiliense, C. carriense) species. The existence of "shade flora" (Karsten, 1907) was stressed by Karsten (1907), Schröder (1906), Jörgensen (1920), Peters (1932), Steemann Nielsen (1934, 1939), and Graham and Bronikovsky (1944) based on material collected by large expeditions which included considerable numbers of Ceratium. Light was regarded as the leading controlling factor of the vertical distribution of representatives of this genus. The species discussed in this paper (except for Ceratium trichoceros)* were classified as surface species by these authors. However, this conclusion might have been drawn, firstly, on account of the collection techniques, which often resulted in catches from the layers 50 — 0, 100 — 0 or 100 — 50, 50 — 0 m and secondly on account of the trophic differences between the regions investigated.

The preference of Pyrocystis pseudonoctiluca for waters deeper than 25 m was mentioned by Semina (1960, 1967) for the Pacific Ocean and by us (Sukhanova, 1962, 1969) for the Indian Ocean.

* Ceratium trichoceros was classified with the "shade flora" by Jörgensen (1920) and Graham and Bronikovsky (1944).

The techniques used for collecting and processing the diurnal series data permitted a fairly clearcut determination of the boundaries of vertical distribution of the species and demonstrated the stratified distribution of Ceratium.

For instance, the population core of the surface species Ceratium deflexum does not descend below 30 m during the diurnal period (the upper boundary of the core lies within the top 10 m). The species exists at minimum concentrations of biogens and at high illuminance (Table 4).

TABLE 4. Temperature, illuminance,* and phosphates at the upper and lower boundary of the population cores of five species of peridinians

Species	Temperature, °C	Illuminance, lux	Phosphates, µg-at P/l
Pyrocystis pseudonoctiluca	29.77–28.14	25,000– 2,000	0.22–0.47
Ceratium carriense	29.75–29.04	55,000– 4,700	0.20–0.36
Ceratium massiliense	29.73–29.04	70,000– 4,800	0.20–0.36
Ceratium trichoceros	29.75–29.04	86,500– 4,500	0.15–0.36
Ceratium deflexum	29.95–29.77	120,000–53,000	0.10–0.20

* Data on illuminance were kindly supplied by O. V. Kopelevich, a member of the Optical Division.

The population core of the deepwater Pyrocystis pseudonoctiluca is situated in the 35–105-m layer, its upper boundary fluctuating from 35 to 65 m over the diurnal period, while its lower boundary is less mobile (maximum shift 15 m). The illuminance at which the core exists is extremely low, so that its lower boundary lies at the depth of the compensation point.

The population cores of the two midwater species Ceratium massiliense and C. carriense occur in the 20–85-m layer throughout the diurnal period. The upper boundary of the population core of Ceratium trichoceros rises to 10 m, with a corresponding expansion of the layer inhabited by the species. These three species are distinguished by considerable diurnal fluctuations of the upper and lower boundaries of the population cores (the maximum shifts are 20–25 m for the upper boundaries, 28–36 m for the lower boundaries). The illuminance range that is favorable for the development of these species is very broad, from 5,000 to 87,000 lux.

These diurnal shifts of the population cores show no time rhythm. The existence of vertical migrations in the species of Ceratium in question may be assumed by analogy with the closely related species Ceratium fusus and C. tripos (Peters, 1929; Tschirn, 1920). However, the migration amplitudes of these species is too small (about 5 m) and such insignificant migrations could not be detected by our broad survey. Even assuming that the migration amplitude of the investigated species is close to the maximum known amplitude (about 16 m), observed in the very mobile species Gonyaulax polyedra and Cachonina niei (Eppley et al., 1968), our collecting techniques would not detect layerwise variation of the abundance resulting from the migrations. Most probably, the variations of the depth inhabited by peridinian

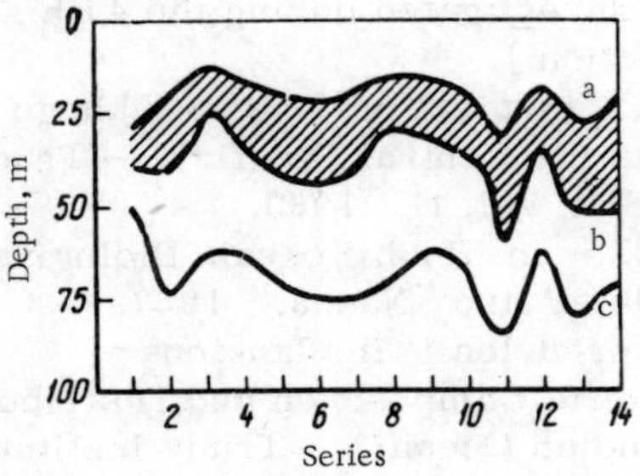

FIGURE 4. Position of the core and 50% level of Ceratium trichoceros population: a — upper boundary of core; b — 50% level; c — lower boundary of core.

populations are due to the continuous exchange of water in the region and to the influx of a new part of the population (observed velocities 5–25 cm/sec — Arsen'ev et al., 1971) and vertical movements of water.

Asymmetric distribution was observed within the layer occupied by the population cores and was manifested most distinctly in Ceratium trichoceros and C. carriense (Figure 4). This means a difference in the spread of the layers included between the 25–50 and 50–75% level of the population's numbers. The cells of Ceratium trichoceros are concentrated near the upper boundary of the population core — a positive asymmetry. The opposite trend was exhibited by the vertical distribution of the population of Ceratium carriense.

We are inclined to interpret the above information concerning the layers inhabited by the species and the asymmetry of the cores of the two midwater species by applying the concept of two leading critical factors in phytoplankton development. The layerwise distribution of species implies different boundary conditions for their existence. Thus, the vertical extent of the population core of Pyrocystis pseudonoctiluca seems to be principally dependent on the concentration of biogenic elements. The species subsists on a minimum of light in order to be as close as possible to the biogen-rich layer. The existence of a population at low illuminance, a part of it even below the compensation point, suggests the possibility of heterotrophic feeding, at any rate in certain stages of development.

Contrary to the preceding species, the vertical distribution boundaries of the population of Ceratium deflexum are primarily controlled by the amount of light. The population core keeps to a thin, well-illumined subsurface layer with slight diurnal oscillations of the upper and lower boundary. The vertical distribution of the midwater species Ceratium carriense exhibits the trend of approaching the layer of concentration of biogenic elements bounded by the lower limit of light required for photosynthesis. This is also indicated by the negative asymmetry of the population core of Ceratium carriense. A similar mechanism of formation for the lower plankton maximum was assumed by Vinogradov, Gitel'zon, and Sorokin (1971). The positive asymmetry displayed by the midwater species Ceratium trichoceros may result from the limiting influence of low concentrations of phosphates in the surface layer.

Bibliography

Publications in Russian

Arsen'ev, V.S. et al. Osnovnye cherty gidrologii i gidrokhimii raiona issledovanii 44-go reisa "Vityazya" (Major Hydrological and

Hydrochemical Features of the Region Investigated during the 44th Cruise of the "Vityaz"). (In this collection.)

Semina, G. I. Raspredelenie fitoplanktona v tsentral'noi chasti Tikhogo okeana (Distribution of Phytoplankton in the Central Pacific). — Trudy Instituta Okeanologii Akademii Nauk SSSR, Vol. 41. 1960.

Semina, G. I. Fitoplankton (Phytoplankton). — In: Tikhii okean, Biologiya Tikhogo okeana, Book 1. Moskva, Izdatel'stvo "Nauka." 1967.

Sukhanova, I. N. O vidovom sostave i raspredelenii fitoplanktona v severnoi chasti Indiiskogo okeana (Specific Composition and Distribution of Phytoplankton in the Northern Indian Ocean). — Trudy Instituta Okeanologii Akademii Nauk SSSR, Vol. 58. 1962.

Sukhanova, I. N. Fitoplankton severo-vostochnoi chasti Indiiskogo okeana (Phytoplankton in the Northeastern Indian Ocean). — Candidate Thesis. Moskva, Institut okeanologii Akademii Nauk SSSR. 1969.

Vinogradov, M. E. and Yu. A. Rudyakov. Sutochnye izmeneniya vertikal'nogo raspredeleniya obshchei biomassy planktona zapadnoi ekvatorial'noi Patsifiki (Diurnal Variation of the Vertical Distribution of the Plankton Biomass in the Equatorial West Pacific). (In this collection.)

Vinogradov, M. E., I. I. Gitel'zon, and Yu. I. Sorokin. O prostranstvennoi strukture soobshchestv evfoticheskoi zony tropicheskikh vod okeana (The Spatial Structure of Communities in the Euphotic Zone of the Ocean Tropics). (In this collection.)

Publications in Other Languages

Eppley, R. W. and others. Some Observations on the Vertical Migration of Dinoflagellates. — Phycologia, Vol. 4. 1968.

Graham, H. W. and N. Bronikovsky. The Genus Ceratium in the Pacific and North Atlantic Oceans. — Sci. Res. Cruise VII "Carnegie", 1928—1929. — Biology, Vol. 5. 1944.

Jørgensen, E. Mediterranean Ceratia. — Rept. Danish Oceanogr. Exp. Mediterranean and Adjacent Seas 1908—1910, Vol. 2. 1920.

Karsten, G. Das Indische Phytoplankton. — Wiss. Ergebn. Dtsch. Tiefsee-Exped "Valdivia," Vol. 2, No. 3. 1907.

Moore, H. B. a. oth. Plankton of the Florida Current. III. The Control of the Vertical Distribution of Zooplankton in the Daytime by Light and Temperature. — Bull. Mar. Sci. Gulf Carrib., Vol. 3, No. 2. 1955.

Peters, N. Über Orts- und Geisselbewegung bei marinen Dinoflagellaten. — Archiv Protistenkunde, Vol. 67, No. 2/3. 1929.

Peters, N. Die Bevölkerung des Südatlantischen Ozeans mit Ceratien. — Wiss. Ergebn. Dtsch. Atlant. Exped. "Meteor," 1925—1927, Vol. 12. 1932.

Schröder, B. Beiträge zur Kenntnis des Phytoplanktons warmer Meere. — Ztschr. naturforsch. Ges. Zürich, Vol. 51. 1906.

Steemann, Nielsen, E. Untersuchungen über die Verbreitung, Biologie und Variation der Ceratien im südlichen Stillen Ozean. — Dana Rept. No. 4. 1934.

S t e e m a n n, N i e l s e n, E. Die Ceratien des Indischen Ozeans und der Ostasiatischen Gewässer. — Oxford Univ. Press. 1939.
T s c h r i n, E. Biologische Studien an Ceratien der Kieler Föhrde. — Dissertation. Kiel. 1920.

POPULATION COMPOSITION AND VERTICAL DISTRIBUTION OF PYROCYSTIS PSEUDONOCTILUCA (W. THOMSON) IN THE WESTERN EQUATORIAL PACIFIC

I. N. Sukhanova and Yu. A. Rudyakov

Shirshov Institute of Oceanology of the Academy of Sciences of the USSR

Planktonologists have recently been paying increasing attention to investigations of the structure and activity of pelagic communities (Vinogradov, Gitel'zon, and Sorokin, 1970). Successful prediction of the production of natural communities depends on the extent of knowledge on the biology of different species of plants and animals. In this connection, the little studied aspects of the vertical distribution of phytoplankton are also of interest.

In this paper we analyze the vertical distribution and certain features of the life cycle of Pyrocystis pseudonoctiluca, one of the most populous species of peridinians in the tropical regions of the ocean (Semina, 1962; Sukhanova, 1962).

The material was collected during the 44th cruise of the "Vityaz" at the diurnal station 6033 (04°10' S, 168°31' E), from 2300 hr on December 21 to 0100 hr on December 23, 1969. The material was collected with a Juday 80/113 net (No. 38 bolting cloth) with meshes of 0.18 mm) from the layers 0—25, 25—50, 50—75, 75—100, 100—150, and 150—200 m. Each layer was investigated at intervals of 2 hours simultaneously with two nets operated synchronously. The cells of P. pseudonoctiluca were counted and measured in a Nauman chamber. In this manner we processed $^1/_{25}$ of each of the 168 subsamples obtained by splitting the original samples into half with a Folsom plankton sample splitter (McEwen et al., 1954).

Little information is found in the literature on the biology and vertical distribution of P. pseudonoctiluca. The species is known to occur mostly below 25 m (Semina, 1960; Sukhanova, 1962, 1969), and it is assumed that after the nucleus and protoplasm have divided into two or four* daughter cells the latter leave the parent cell and proceed to grow until the next division (Schiller, 1937).

The specimens of P. pseudonoctiluca examined differed in size, from the cells and stages of life cycle. The diameter of the cells varied from 70 to 900 μ. To save time, the cells were differentiated only roughly according to size during the processing of the material, by dividing them into two size groups, small, with a diameter of less than 250 μ, and large,

* Cases of division into four daughter cells were not observed by us.

with a diameter of more than 250 μ. The following three stages of the life cycle were distinguished in accordance with the state of the nucleus and cytoplasm: vegetating cells, cells in the initial stages of division and cells in the final stages of division (Figure 1). Cells of both size groups and cells in different stages of division occurred in the typical spherical form (Blackman, 1902; Schiller, 1937) and also in modified forms approaching the ovoid and rhomboid in plan with gradual transitions from the typical to the two modified forms (Figure 2). Accordingly, all the specimens were divided into two groups with respect to the form of the cells, typical (spherical) and atypical.

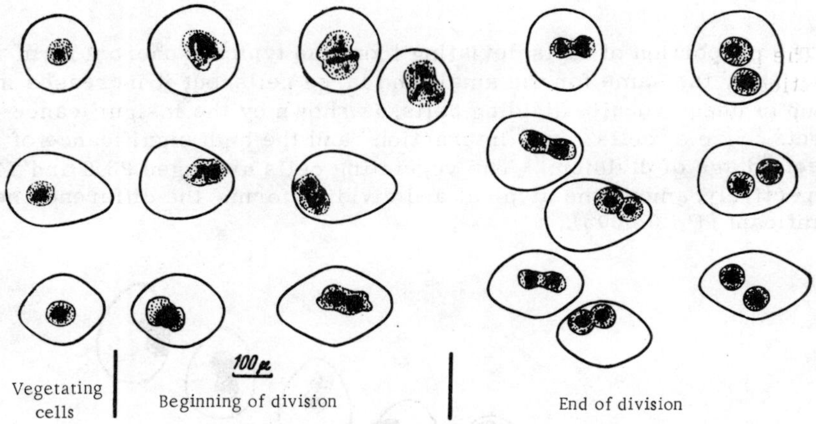

FIGURE 1. Grouping of cells of P. pseudonoctiluca by stages of the life cycle

Table 1 gives information on the numbers of cells of the typical and atypical forms grouped by size groups and stages of division.

TABLE 1. Distribution of typical and atypical cells of P. pseudonoctiluca from the 0–150-m layer by size groups and stages of division

Stage of division	Small cells			Large cells		
	typical	atypical	percentage of atypical cells	typical	atypical	percentage of atypical cells
Vegetating	829	262	24.0	998	302	23.2
Beginning of division . .	231	45	16.3	808	303	27.2
End of division	48	116	70.7	144	458	76.1

The group of cells in the initial stage of division was temporarily excluded from consideration in order to compare the cells that were unequivocally either vegetating or dividing. Transformation of the percentages of atypical cells into angles* and variance analysis (Snedecor, 1961, pp. 357 – 358) yielded the results shown in Table 2.

TABLE 2. Results of variance analysis

Source of variation	Number of degrees of freedom	Mean square	Probability
Size of cells	1	45	
Stage of division	1	529,700	< 0.0005
Interaction	1	1,863	> 0.05
Error	—	821	

The proportion of cells deviating from the typical spheroid form was practically the same for the small and large cells, but it increased in the group of unequivocally dividing cells, as shown by the insignificance of the effects "Size of cells" and "Interaction" and the high significance of the effect "Stage of division." The vegetating cells averaged 23.6 and 74.9%, respectively, among the atypical and dividing forms, the difference being significant ($P < 0.0005$).

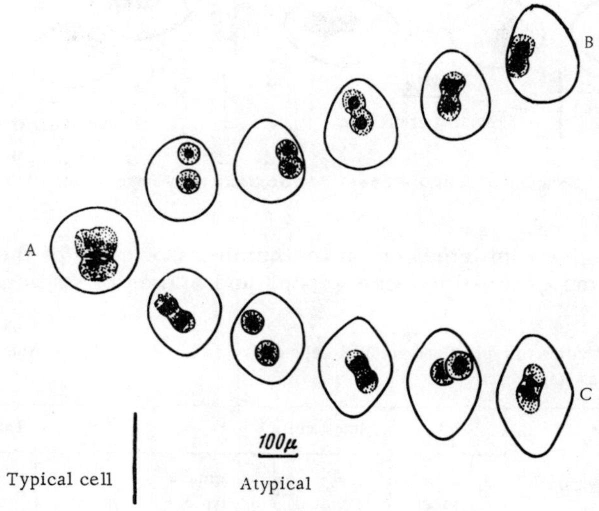

FIGURE 2. Variations in the form of cells of P. pseudonoctiluca from the typical spheroid (A) to atypical ovoid (B) and rhomboid (C) form

This result may be interpreted in a variety of ways. Firstly, the change in the shape of the cell wall may be directly related to the division of the nucleus and cytoplasm. Secondly, the division of ovoid and rhomboid cells may be more protracted than that of the spheroid cells. Thirdly, in spite

* Transformation through arcsines is necessary, when applying statistical methods based on the normal distribution hypothesis, in the processing of data which have binomial distribution and are stated in the form of fractions or percentages (Steel and Torrie, 1960; Snedecor, 1956).

of the continuous transition from the spheroid to ovoid and rhomboid cells, there are some doubts as to their belonging to a single population. In particular, as seen from Table 3, the mean depth of occurrence (Rudyakov 1971) of atypical cells is significantly greater than that of the spheroid cells. However, the first interpretation appears to be the most plausible one, since the distribution of cells in the initial stage of division (see Table 1) appears to indicate that the proportion of atypical cells increases only toward the end of division of the cytoplasm.

TABLE 3. 95% confidence interval of the mean depth of occurrence of different groups of cells of P. pseudonoctiluca (the figures in parentheses indicate the abundance of each group)

	Vegetating		Beginning of division		End of division	
	Small	Large	Small	Large	Small	Large
Typical cells	71.2 ± 2.4 (837)	64.9 ± 2.2 (1012)	75.9 ± 3.8 (231)	75.1 ± 2.4 (808)	101.0 ± 9.0 (48)	110.7 ± 6.8 (149)
Atypical cells	77.5 ± 3.2 (262)	76.0 ± 3.6 (305)	80.6 ± 9.7 (45)	80.8 ± 3.3 (303)	116.0 ± 3.7 (117)	120.4 ± 1.5 (460)
Probability of the greater value of Student's criterion for the difference of the depth of occurrence of typical and atypical cells	< 0.01	< 0.001	> 0.05	< 0.01	< 0.001	< 0.001

Let us examine Table 4 for a tentative comparison of the sizes of dividing and nondividing cells.

TABLE 4. Size distribution of vegetating and dividing cells of P. pseudonoctiluca from the 0–150-m layer

State	Small	Large	Total
Vegetating	1091 (45.6%)	1300 (54.4%)	2391
End of division	164 (21.6%)	602 (78.4%)	766
Total ...	1255	1902	3157

For this table of the type 2 × 2, $\chi^2 = 141.1$ and the probability of obtaining the larger value is less than 0.0005. This means that the mean size of cells in the final stage of division is significantly greater than that of the vegetating cells. However, we still do not know whether the cells grow in the course of division or whether division predominantly occurs in the large cells. The latter assumption is borne out by data on a lowered metabolism of dividing cells and by the size distribution of cells in the initial stage of division (276, or 19.9% small cells, 1108, or 80.1% large cells), which does not significantly differ from the distribution of cells in the final stage of division ($\chi^2 = 0.94$; $p > 30\%$). Hence, division apparently begins after the growing cell has reached a certain limiting size. This conforms to the hypothesis on the existence of a critical ratio of the volumes of nucleus and protoplasm on reaching which a cell begins to divide. This view was advanced by Hertwig in 1908 and is upheld, in general, by modern cytologists (Mazia, 1963).

In embarking on an examination of the vertical distribution of cells, one should have a clear idea of the limitations that may be imposed on the correctness of judgments and conclusions by the techniques used for collecting material. Since it is mainly large cells which are dividing, it may be hoped that the material investigated furnishes an undistorted idea of the vertical distribution patterns and the depth of occurrence of dividing cells. With respect to the group of cells designated "vegetating" in this paper, one must remember that it does not include cells smaller than the meshes of the nets used. We therefore could not examine the distribution patterns of all the cells in the interphase stage. In view of this, we have to abandon attempts to determine the depth and other attendant conditions under which the growth rate of P. pseudonoctiluca cells is most intensive.

The averaged picture of the vertical distribution of different stages and the variation of their mean depth of occurrence over the diurnal period are shown in Figure 3.* The diurnal variations of the mean depths are fairly large (amplitudes up to 20 – 25 m) and are synchronous for all stages. These fluctuations may be caused by a combination of such factors as inhomogeneity and vertical movements of water in the region, variations in the specific gravity of the algae, and finally variations in the activity and vertical distribution of phytophagous animals.

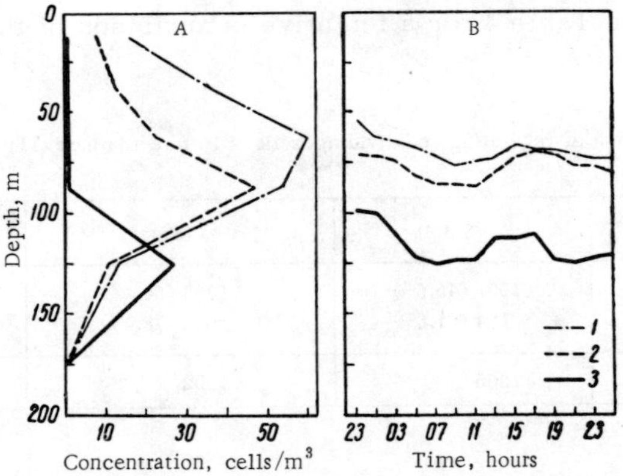

FIGURE 3. Vertical distribution (A) and diurnal variations of the mean depth of occurrence (B) of the different stages of life cycle of P. pseudonoctiluca:

1 – vegetating cells; 2 – cells at the beginning of division; 3 – cells at the end of division.

* Fitted mean depth values are shown in Figure 3B. The fitting was done by linear interpolation from three points (Urbakh, 1963), taking into account the cell count in each series of catches.

In our view, the most intriguing conclusion from examination of Figure 3 is that dividing cells are situated deeper than vegetating cells, the cells in the final stage of division being located deeper than those showing slight changes in the nucleus and cytoplasm. If we summarize all the information, the estimated mean diurnal depth of occurrence is 69.8 m for vegetating cells, 76.6 m for cells at the beginning of division, and 117.7 m for cells at the end of division. All the differences between these mean values are significant, $P < 0.0001$.

In our opinion, the only possible cause of the different depths of occurrence of cells in the different phases of the life cycle is the increased specific gravity and consequently higher settling rate of dividing cells.

As early as 1915, Gran (1915) cited in Steele (1959) pointed out the impossibility of explaining the features of vertical distribution of algae without resorting to the hypothesis of their sinking. The mean density of living protoplasm is known to be 1.045 — 1.050 (Jacobs, 1935) and to exceed even the maximum density of seawater, which is 1.032 at a temperature of −2° and salinity of 40% (Zubov, 1957). The density of the protoplasm of *Scenedesmus* sp. is 1.07—1.15 (Jacobs, 1935), that of *Chlamydomonas* Dang. 1.077—1.091 (Leontjew, 1926). The nuclear densities are still higher. For instance, *Spirogyr majuscula* has a cytoplasm density of 1.076 and a nuclear density of 1.094 (Milovidov, 1949). The specific gravity is especially high in organisms with skeletal formations, for instance 2.071 in diatomaceous algae (Jacobs, 1935). A cursory review of the results of painstaking investigations of the physical properties of cytoplasm carried out mainly at the beginning of this century leads to the conclusion that planktonic algae must have negative buoyancy unless they possess mechanisms for its compensation. One of such mechanisms consists in lowering the density of the cell sap by active elimination of heavy bivalent ions (Gross and Zeuthen, 1948; Beklemishev, Petrikova, and Semina, 1961; Braarud, 1962; Kiselev, 1969). However, neutral buoyancy of algae that are incapable of active movement, such as the diatoms and species of the genus *Pyrocystis*, may apparently be achieved only in water that is very rich in biogenic elements. This, as shown by the theoretical studies carried out by Munk and Riley (1952), is because for diatoms to grow, there must be a certain minimal rate of incorporation of biogenic elements into the cell, and this rate can only be achieved if the cells move at a certain speed with respect to the water. This speed of displacement is ensured by the negative buoyancy causing their sinking. The sinking rate of planktonic algae, determined experimentally and also calculated from data on the vertical distribution of phytoplankton or chlorophyll, varies within a range of 0—30 m/day (Steele, 1959; Eppley et al., 1967; Smayda, 1969).*

Even estimating the rate of sinking of dividing cells of *P. pseudonoctiluca* at 30 m/day, one would have to conclude that cell division in this

* According to Vinogradov (1961), the rate of sinking of the diatom *Ethmodiscus rex* reaches 19—21 m/hour. This result can hardly be regarded as representative for phytoplankton as a whole in view of the large size of the cells of this species, 1.50—1.75 mm (Smayda, 1969), but it is fairly close to that obtained by extrapolation of the graph constructed by Eppley et al., (1967). By extrapolation of the same relationship, the rate of sinking of nongrowing cells of 600 μ diameter (which is about the size of dividing *P. pseudonoctiluca* cells) is estimated at 8.3 m/hr. However, extrapolated estimates may prove very far from the real values.

species takes longer than 24 hr. Mazia (1963), one of the most prominent experts on mitotic division of cells, believes such a low rate of division to be hardly likely. For instance, the extensive compendium of data on the different duration of mitosis of animal and plant cells compiled by Milovidov (1949) contains only five cases of mitosis taking longer than three hours. The most prolonged mitosis occurs in the diatom Surirella calcarata, 5–5.5 hr (Milovidov, 1949), in rhizoid cells of Chara 5–6 hr (Pekarek, 1932), in Noctiluca 2–7 hr (see Milovidov, 1949) and Spirogyra, in which mitosis takes 45 min at the optimal temperature 12°, 11 hr at 16–18° and 14 hr at 3–4° (Milovidov, 1949).

Hence, even the maximum known sinking rate is insufficient to allow the duration of division in P. pseudonoctiluca to be included in the range of the observed values. Moreover, these algae have a very efficient mechanism for preserving zero buoyancy, according to Blackman's observations (1902).

However, as far as we know, the determinations of the rate of sinking of algae did not include an attempt at a separate consideration of vegetating and dividing cells, the specific gravities of which may be quite different; it is known that the sinking rate, minimal in intensively growing algae (Smayda and Boleyn, 1966), increases approximately fourfold as growth ceases (Eppley et al., 1967). Moreover, numerous investigations reviewed by Mazia (1963) show that a dividing cell is similar to an anuclear cell; the intensity of respiration, which may be considered a measure of the synthesis of the respiratory enzymes, is lowered, the movement of protoplasm ceases, the rate of incorporation of phosphates decreases, and at least some of the biosynthetic processes come to a halt. In this way, many of the observed manifestations of activity of cells suggest that the period of division is one of low physiological activity as if the cells were concentrating on reproduction, neglecting its ordinary "domestic chores" (Mazia, 1963). The mechanism for maintaining zero buoyancy postulated by Gross and Zeuthen (1948) calls for continuous expenditure of energy and it may be assumed that this mechanism is similarly blocked during the period of division, when the cell is in a state of physiological depression. There are also experimental proofs of an increasing density of cells (in Schizosaccharomyces pombe) in the course of their division (Mitchison, 1957). Hence, an increase in the specific gravity and consequently a higher rate of sinking of dividing cells appear to be quite probable, but nevertheless remain to be proved.

Young cells of P. pseudonoctiluca can return to the euphotic layer either by achieving positive buoyancy or by active migrations of the hypothetical mobile stage of the life cycle. Such a flagellate form (the Gymnodinium stage) is known for the closely related species P. lunula (Dogiel, 1906; Schiller, 1937; Graham, 1951). Moreover, we found a cell membrane belonging to P. pseudonoctiluca with two inclusions resembling the Gymnodinium stage in form in one of the samples taken at the station (from the 200–300-m layer) and in one plankton sample from the Indian Ocean.

It should be noted that measurements of the specific gravities and sinking rates of planktonic algae and also the mechanisms responsible for maintaining the algae in the suspended state may explain many features of their distribution. We shall illustrate this thesis with several examples.

One of the conclusions from the extensive studies carried out by Riley, Stommel, and Bumpus (1949) was that a stable vertical distribution of a population of algae cannot be preserved if $V^2 > 4A\omega$, where V signifies the rate of sinking of the algae, A the turbulent diffusion coefficient and ω, the mean production coefficient in the euphotic zone. An important consequence of this inequality is the conclusion that artificial enrichment of the euphotic layer with biogenic elements in regions with low turbulent diffusion coefficient values does not necessarily increase the biomass. The copious development of algae in regions of upwelling appears to be due not solely to the high concentration of biogenic elements but also to intensification of turbulent processes. Cushing (1968) greatly doubts that the production of natural water bodies is limited by a deficiency of biogenic elements; the distribution of dead suspended matter in the ocean is similar to that of phytoplankton in general features (Riley, 1965).

The effect of turbulent movements of water on the settling of suspended particles consists in rendering the settling laws stochastic, i. e. expressing probabilities. For instance, Velikanov (1949) demonstrated that if a particle suspended in a laminar flow reaches the bottom at time t depending upon its diameter of fall (i. e., its rate of settling in still water), the same particle suspended in a turbulent flow will reach the bottom in the same time with the probability of only 50%, and theoretically this probability can become 100% only if $t=\infty$. This means that in the presence of a large number of such particles, half of them will reach the bottom in time t in a turbulent flow, and all the suspended matter will settle on the bottom only (theoretically) after an infinitely long period of time. This is why plankton can include organisms with considerable residual weight which are incapable of active migrations, as exemplified by the diatomaceous algae. However, this can only be if the rate of reproduction is high enough to compensate for the continuous loss of specimens to the lower boundary of their habitat layer. It follows that the higher the sinking rate and the weaker the mechanisms maintaining the individuals of species in the suspended state (upwelling, pronounced turbulent processes and capacity for active orientated migrations), the higher must be the rate of reproduction of individuals within the stage of the independent population. A weakening of the maintaining mechanisms would endanger or even doom the existence of species with a high residual weight and low rate of reproduction. In this connection, it is of considerable interest that the abundance of diatomaceous algae is much higher in regions with intensive upwelling, whereas the actively swimming peridinians are distributed more uniformly (Sukhanova, 1962; Semina, 1967). This difference in the distribution of the two main groups of phytoplanktons can be attributed (Hulburt, 1962) to the capacity of peridinians for active movement, as well as to the higher specific gravity of diatoms and apparently their faster rate of division.

In accordance with the principles worked out by Munk and Riley (1952), in order for small and large cells to have the same rate of division, the large cells should have a higher sinking rate. Small cells always have this advantage over large ones, and the advantage increases, the lower the concentration of biogenic elements and the lower the turbulent diffusion coefficient. This fully explains the diminishing mean size of cells with the decreasing intensity of upwelling, as reported by Semina (1968, 1969).

The effect of the rate of sinking of algae, upwelling, and the phosphorus concentration on the abundance of phyto- and zooplankton was studied theoretically by Riley (1965). However, there are still no factual data on the sinking rate, and this gap must be filled by thorough investigations.

Bibliography

Publications in Russian

Beklemishev, K. V., M. N. Petrikova, and G. I. Semina. O prichine plavuchesti planktonnykh diatomovykh (On the Cause of Buoyancy of Planktonic Diatoms). — Trudy Instituta Okeanologii Akademii Nauk SSSR, Vol. 51. 1961.

Kiselev, I. A. Plankton morei i kontinental'nykh vodoemov (Plankton of Seas and Continental Water Bodies). Vol. 1. Moskva — Leningrad, Izdatel'stvo "Nauka." 1969.

Kushing, D. G. (Cushing, D. H.) Modelirovanie produktsionnogo tsikla v more (Simulation of the Production Cycle in the Sea). — In Sbornik: "Osnovnye problemy okeanologii." Moskva, Izdatel'stvo "Nauka." 1968.

Rudyakov, Yu. A. Detali gorizontal'nogo raspredeleniya i sutochnye verticakl'nye migratsii Cypridina (Pyrocypris) sinuosa (G. W. Müller) (Crustacea, Ostracoda) v zapadnoi chasti ekvatorial'noi Patsifiki (Details of the Horizontal Distribution and Diurnal Vertical Migrations of Cypridina (Pyrocypris) sinuosa (G. W. Müller) in the Western Equatorial Pacific). (In this collection.)

Semina, G. I. Raspredelenie fitoplanktona v tsentral'noi chasti Tikhogo okeana (Distribution of Phytoplankton in the Central Pacific). — Trudy Instituta Okeanologii Akademii Nauk SSSR, Vol. 41. 1960.

Semina, G. I. Fitoplankton tsentral'noi chasti Tikhogo okeana na razreze po 174° z. d. (Phytoplankton in the Central Pacific over the Section along 174° W). Part 1. — Trudy Instituta Okeanologii Akademii Nauk SSSR, Vol. 58. 1962.

Semina, G. I. Fitoplankton (Phytoplankton). — In: Tikhii okean, Vol. 7. Biologiya Tikhogo okeana, Book I. Moskva, Izdatel'stvo "Nauka." 1967.

Semina, G. I. Razmery kletok fitoplanktona na razreze po 174° z. d. v Tikhom okeane (Size of Phytoplankton Cells over the Section along 174°W in the Pacific). — Okeanologiya, Vol. 9, No. 3. 1969.

Sukhanova, I. N. O vidovom sostave i raspredelenii fitoplanktona v severnoi chasti Indiiskogo okeana (On the Specific Composition and Distribution of Phytoplankton in the Northern Indian Ocean). — Trudy Instituta Okeanologii Akademii Nauk SSSR, Vol. 58. 1962.

Sukhanova, I. N. Fitoplankton severo-vostochnoi chasti Indiiskogo okeana (Phytoplankton in the Northeastern Indian Ocean). Candidate thesis. Moskva, Institut Okeanologii Akademii Nauk SSSR. 1969.

Urbakh, V. Yu. Matematicheskaya statistika dlya biologov i medikov (Mathematical Statistics for Biologists and Medical Personnel). Moskva, Izdatel'stvo Akademii Nauk SSSR. 1963.

Velikanov, M. A. Dinamika ruslovykh potokov (Dynamics of Channel Flows). Leningrad, Gidrometeoizdat. 1949.

Vinogradov, M. E. Istochniki pitaniya glubokovodnoi fauny. O skorosti razlozheniya otmerskikh Pteropoda (Forage Sources of Deepwater Fauna. On the Rate of Decomposition of Dead Pteropoda). — Doklady Akademii Nauk SSSR, Vol. 138, No. 6. 1961.

Zubov, N. N. Okeanologicheskie tablitsy (Oceanological Tables). Leningrad, Gidrometeoizdat. 1957.

Publications in Other Languages

Blackman, V. H. Observations on the Pyrocystae. — New Phytologist, Vol. 1, No. 8. 1902.

Braarud, T. Species Distribution in Marine Phytoplankton. — J. Oceanogr. Soc. Japan, Vol. 20. 1962.

Dogiel, V. Beiträge zur Kenntnis der Peridineen. — Mitt. Zool. Stat. Neapel, Vol. 18, pt. 1, 2. 1906.

Eppley, R. W., R. W. Holmes and J. D. H. Strickland. Sinking Rates of Marine Phytoplankton Measured with a Fluorometer. — J. exp. Biol. Ecol., Vol. 1. 1967.

Graham, H. W. Pyrrophyta. — Manual of Phycology, ed. by G. M. Smith. A New Series of Plant Science Books ed. by F. Verdoorn, Vol. 27. 1951.

Gross, F. and E. Zeuthen. The Buoyancy of Plankton Diatoms: A Problem of Cell Physiology. — Proc. Roy. Soc., B. Vol. 135. 1948.

Hulburt, E. M. Phytoplankton in the Southern Sargasso Sea and North Equatorial Current, February 1961. — Limnol. and Oceanogr., Vol. 7, No. 3. 1962.

Jacobs, W. Das Schweben der Wasserorganismen. — Ergebn Biol., Vol. 11. 1935.

Leontjew, H. Zur Biophysik der niederen Organismen. 2. Die Bestimmung des spezifischen Gewichts und der Masse von Naegleria sp. Pflügers Arch., Vol. 213, No. 1. 1926.

Mazia, D. Mitosis and Cell Fission Physiology. [Russian translation. 1963.]

McEwen, G. E., M. M. Johnson and T. R. Folsom. A Statistical Analysis of the Performance of the Folsom Plankton Sample Splitter, Based upon Test Observations. — Arch. Meteorol., Geophysik u. Bioklimatol., S. A. Meteorol. u. Geophysik, Vol. 7. 1954.

Milovidov, P. Physik und Chemie des Zellkernes. — Protoplasma Monographien, Vol. 20, No. 1. 1949.

Mitchison, J. M. The Growth of Single Cells. — Exptl. Cell Research, Vol. 13. 1957.

Munk, W. H. and G. A. Riley. Absorption of Nutrients by Aquatic Plants. — J. Mar. Res., Vol. 11, No. 2. 1952.

Pekarek, J. Ein vergessenes Objekt für das Studium der Kern- und Zellteilungsvorgänge im Leben. — Planta, Vol. 16. 1932.

Riley, G. A. A Mathematical Model of Regional Variations in Plankton. — Limnol. and Oceanogr., suppl. Vol. 10. 1965.

Riley, G. A., H. Stommel, and D. F. Bumpus. Quantitative Ecology of the Plankton of the Western North Atlantic. — Bull. Bingham Oceanogr. Coll. Vol. 12, No. 3. 1949.

Schiller, Jos. Dinoflagellatae, No. 2. — Kryptogamen-Flora, Vol. 10, pt. 3. 1937.

Semina, H. I. Movements of Water and the Size-groups of Phytoplankton. — Sarsia, Vol. 34. 1968.

Smayda, T. J. Some Measurements of the Sinking Rate of Fecal Pellets. — Limnol. and Oceanogr., Vol. 14, No. 4. 1969.

Smayda, T. J. and B. J. Boleyn. Experimental Observations on the Flotation of Marine Diatoms. 2. Skeletonema costatum and Rhizosolenia setigera. 3. Bacteriastrum hyalinum and Chaetoceros lauderi. — Limnol. and Oceanogr. Vol. 11, No. 1. 1966.

Snedicor, G. W. Statistical Methods Applied to Experiments in Agriculture and Biology. Ames, Iowa State College Press. 1956.

Steele, J. H. The Quantitative Ecology of Marine Phytoplankton. — Biol. Rev. Vol. 34, No. 2. 1959.

Steel, R. G. D. and J. H. Torrie. Principles and Procedures of Statistics with Special Reference to the Biological Sciences. McGraw-Hill, New York. 1960.

Vinogradov, M. E., I. I. Gitel'zon, and Yu. I. Sorokin. The Vertical Structure of a Pelagic Community in the Tropical Ocean. — Marine Biology, Vol. 6, No. 3. 1970.

VERTICAL DISTRIBUTION AND DIURNAL MIGRATIONS OF MASS SPECIES OF CYCLOPOIDA (COPEPODA) IN THE WESTERN EQUATORIAL PACIFIC

A. V. Tsalkina
Shirshov Institute of Oceanology of the Academy of Sciences of the USSR

Cyclopoida (Copepoda) constitute one of the most important groups of surface plankton in the tropical zone of the ocean. Their ecological significance in the tropics is high, as they are represented by numerous species and are among the most populous groups in general.

The material was collected during the 44th cruise of the "Vityaz" in the western equatorial Pacific at the diurnal station 6033 (4°09.2'S, 168°31.0'E) on December 21–23, 1968.

The collecting techniques and the hydrological conditions in the region are described by Vinogradov and Rudyakov (1971) in this collection.

Catches were made down to 300 m by the layers 300–200, 200–150, 150–100, 100–75, 75–50, 50–25, and 25–0 m. A sample obtained from each layer was divided into two equal parts in the sample splitter (McEwen et al., 1954), and only one half was processed. The chamber processing was done in the following manner. A sample volume was brought up to 100 ml, 10 cm^3 was drawn off with a plunger pipet and the animals were counted under the binocular microscope in a Bogorov chamber. Separate counts were made of males, females, and copepodite stages of different species. After the processing, the results from the two samples taken simultaneously from the same horizon were averaged.

Seventeen species of Cyclopoida belonging to three families, Oithonidae, Oncaeidae, and Corycaeidae, were chosen for the investigation, since their abundance permitted their distribution patterns to be traced,* as follows: Oithona plumifera*, O. setigera*, O. tenuis*, O. robusta*, O. vivida, Oncaea venusta*, O. mediterranea*, O. media, O. conifera*, O. clevei, Corycaeus speciosus*, C. crassiusculus*, Agetus typicus*, Onychocorycaeus agilis, Farranula carinata, F. concinna, F. gibbula.

* Only the mature stages were counted for the majority of species, and in addition copepodites IV and V in the species marked with an asterisk. The mesh size of the net does not make it possible to take into account the copepodite stages of all the species investigated, because the young Cyclopoida pass through the net.

DIURNAL VARIATIONS OF THE ABUNDANCE OF CYCLOPOIDA

The region studied was poor in plankton. The biomass of zooplankton in the 0 — 100-m layer averaged 24 mm^3/m^3 over the diurnal period (Vinogradov and Rudyakov, 1971).

The number of Cyclopoida in the material was small. The following species had the highest average abundance throughout the layer investigated (diurnal averages, specimens/m^3):

Oitnona plumifera	11.5	O. conifera	3.0
O. robusta	1.5	O. mediterranea	2.5
O. setigera	1.3	Farranula carinata	3.0
Oncaea media	12.7	F. concinna	1.9
O. venusta	4.2	F. gibbula	1.5
O. clevei	3.7		

The average abundances were 16, 41, and 31 specimens/m^3 for the Corycaeidae, Oncaeidae and Oithonidae, respectively.

For comparison it may be pointed out that the abundances of the same families in the 0 — 500-m layer in the central waters of the Indian Ocean were 28, 70 — 80 and 50 — 60 specimens/m^3, respectively (our data).

The diurnal variations of the total abundance of Cyclopoida in the water column were small and irregular (Figure 1).

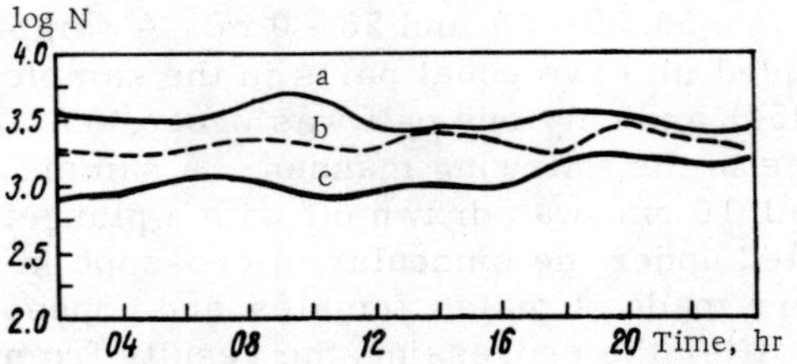

FIGURE 1. Diurnal variations of absolute abundance of Oncaeidae (a), Oithonidae (b), and Corycaeidae (c) in the 0—300-m layer

We also examined the quantitative variations of three mass species, Oithona plumifera, Oncaea media, and Farranula carinata (Figure 2), which account for over 50% of the abundance of the families to which they belong. The variations of their own numbers are also small and apparently random. In order to make sure that the described pattern did not result from a wide scatter of the results from two parallel samples, we provide data on Oncaea media and Farranula carinata for the two series in which these species were the most populous (Figure 3). It is seen that the scatter of data is small and can be neglected.

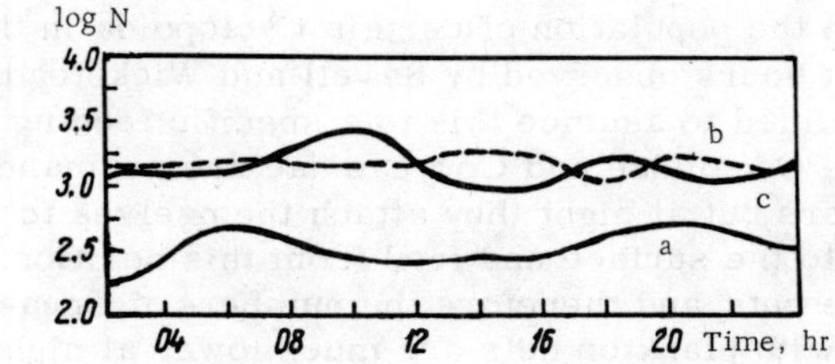

FIGURE 2. Diurnal variations of absolute abundance of some mass species of Cyclopoida in the 0—300-m layer:

a — Farranula carinata; b — Oithona plumifera; c — Oncaea media.

More detailed and frequent hauls did not disclose any considerable increase of numerical abundance in the daytime or a decrease at night as reported by Wickstead (1961, 1962) and Sewell (1947—1948).

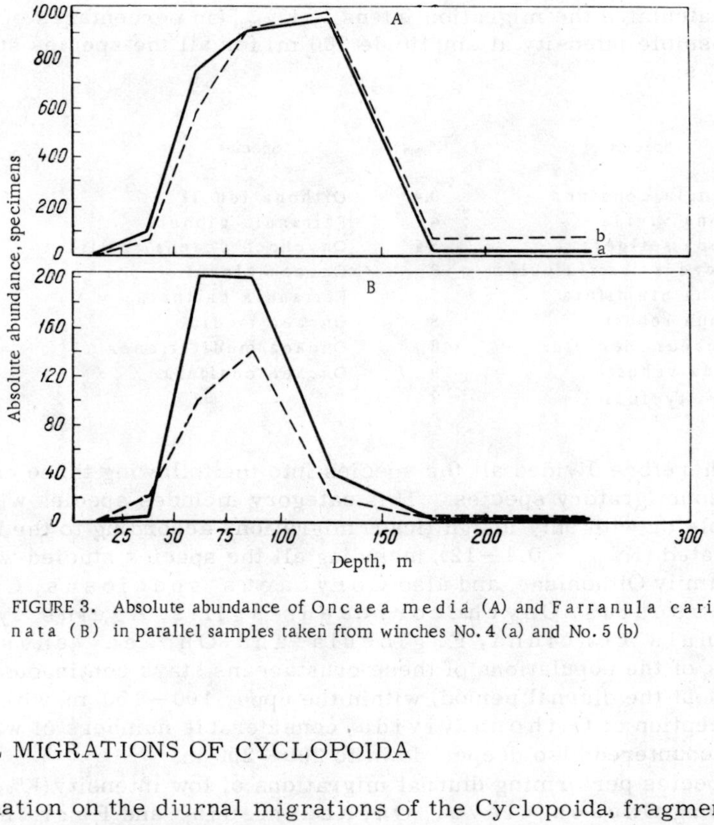

FIGURE 3. Absolute abundance of Oncaea media (A) and Farranula carinata (B) in parallel samples taken from winches No. 4 (a) and No. 5 (b)

DIURNAL MIGRATIONS OF CYCLOPOIDA

Information on the diurnal migrations of the Cyclopoida, fragmentary and often contradictory, has been reported by Moore et al., (1953, 1957, 1963), Motoda and Anraku (1956), Sewell (1947—1948), and Wickstead (1961, 1962).

The increase in the population of certain Cyclopoida in the surface layers during the daylight hours observed by Sewell and Wickstead is of interest. Wickstead was inclined to ascribe this to a specific feeding pattern. According to him, the Oncaeidae and Corycaeidae are permanent inhabitants of the surface layers but at night they attach themselves to macroplanktonic organisms rising to the surface and feed from this position. Macroplankton actively avoids the nets, and therefore the numbers of Oncaeida and Corycaeida in catches with plankton nets are much lower at night. In contrast, plankton catches taken in the daytime, when the Oncaeidae and Corycaeidae do not feed and lead a free pelagic life, reflect the true abundance.

Moore et al., (1957, 1963, 1966) divided the Copepoda in his study into surface, intermediate, and deepwater forms and found that the surface species scarcely migrate over the diurnal period or even perform a reversed migration. His group of surface species included some species of Cyclopoida that also figures in our material.

All the species of Cyclopoida that we investigated are characteristic inhabitants of the surface waters. Judging from Moore's conclusions, they should not perform any considerable vertical migrations, but this was not confirmed by our data.

We calculated the migration intensity $K\%_{300}$ (in percentages of the maximum possible intensity at amplitude 300 m) for all the species studied* as follows:

Species	$K\%_{300}$	Species	$K\%_{300}$
Farranula concinna	0.4	Oithona tenuis	9
Oithona vivida	4	Farranula gibbula	9
Oithona setigera	5	Onychocorycaeus agilis	12
Corycaeus crassiusculus	6	Oncaea clevei	17
Oithona plumifera	7	Farranula carinata	20
Oithona robusta	8	Oncaea media	23
Corycaeus speciosus	8	Oncaea mediterranea	32
Oncaea venusta	9	Oncaea conifera	62
Agetus typicus	9		

We therefore divided all the species into the following three categories.

1. Nonmigratory species. This category includes species without detectable or with only insignificant migrations according to the horizons investigated ($K\%_{300}$ = 0.4 — 12), including all the species studied which are of the family Oithonidae, and also Corycaeus speciosus, C. crassiusculus, Onychocorycaeus agilis, Agetus typicus, Farranula concinna, F. gibbula, and Oncaea venusta. The bulk of the populations of these crustaceans stays continuously (throughout the diurnal period) within the upper 100 — 150 m, with the sole exception of Oithona vivida, considerable numbers of which were encountered also deeper, down to 200 — 250 m.

2. Species performing diurnal migrations of low intensity ($K\%_{300}$ = 17 — 23), comprising Oncaea clevei, Oncaea media, and Farranula carinata. All perform migrations of the usual type, descending deeper in the daytime and rising to the surface at night. Most of these species inhabit

* The migration intensities were calculated by Vinogradov's method (1968).

depths down to 150—200 m. The migration amplitude of their abundance maximum is small, ranging from 50 m in Oncaea clevei and 70 m in F. carinata to 80—85 m in O. media.

3. Species performing diurnal migrations of medium intensity ($K\%_{300} = 32-62$). Oncaea mediterranea and Oncaea conifera were classified in this category. The migration amplitude of the abundance maxima of these species reaches 125 m; for the most part they remain within the upper 300 m throughout the diurnal period.

The nighttime ascent to the surface layers does not occur simultaneously in the different migratory species of Cyclopoida. Thus, maximum abundance in the 0—25-m layer occurs at 2000 hr in the case of O. clevei, 2200 hr in F. carinata, and 0200 hr in O. media.

During our work at the diurnal station, maximum illuminance occurred at noon, while values approaching the maximum were observed by 1000 hr. and persisted until approximately 1400 hr; from this time on, illuminance declined rapidly.

The water temperature at depths inhabited by the nonmigratory species (the upper 150 m) remained practically constant throughout the diurnal period, the depth interval occupied by waters at about 29° being approximately 100 m round the clock. The migrations of these species were apparently not affected to any considerable extent by variations of illuminance.

The majority of individuals belonging to the weakly migrating species (over 75% of the total population) occur in the gradient layer (150—200 m), where the temperature drops from 26 to 20°. These species begin to rise to the surface at 1400 hr, that is, just when illuminance starts to decline.

The bulk of Oncaea mediterranea and O. conifera (over 75% of the total population) belonging to the group of species with diurnal migrations of medium intensity occurs in the 200—300-m layer. In this layer, however, there were also considerable numbers of the nonmigratory Oithona vivida. The temperature gradient here is considerable, from 20° at 200 m to 11° at 300 m. The rise to the surface of species in this group likewise starts at 1400 hr.

Our catches, performed by very detailed horizons at two-hour intervals, did not reveal any signs of reversed vertical migrations in the Cyclopoida.

The migrations of females and males were identical in the investigated species, and therefore we combined their numerical abundances for calculation purposes. The sex ratio varies in species belonging to the different families. Thus, in the Oithonidae, over 99% of all specimens were females, males occurring sporadically. The ratios of females to males among the Oncaeidae was 4:1 in the mass species Oncaea venusta and O. clevei and 5:1 in O. media. In the nonpopulous species O. mediterranea and O. conifera the sex ratio was 1:1. Males were slightly more numerous than females in all the species of Corycaeidae.

VERTICAL DISTRIBUTION OF THE CYCLOPOIDA

The distribution of populations is very drawn out vertically, specimens occurring sporadically at depths that are uncharacteristic of the population as a whole. A fairly clear idea of the vertical distribution of species can be gained by examining the core of the population, i. e. the part comprised between the 25 and 75% levels (Vinogradov, 1968; Moore et al., 1953, 1957, 1963).

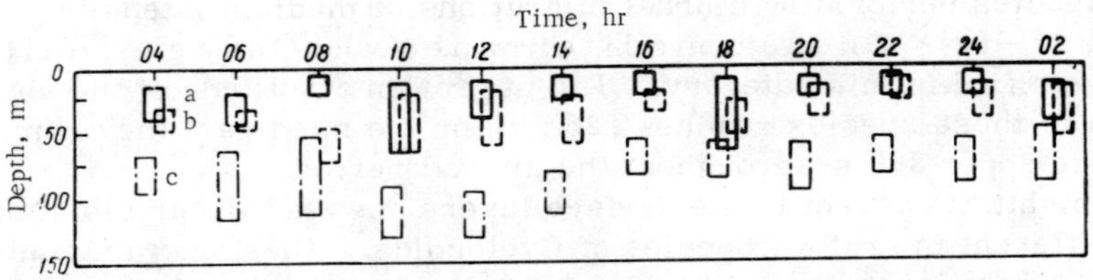

FIGURE 4. Vertical distribution of the "combined cores" of the populations of species of the families Corycaeidae (a), Oncaeidae (b), and Oithonidae (c) over the diurnal period

Figure 4 shows the depth of occurrence of the "combined cores" of the populations of all the investigated species of the families Oithonidae, Oncaeidae, and Corycaeidae over the diurnal period.* It is immediately seen that the vertical divergence of the "combined cores" of the populations of species belonging to the three families, forms a stepwise distribution which persists throughout the diurnal period. The "combined core" of the populations of Corycaeidae species is fixed in the layer 20(12) — 70(120) m deep** throughout the diurnal period. The next step is occupied by the "combined core" of the populations of Oncaeidae species occupying the layer 52 (25) — 102 (145) m deep; this core shifts toward the surface at night. The "combined core" of the populations of Oithanidae species occupies the lowest step, 80(70) — 130 (140) m.

Different representatives of each family exhibit the same stepwise distribution of population cores, at any rate their upper boundaries (25% levels). Among the species of Oithonidae (Figure 5a, b) the core of Oithona tenuis is located the highest, in the 68(55) — 98 (125)-m layer throughout the diurnal period; the core of O. plumifera lies deeper down, 78 (65) — 134 (145)m and that of O. robusta still deeper 105 (87) — 122 (135) m. The lowest level is occupied by the population core of O. vivida, 135(104) — 203 (255) m. It is of interest to note, as reported by Moore et al., (1953, 1957), that the vertical extent of a core is greater for deeper-dwelling species. The vertical extent of the core of O. tenuis, which is the most surface-inhabiting species of the Oithonidae, averages 24(40) m, that of O. robusta 35 (40) m, and of O. plumifera and O. vivida 54 (65) and 99 (150) m,† respectively.

The stepwise pattern is also readily observable among the species of the Oncaeidae (Figure 6, a, b). The core of Oncaea media is situated at depths of 46 (15) — 101 (120)m, that of O. venusta in the 61 (20) — 119 (130)-m layer, and that of O. clevei deeper still, at 71 (55) — 100 (125) m. Lower levels are occupied by the population cores of O. mediterranea, 110 (80) — 165 (230) m, and O. conifera, 121 (210) — 198 (270) m. Populations of the Oncaeidae species perform small vertical migrations, rising

* The term "combined core" in this context is applied to the total numerical abundance of all the species in the family comprised between the 25 and 75% levels.

** The figures in parentheses denote the extreme depths of the levels.

† The figures in parentheses denote the maximum extents of the population cores.

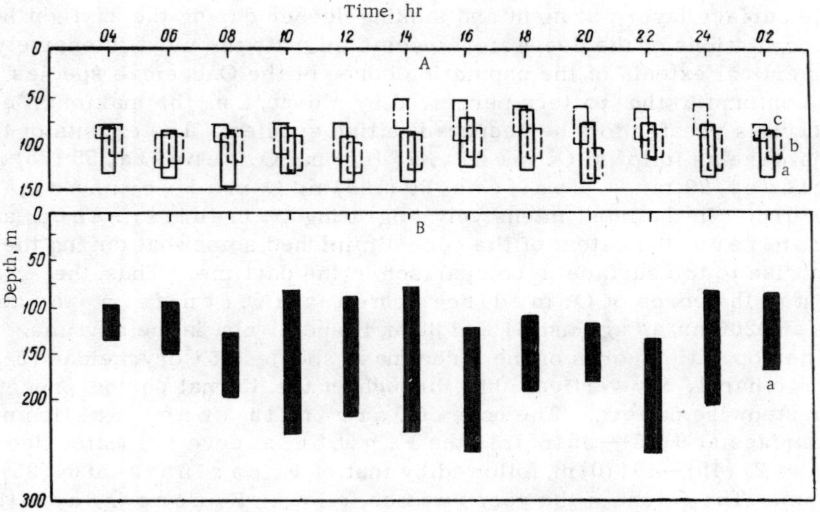

FIGURE 5. Vertical distribution of the population cores of species of Oithonidae over the diurnal period:

A — surface species: a — Oithona plumifera; b — O robusta; c — O. tenuis; B — deepwater species: a — Oithona vivida.

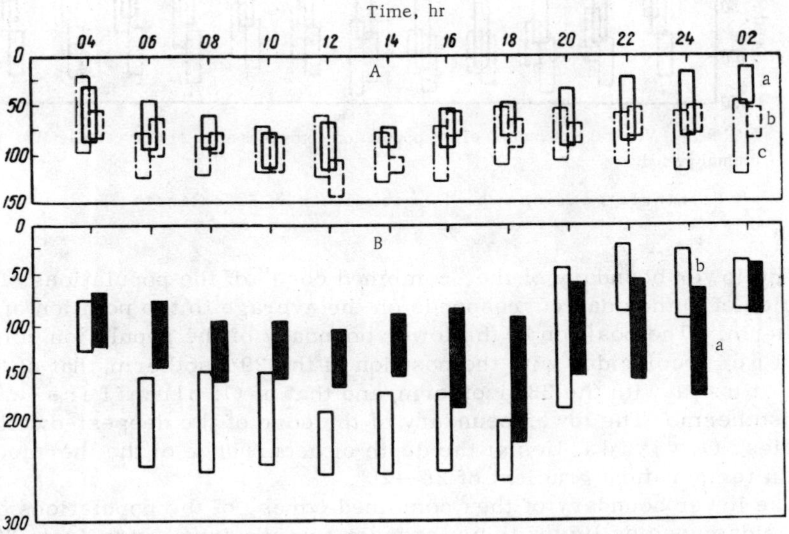

FIGURE 6. Vertical distribution of the population cores of species of Oncaeidae over the diurnal period:

A — surface species: a — Oncaea media; b — O. clevei; c — O venusta; B — deepwater species: a — Oncaea mediterranea; b — O. conifera.

to the surface layers at night and sinking deeper during the daylight hours. The populations of the deepwater species migrate the most intensively. The vertical extents of the population cores of the Oncaeidae species also conform to the pattern perceived by Moore, i. e. the maximum extent of cores is greater for the deeper-dwelling species. The extents of the cores were as follows: O. media, 45 (65) m; O. venusta, 55 (75) m; O. clevei, 29 m; O. conifera, 80 (125) m; O. mediterranea, 84 (130) m. In the most intensively migrating O. mediterranea and O. conifera the extent of the core diminished somewhat during the night rise to the surface in comparison to the daytime. Thus, the extent of the cores of O. mediterranea and O. conifera was about 50 m at 0200 hr, as against 84 and 80 m, respectively, in the daytime.

The population cores of the three mass species of Corycaeidae (Figure 7) undergo hardly any vertical shift throughout the diurnal period, preserving their stepwise pattern. The core of Farranula concinna lies nearest the surface at 9 (15) — 35 (65) m; the F. gibbula core is located deeper down at 21 (45) — 49 (70) m, followed by that of F. carinata at 64 (95) — 102 (135) m. The extent of the cores was as follows: F. concinna, 24 (50) m; F. gibbula, 27 (45) m; F. carinata, 37 (65) m. The population cores of the first two species remain practically at the same level throughout the diurnal period, while the core of F. carinata rises somewhat toward the surface at night but still without disturbing the stepwise distribution.

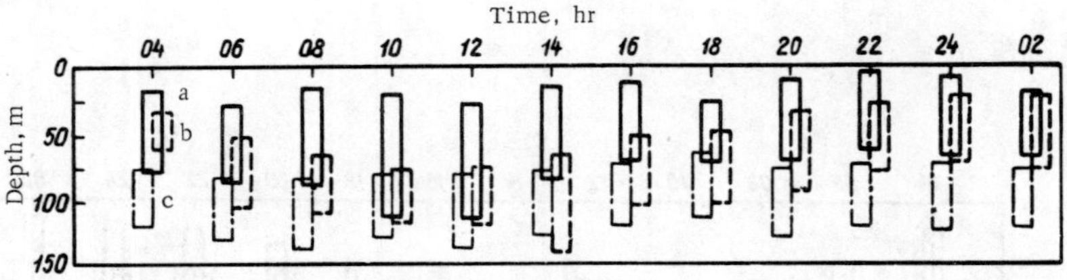

FIGURE 7. Vertical distribution of the population cores of species of Corycaeidae over the diurnal period:

a — Farranula concinna; b — F. gibbula; c — F. carinata.

The lower boundary of the "combined core" of the populations of all the species of Oithonidae corresponds on the average to the position of the 26° isotherm. The position of the lower boundary of the population core of O. tenuis coincides with the position of the 29° isotherm, that of the O. robusta with the 28° isotherm, and that of O. plumifera with the 27° isotherm. The lower boundary of the core of the deepest-dwelling species, O. vivida, lies at the depth of occurrence of the thermocline, with a temperature gradient of 26 — 20°.

The lower boundary of the "combined cores" of the populations of Oncaeidae species lies within a broader temperature range, from 29 to 26°. The lower boundary of the core of O. media coincides with the position of the 29° isotherm, while that of O. clevei and O. venusta coincides with the 27 — 29° and 28 — 29° isotherms, respectively. The temperature of the layers containing the lower boundary of the population cores varies considerably in the species performing the most intensive vertical migrations and passing through the thermocline, 22 – 27° in O. mediterranea and 15 – 29° in O. conifera.

The lower boundary of the "combined core" of the populations of all the Corycaeidae species is situated in layers with temperatures of 28 – 29°.

DISCUSSION

The most striking feature of the distribution of the Cyclopoida is the stepwise arrangement of the population cores of the different species, which is rigidly preserved throughout the diurnal period and is equally characteristic of the vertically migratory and nonmigratory species. A similar distribution of copepods was described by Vinogradov (1956) for the Bering Sea and by Geinrikh (1957) and Vinogradov and Voronina (1964) for tropical waters. Alternation of dominance layers is also known for deep-sea plankton (Vinogradov, 1968; Cushing, 1951), freshwater Cyclops (Frayer, 1957), and some tropical fishes (Paxton, 1967).

The ecological differences in the marine pelagic biocenoses are primarily related to the mode of feeding of their components. Utilization of food resources is the fundamental aspect of competition (Vinogradov, 1968). Practically all Cyclopoida have a similar type of feeding (Wickstead, 1962), in other words, all the species of this group are potential food competitors. According to the principle expounded by Gauze (1936), species occupying the same ecological niche cannot maintain a stable coexistence. Alternation of dominance layers of similarly feeding species ensures that ecologically similar, systematically related, and identically feeding species (Vinogradov, 1968) can exist in the same oceanic region, and it has been demonstrated that the strictness of differentiation of the dominance layers is higher as the food resources are more limited (Vinogradov, 1956; Geinrikh, 1957; Vinogradov and Voronina, 1964).

Our investigations, conducted in a region that is very poor in plankton, revealed a distinct alternation of dominance layers. Vinogradov (1968) pointed out that the dominance layers of tropical plankton have a very small vertical extent and are therefore obliterated in broad catches. We succeeded in pinpointing the dominance layers and their stepwise arrangement by performing very detailed catches at small depth and time intervals.

A noteworthy feature is the divergence of migrating species of Cyclopoida as to the time of nocturnal ascent into the surface layers, i. e. the divergence of population cores with respect not only to the depth of habitat but also to the duration of their stay in the principal foraging area; this undoubtedly reduces the competition for food among similarly feeding species.

The data presented in this paper suggest that as with all planktonic animals, the distribution of the Cyclopoida obeys the principle of elimination of competition. However, the scatter of species with respect to depth is incomplete because of the small extent of their vertical zones of habitat, high mobility of the animals and transport of waters. The principle of elimination of competition is manifested in the diversification of the habitats of population cores with respect to depth and also in the different times of ascent of migratory species to the surface layers at night.

In conclusion I wish to express my deep gratitude to M. E. Vinogradov and Yu. A. Rudyakov for their unfailing assistance and valuable advice.

CONCLUSIONS

1. Studies were made of the vertical distribution and diurnal migrations of 17 species of Cyclopoida (Copepoda) in the Western Equatorial Pacific.
2. The abundance of Cyclopoida in the waters investigated was practically constant throughout the diurnal period.
3. None of the investigated species of the family Oithonidae, or of the Corycaeidae (with the exception of one species), performs diurnal vertical migrations. The investigated species of the family Oncaeidae (again with the exception of one), and also one species of Corycaeidae perform diurnal migrations of the usual type, rising to the surface layers at night and descending to deeper layers in the daytime.
4. Reversed diurnal vertical migrations were not discovered in any of the species, contrary to the view advanced by Moore, Wickstead, and Sewell.
5. The population cores of the different Cyclopoida species are located at different depths over the diurnal period, exhibiting a stepwise distribution which is also preserved by the species performing diurnal migrations.
6. The different mass migratory species rise to the surface layer at different times.
7. This distribution pattern apparently tends to reduce food competition among the species of Cyclopoida with similar feeding patterns.

Bibliography

Publications in Russian

Gauze, G. F. O nekotorykh osnovnykh problemakh biotsenologii (On Some Fundamental Problems of Biocenology). — Zoologicheskii Zhurnal, Vol. 15, No. 3. 1936.

Geinrikh, A. K. Vertikal'noe raspredelenie planktona v raione k yugo-vostoku ot Boninskikh ostrovov (Vertical Distribution of Plankton in the Region Southeast of the Bonin Islands). — Doklady Akademii Nauk SSSR, Vol. 117, No. 2. 1957.

Vinogradov, M. E. Raspredelenie zooplanktona v zapadnykh raionakh Beringova morya (Distribution of Zooplankton in the Western Bering Sea). — Trudy Vsesoyuznogo gidrobiologicheskogo obshchestva, Vol. 7. 1956.

Vinogradov, M. E. Vertikal'noe raspredelenie okeanicheskogo zooplanktona (Vertical Distribution of Oceanic Zooplankton). Moskva. 1968.

Vinogradov, M. E. and N. M. Voronina. Raspredelenie biomassy planktona v vodakh ekvatorial'noi chasti Tikhogo okeana. II. Vertikal'noe raspredelenie otdel'nykh vidov (Distribution of the Plankton Biomass in Waters of the Equatorial Pacific. II. Vertical Distribution of Individual Species). — Trudy Instituta Okeanologii Akademii Nauk SSSR, Vol. 65. 1964.

Vinogradov, M. E. and Yu. A. Rudyakov. Sutochnye izmeneniya vertikal'nogo raspredeleniya obshchei biomassy planktona v zapadnoi ekvatorial'noi Patsifike (Diurnal Variations of the Vertical Distribution of the Plankton Biomass in the Equatorial West Pacific). (In this collection.)

Publications in Other Languages

Cushing, D. H. The Vertical Migration of Planktonic Crustacea. — Biol. Rev. Vol. 26, No. 2. 1951.

Frayer, G. 1957. The Food of Some Freshwater Cyclopoid Copepods and its Ecological Significance. — J. Anim. Ecol. Vol. 26. 1957.

McEwen, G. E., M. M. Johnson, and T. R. Folsom. A Statistical Analysis of the Performance of the Folsom Plankton Sample Splitter, Based Upon Test Observations. — Arch. Meteorol. Geophysik u. Bioklimatol., Ser. A: Meteorologie u. Geophysik, Vol. 7. 1954.

Moore, H. B. and D. L. O'Berry. Plankton of the Florida Current. IV. Factors Influencing the Vertical Distribution of Some Common Copepods. — Bull. Mar. Sci. Gulf and Caribbean, Vol. 7, No. 4. 1957.

Moore, H. B. and M. Foyo. A Study of the Temperature Factor in Twelve Species of Oceanic Copepods. — Bull. Mar. Sci. Gulf and Caribbean, Vol. 13, No. 4. 1963.

Moore, H. B., R. Owre, E. C. Jones, and T. Dow. Plankton of the Florida Current. III. The Control of the Vertical Distribution of Zooplankton in the Daytime by Light and Temperature. — Bull. Mar. Sci. Gulf and Caribbean, Vol. 3, No. 2. 1953.

Moore, H. B. and M. G. Roehr. Nul Values of Temperature, Illumination and Pressure for a Series of Oceanic Copepods. — Bull. Mar. Sci., Vol. 16, No. 2. 1966.

Motoda, S. and M. Anraku. Marine Plankton Copepods from the Machijo Island, Idzu Island, with Special Reference to Their Vertical Distribution. — Rec. Oceanogr. Japan., Vol. 2, No. 2. 1956.

Paxton, J. L. A Distributional Analysis for the Lantern-Fishes (Family Myctophidae) of the San Pedro Basin, California. — Copeia, No. 2. 1967.

Sewell, R. B. S. The Free-Swimming Planktonic Copepoda. Systematical Account. Geographical Distribution. — John Murray Exp. 1933—1934, Sci. Repts., Vol. 8, No. 1/3. 1947—1948.

Wickstead, J. H. A Quantitative and Qualitative Study of Some Indo-West-Pacific Plankton. — Fishery Publ., No. 16. London. 1961.

Wickstead, J. H. Food and Feeding in Pelagic Copepods. — Proc. Zool. Soc., Vol. 139, No. 4. London. 1962.

DETAILS OF THE HORIZONTAL DISTRIBUTION AND DIURNAL VERTICAL MIGRATIONS OF CYPRIDINA (PYROCYPRIS) SINUOSA (G.W. MÜLLER) (CRUSTACEA, OSTRACODA) IN THE WESTERN EQUATORIAL PACIFIC

Yu. A. Rudyakov
Shirshov Institute of Oceanology of the Academy of Sciences of the USSR

The overall quantitative distribution pattern of plankton in the World Ocean has been established (Bogorov et al., 1968), and planktonologists are now turning away from the field of general evaluation of the forage resources of different waters and are going over to studies of the structure and activity of pelagic communities, which involve research into the horizontal and vertical distributions of plankton on the scale of hundreds and tens of meters and even centimeters. Attention is becoming more and more centered on the distribution patterns of individual species and coeval groups of individuals of the same species. Separate consideration of coeval groups within populations shows signs of becoming the rule in the future, as it is known that organisms of the same species but of different age eat different foods (Petipa, 1965) and consequently play quite different parts in the trophic web of a community.

Plankton collections obtained from the 0—25, 25—50, 50—75, 75—100, 100—150, 150—200, and 200—300-m layers at diurnal stations during the 44th cruise of the "Vityaz" provide excellent material for studies of the details of horizontal distribution and diurnal vertical migrations, since the use of parallel catches makes it possible to apply mathematical statistical methods when processing the material.

Large numbers of pelagic Ostracoda, Cypridina sinuosa (G. W. Müller), were found at the diurnal station 6033 executed at the northern periphery of the South Equatorial Current (04°10' S, 168°31' E). Since it lent itself easily to counts and to approximate determination of age, this species was chosen as the object of study. Work at the station was conducted from 2300 hr on December 21 to 0200 hr on December 23, 1968 (zonal time is used in this paper). Sunrise is at 0530 hr, sunset at 1800 hr, and noon at 1145 hr. For a detailed description of the techniques of collection and preliminary processing of the material the reader is referred to the paper by Vinogradov and Rudyakov (1971).

The laboratory processing included counts of the ostracods in half of each sample, measurement of the length of the shell of each specimen, and determination of the sex of mature specimens.

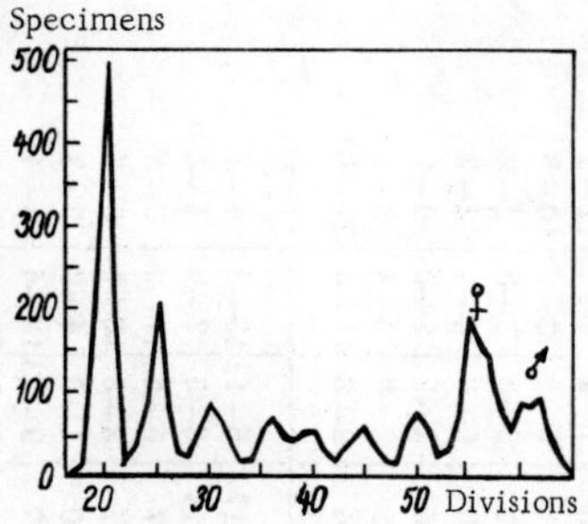

FIGURE 1. Numerical distribution of C. sinuosa specimens according to the length of the shell stated in scale divisions of the ocular micrometer (scale value 0.025 mm)

About 3,600 specimens of Cypridina sinuosa were counted and measured; their distribution with respect to the shell length is shown in Figure 1.

The length differences among mature females and males suggest that they hold also for immature individuals, being smaller the younger the animals. Apparently the maxima at 50 and 40 divisions of the ocular micrometer* correspond to males, those at 45 and 36 divisions to females of two consecutive age stages. The differences in the length of younger females and males are probably unnoticeable, and the maxima at 30, 25, and 20 divisions correspond to consecutive age stages of females and males together. Hence, in addition to mature individuals of both sexes we are apparently dealing with five larval stages with arbitrary boundaries of shell length 17—22, 23—27, 28—33, 34—42, and > 42 divisions of the ocular micrometer. The seven differentiated groups of specimens were considered separately (Table 1).

HORIZONTAL DISTRIBUTION

The inhomogeneities of horizontal distribution were studied by using information on the number of individuals in the different layers totaled for the entire water column investigated, 0—300 m, separately for each of the two nets working in parallel.

The time variation of the logarithmic abundances of the differentiated groups in the 0—300-m layer is shown in Figure 2. Comparison of the experimental curves reveals, along with certain common features, their specific characteristics. For a more precise analysis of the abundance variations we used successive fitting of approximating parabolas of increasing degrees by the method of orthogonal Chebyshev polynomials the principles of which were described by Snedecor (1956).

In the first stage of the analysis, the sum total of squares of central deviations was divided into four parts related to the differentiated "groups," "time" moments, their "interaction" and experimental "error." Then parts related to parabolic regression of steadily increasing degree were successively differentiated from the sums of squares related to "time" and "interaction," until the mean square "residue" from "interaction" became comparable with the mean square of "error." This occurred (Table 2) when the experimental data were approximated by a fourth-degree polynomial. Hence, a fourth-degree parabola provides the best description of the differences in the general trends of variation of the abundance of the differentiated groups. However, the significance of "fourth-degree

* One division of the ocular micrometer is equivalent to 0.025 mm.

TABLE 1. Numerical abundance of differentiated groups of C. sinuosa specimens in samples obtained with Juday 80/113 nets operated from winches No.4 (left-hand figure) and No.5 (simultaneous catches for each layer)

Layer m	Sex and size (micrometer scale divisions)	Time, hr													
		23	01	03	05	07	09	11	13	15	17	19	21	23	01
0—25	♀	2—1	4—2	0—0	1—0	0—0	0—0	0—0	0—0	0—0	9—15	5—7	4—7	2—2	6—6
	♂	4—2	0—0	0—0	0—0	0—0	0—0	0—0	0—0	0—0	6—7	0—1	1—1	0—3	2—3
	>42	1—0	0—1	0—1	0—0	0—0	0—0	0—0	0—0	0—0	1—1	1—0	0—1	0—0	0—0
	34—42	2—0	2—0	2—1	0—0	0—0	0—0	0—1	0—0	0—0	0—0	1—0	1—4	1—0	0—1
	28—33	1—1	1—1	1—0	0—0	0—0	0—0	0—0	0—0	0—0	0—0	0—0	1—0	1—0	0—1
	23—27	0—2	1—0	0—0	0—0	0—0	0—0	0—0	0—0	0—0	0—0	1—0	2—6	4—0	0—3
	17—22	8—8	1—0	1—1	0—0	0—0	0—0	0—0	0—0	0—0	0—0	1—1	5—8	4—0	2—6
25—50	♀	3—0	0—2	3—2	2—0	0—0	0—0	0—0	0—0	0—1	13—16	10—3	17—9	9—4	12—9
	♂	5—2	1—0	5—3	0—0	1—0	0—0	0—0	0—0	0—0	7—7	3—1	5—6	2—2	10—3
	>42	1—0	0—2	1—2	0—0	0—0	0—0	0—0	0—0	0—0	2—5	5—1	3—5	3—0	2—2
	34—42	1—1	2—1	6—0	0—0	0—0	0—1	0—0	0—0	0—0	0—3	2—0	6—6	4—1	2—2
	28—33	2—0	0—0	6—1	0—0	0—0	0—1	0—0	0—0	0—0	0—1	3—0	2—2	1—0	0—1
	23—27	1—1	0—3	3—0	1—0	0—0	0—0	0—0	0—0	0—1	1—1	2—1	3—6	3—4	3—1
	17—22	1—2	1—1	3—4	0—0	0—0	0—0	0—0	0—0	0—0	2—0	8—4	8—8	11—6	27—10
50—75	♀	0—1	0—0	1—0	0—1	0—0	0—0	0—0	0—0	2—0	14—1	14—17	5—14	13—2	4—7
	♂	0—1	2—1	1—0	0—0	0—0	0—0	0—0	0—0	0—0	4—2	11—14	6—3	7—0	4—4
	>42	0—1	0—0	0—2	0—1	0—0	0—0	0—0	0—0	1—0	4—1	8—7	5—5	17—9	3—5
	34—42	2—2	2—2	1—1	0—0	2—0	0—0	0—0	0—0	0—0	0—1	6—7	3—8	10—5	6—5
	28—33	2—3	3—1	2—1	0—0	0—0	0—0	1—0	0—0	0—0	0—0	3—0	1—6	4—0	3—3
	23—27	0—1	5—0	3—6	0—2	0—0	0—0	1—0	1—0	0—0	1—5	2—5	9—13	12—4	4—4
	17—22	0—1	2—1	7—2	0—0	3—0	1—0	1—0	0—0	0—0	8—7	26—17	16—17	22—10	11—11

75—100	♀	0—0	0—0	0—0	1—2	0—0	1—0	0—0	0—0	3—9	6—4	11—6	9—5	6—5	3—2
	♂	0—0	0—0	0—0	1—0	0—0	0—0	1—0	0—0	0—0	6—1	12—12	3—5	4—1	4—0
	>42	0—0	0—1	0—0	0—1	0—1	1—0	0—0	0—0	2—1	4—1	3—7	9—8	9—12	9—6
	34—42	1—0	1—1	1—1	2—1	0—0	0—0	0—0	0—0	0—0	2—5	4—5	13—11	18—11	10—7
	23—33	2—2	1—0	3—1	2—2	0—0	0—0	0—0	0—0	0—0	6—5	2—0	6—5	6—6	4—6
	23—27	1—1	0—0	2—0	6—4	0—3	0—0	0—1	1—0	1—0	6—6	8—5	12—17	11—12	20—13
	17—22	0—0	0—0	0—1	0—5	1—1	2—2	0—0	1—0	1—0	28—21	23—16	24—19	12—15	14—10
100—150	♀	1—0	0—0	0—0	9—12	0—2	0—1	0—0	1—0	22—19	6—3	0—0	0—1	1—0	1—0
	♂	0—1	0—0	0—0	3—5	0—2	0—1	0—1	0—0	5—6	0—0	1—0	1—3	1—1	0—0
	>42	0—0	0—0	0—0	5—13	2—3	0—0	2—0	0—0	6—7	7—3	1—1	2—3	4—7	1—2
	34—42	0—0	0—0	0—0	6—17	1—3	0—2	1—0	0—0	2—0	1—2	1—1	2—2	6—4	4—1
	23—33	0—0	0—0	1—0	11—27	2—3	0—1	0—0	0—1	0—1	5—0	1—1	0—0	4—6	0—0
	23—27	1—0	0—0	0—0	9—16	5—8	0—4	0—1	0—1	4—2	3—2	0—1	1—1	4—5	0—1
	17—22	0—1	0—0	0—1	5—12	20—25	14—11	4—3	3—2	24—8	21—16	1—0	1—2	2—3	3—1
150—200	♀	0—0	0—0	1—0	0—1	20—10	18—7	11—6	2—4	24—19	0—0	0—1	0—0	0—1	0—1
	♂	0—0	0—0	0—0	0—0	9—5	8—7	3—4	1—2	15—15	2—1	0—1	0—0	0—0	0—0
	>42	0—0	0—0	0—0	0—0	18—12	14—12	16—8	1—1	10—12	0—0	0—1	0—0	0—0	0—1
	34—42	0—0	0—0	0—0	2—0	24—19	9—14	11—6	1—0	6—9	0—0	0—1	0—0	0—0	0—0
	23—33	0—0	0—0	0—0	0—0	26—27	6—10	8—12	2—1	3—6	1—0	0—1	0—1	0—0	0—0
	23—27	0—0	0—0	0—0	0—0	20—23	13—10	15—9	2—4	10—14	1—0	0—1	0—0	0—0	0—1
	17—22	0—0	0—0	0—0	1—1	12—9	8—30	45—23	4—35	29—39	0—0	0—1	0—0	0—2	0—3
200—300	♀	0—0	0—0	0—0	0—1	10—12	10—13	38—37	28—26	2—4	0—0	0—0	0—0	0—0	0—0
	♂	0—0	0—0	0—0	0—0	5—8	2—12	19—21	10—8	3—7	0—0	0—0	0—0	0—0	0—0
	>42	0—0	0—0	0—0	0—0	4—2	0—5	10—15	13—8	4—3	0—0	0—0	0—0	0—0	0—0
	34—42	0—0	0—0	0—0	0—0	3—5	3—3	9—10	6—7	3—3	0—0	0—0	0—0	0—0	0—0
	23—33	0—0	0—0	0—0	0—0	2—3	0—1	7—5	2—1	2—3	0—0	0—0	0—0	0—0	0—1
	23—27	0—0	0—0	0—0	0—0	0—0	0—1	0—0	0—1	1—0	0—0	0—0	0—1	0—0	0—0
	17—22	0—0	0—0	0—0	0—0	0—1	0—0	0—0	0—0	0—3	0—0	0—1	0—0	0—0	0—0

interaction" is small, and that of the "fourth power of time" is nil. It was therefore decided to limit the approximation of empirical data to a third-degree polynomial.

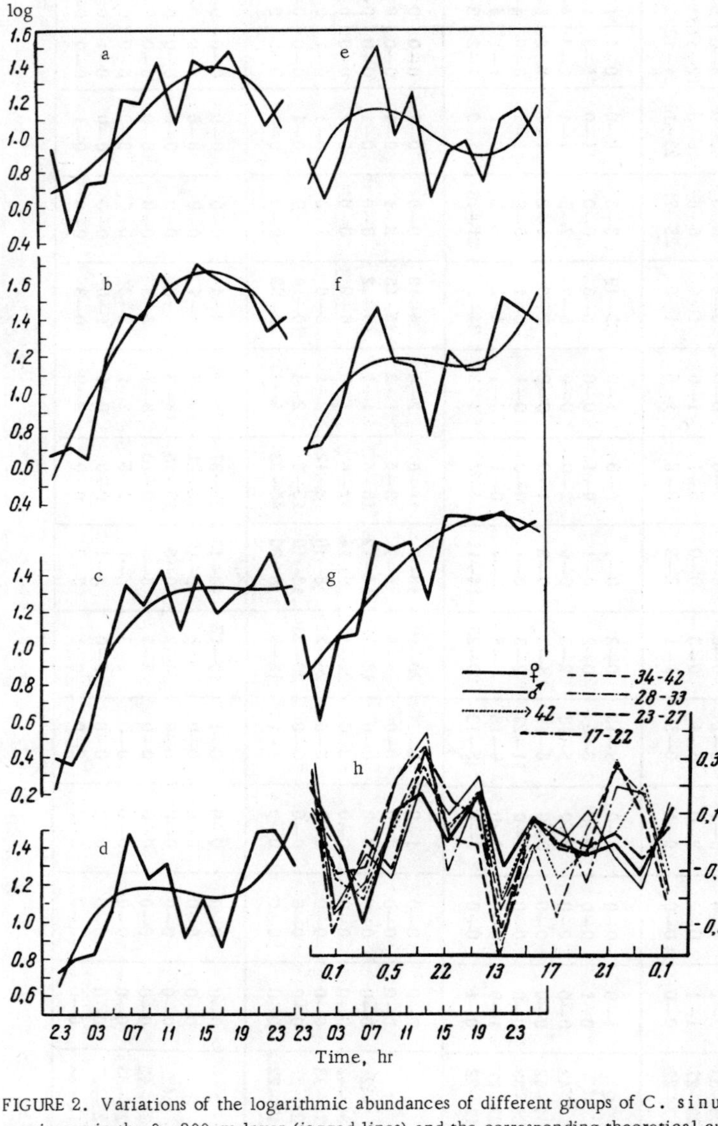

FIGURE 2. Variations of the logarithmic abundances of different groups of C. sinuosa specimens in the 0–300-m layer (jagged lines) and the corresponding theoretical curves:

a – males; b – females; c – g – immature specimens of the following size: c, > 42; d, 34–42; e, 28–33; f, 23–27, g, 17–22; h, deviations of the empirical graphs from the theoretical curves.

TABLE 2. Complete analysis of variance for data on variations of logarithmic abundances of different groups of C. sinuosa

Source of variation	Number of degrees of freedom	Sum of squares	Mean square	Probability, %
Total	195	24.8693		
Groups	6	3.6035		
Time	13	13.5698		
linear	1	3.7016	3.7016	< 0.05
square	1	1.0830	1.0830	< 0.05
cube	1	0.1440	0.1440	< 2.5
4th power	1	0.0008	0.0008	> 50
residue	9	8.6404	0.9600	< 0.05
Interaction	78	5.2901		
linear	6	1.4680	0.2447	< 0.05
square	6	0.9896	0.1649	< 0.05
cube	6	0.9450	0.1575	< 0.05
4th power	6	0.3899	0.0650	< 2.5
residue	54	1.4976	0.0277	< 25
Error	98	2.4059	0.0245	

From Figure 2 it follows that as the specimens are smaller, the theoretical shape of the abundance variation displays a trend toward bimodality, disappearing in the smallest specimens. The variations of abundance of females, males, and larvae of size 17 — 22 are practically the same.

As will be shown, these differences can be ascribed to the significant difference in the amplitudes of the diurnal vertical migrations of individuals belonging to the different size groups. Therefore, in accordance with the principles elaborated by Hardy (1965), differences in the velocity and direction of currents at different depths in the region investigated (Arsen'ev et al., 1971) inevitably lead to a separation of the aggregations of different-aged individuals. This assumption is borne out by the identical course of variation in the numerical abundance of females and males with the same migration amplitude. As for the relatively weakly migrating larvae of size 17 — 22, their accumulations may coincide with those of the mature individuals simply because they have only recently left the maternal brood chamber.

Similar features of the abundance variation of differentiated groups of individuals can be most conveniently examined from deviations of the empirical graphs from the theoretical curves (see Figure 2h). On the whole, these deviations are statistically significant for all the groups, the mean square of "residue" from "time" significantly exceeding the mean square of "error" (see Table 2), and apparently result from the influence of certain factors which equally affect the abundance of all the age groups. Identification of these factors can only be presumptive.

The decreased abundance around 0100 hr at the beginning of work might have been caused by the fact that ostracods were eaten by hyperiids Anchylomera blossevillei, a large concentration of which was recorded at that time (Vinogradov and Rudyakov, 1971). The stomachs of these hyperiids were found to contain limbs of Cypridina (oral communication from E. G. Arashkevich). The drop in abundance around 1400 hr cannot be explained either by some methodological error (for instance, premature closing of the nets or loss of part of the catch in preliminary processing) or by the ostracods' avoidance of the plankton nets in the daytime hours. The first of these explanations is contradicted by there being no such decrease in the variation of the total abundance of phytoplankton and Cyclopoida in the same samples (oral communications from I. N. Sukhanova and A. V. Tsalkina), and the second by the overall shape of the empirical curves depicting the variation of abundance of older age groups, in which the capacity for avoiding the net would be the most pronounced. One reason for the observed decrease could be the daytime descent of a part of the population below 300 m. The results of processing of the sample taken from the 300—500-m horizon around noon on December 21 did suggest that such a descent takes place. But in this case the decrease of abundance in the daytime should be the most pronounced in mature Cypridina performing migrations of the largest amplitude, and this assumption is not supported by experimental data. Therefore the most probable cause of the observed variations is the complex structure of the waters in the sampling region (Arsen'ev et al., 1971). During work at the station the ship may, under the action of wind, drift through areas that are occupied by waters with the same temperature and salinity but of different origin and consequently also with different plankton populations. According to Fager and McGowan (1963), such occurrences are quite possible.

VERTICAL DISTRIBUTION AND DIURNAL MIGRATIONS

One of our principal aims was to estimate the speed of migration of different age groups. This is usually determined from the path traveled by the object over measured periods. However, in most cases the planktonologist does not have the opportunity of studying the migrations of individual organisms and only possesses information concerning the depth variation of their abundance. Therefore his object is to study the distribution of individuals according to the depth of their occurrence and also the time variations of this distribution. The task of determining the path traveled by a group of individuals over a certain period is reduced to comparison of the central trends of two distributions with respect to depths, i. e. the problem of comparing two random groups, which is solved by standard statistical methods.

Either the arithmetic mean or the median can play the role of the central trend of vertical distribution characterizing the depth of occurrence of the group of individuals studied. The use of the median offers some advantages for distributions differing markedly from the normal. This method for describing vertical distribution (see Moore et al., 1953) uses the interquartile range (depths of the 25% and 75% levels are the first and third quartiles, while the 50% level represents the median, or the second quartile) as a measure of dispersion. However, only some of the initial data are utilized in calculation of the median quartiles, while the confidence interval for the median is in most cases considerably wider than for the arithmetic mean (Van der Waerden, 1957). Therefore, as with the large majority of cases of applying the tools of mathematical statistics, it is preferable to use the sample arithmetic mean to characterize the depth of occurrence of a group of animals, while the sample variance or standard deviation is used to characterize the dispersion.

Calculation of the mean habitation depth, or the "center of gravity" of vertical distribution was first used, apparently, by Worthington (1931) and Gardiner (1933). The calculations follow the usual schemes described in manuals of statistics.

For instance, if at the initial moment the layers 0—50, 50—100, 100—150, 150—200, and 200—250 m were found to contain 2, 5, 3, 0, and 0 individuals, respectively, then the arithmetic mean ($\bar{x}_1$) of this distribution is

$$\frac{25\cdot 2+75\cdot 5+125\cdot 3+175\cdot 0+225\cdot 0}{2+5+3+0+0}=\frac{800}{10}=80 \text{ m},$$

and the variance

$$\sigma_1^2=\frac{25^2\cdot 2+75^2\cdot 5+125^2\cdot 3+175^2\cdot 0+225^2\cdot 0-\dfrac{800^2}{10}}{10-1}=\frac{12250}{9}=1361 \text{ m}^2.$$

Let us suppose that, two hours later, the distribution of individuals in the same layers becomes 0, 1, 8, 2, and 0. For this distribution,

$$\bar{x}_2=\frac{1425}{11}=129.5 \text{ m and } \sigma_2^2=\frac{7273}{10}=727.3 \text{ m}^2.$$

Consequently, the path traveled by the group in two hours is $\bar{x}_1-\bar{x}_2=-49.5$ m (the minus sign indicates downward migration). In order to decide whether this figure is statistically significant, we must calculate the error in the difference m_{1-2}, which in our example is

$$\sqrt{\frac{12250+7273}{9+10}\cdot\frac{10+11}{10\cdot 11}}=13.97 \text{ m}.$$

The migration is 3.54 times larger than its error, thereby exceeding the limiting value of Student's test, which is $t=2.86$ when the number of degrees of freedom is $10+11-2=19$ and the significance level 1%. Therefore, it may be concluded with error probability less than 1% that migration indeed occurs, and it is possible to determine the corresponding confidence interval: $X_1-X_2=(\bar{x}_1-\bar{x}_2)\pm t\cdot m_{1-2}$. In our example, with significance level 5%, this interval is $-49.5\pm 2.09\cdot 13.97=-49.5\pm 29.2$, i. e. from -20.3 to -78.7 m. Therefore, the mean speed of migration over the 2-hour period for significance level 5% lies within the range -10.2 to -39.4 m/hr.

If there is a marked difference between the variances of the distributions being compared, which can be judged by the two-sided (!) Fisher test, Cochrane's approximate method is convenient (Snedecor, 1956). As for the applicability of Student's test to construction of confidence intervals of parameters of distributions deviating from normal, the central limit theorem is valid also in such cases. It is true that the power of the t test decreases in the case of deviations from normality (Van der Waerden, 1957), but the only result of this decrease is that the probability of the parameter under evaluation leaving the constructed confidence interval becomes less than the adopted significance level, i. e. it leads to more cautious interval estimates.

The velocity calculated by the above method is the velocity of synchronous vertical migrations of individuals in the investigated group or, which is the same thing, the vertical component of the velocity of the "mean individual." Naturally, such estimates can be used in energy calculations only as minimal ones.

In practice, with the existing techniques of plankton research, it is difficult to use this calculation method because the unequal thicknesses of the layers studied affects the estimates of the arithmetic mean and especially of the variance and makes it impossible to establish the true shape of the distribution. Therefore, if it is impossible to organize the operations in a manner ensuring uniform thickness of layers (or if this is not worthwhile for some other reasons), the layer thicknesses should be multiples of each other, as was the case in the 44th cruise. It is possible thus to achieve equal class intervals by summation of data elicited by catches taken from thinner layers.

Variations of the mean depth of vertical distribution of isolated groups of C. sinuosa are shown in Table 3. The entire population is situated near the surface in the dark part of the diurnal period. It begins to descend at least 2 hours before sunrise and continues to sink approximately to noon. This is followed by a rising phase which is completed toward 1700—1900 hr, or around sunset.

TABLE 3. Variation of the mean depth (m) of vertical distribution of different groups of C. sinuosa

Time, hr	Group							
	♀	♂	♀+♂	>42	34–42	28–33	23–27	17–22
23	44	35	38	38	49	59	56	32
01	19	56	31	44	51	52	49	45
03	54	40	41	42	43	53	61	55
05	130	119	132	120	123	120	107	119
07	⩾191	⩾190	⩾190	162	161	158	159	144
09	⩾194	⩾197	⩾195	162	166	150	154	149
11	⩾232	⩾234	⩾231	⩾199	⩾197	⩾183	166	169
13	⩾239	⩾236	⩾238	⩾241	⩾243	⩾193	140	167
15	141	170	151	156	176	183	147	152
17	50	51	50	77	75	92	85	92
19	60	73	65	71	77	80	73	54
21	50	62	54	70	63	69	56	63
23	59	63	61	82	78	89	73	65
01	46	47	45	77	75	60	71	57

A comparison of the depth variations according to the rows in Table 3 shows them to be quite regular. These regularities can be seen especially clearly from Figure 3. Mature individuals and larvae of size 17—22 come closest to the surface in the darkness hours. The population reaches a depth of 110—130 m towards 0500 hr (at sunrise), and by this time the mature individuals are located deeper than the larvae. In the course of the following six hours, large mature individuals descend increasingly faster than larvae, so that the interval separating them reaches at the most

60 m at the end of the descent.* However, already at around 1500 hr the mature individuals rising toward the surface overtake the larvae and reach their nighttime depth toward 1700 hr, whereas the larvae continue to rise until 1900 hr. On the second night, as on the first, the mature individuals and larvae of size 17—22 came closest to the surface, but rising ceased in deeper waters. As the result of these movements, the amplitude of diurnal vertical migrations amounts to some 110—130 m in individuals of size 17—22 and, increasing nonlinearly with age, becomes at least 200 m in the case of mature specimens.

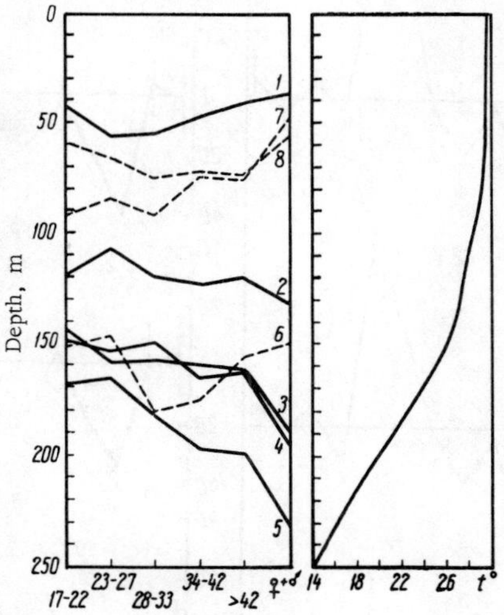

FIGURE 3. Right, vertical temperature variation; left, variation of the mean depth of vertical distribution of the differentiated groups during descent (dotted lines) and ascent (dashed lines)

1 — 2300 to 0300 hr; 2 — at 0500 hr; 3 — at 0700 hr; 4 — at 0900 hr; 5 — at 1100 hr; 6 — at 1500 hr; 7 — at 1700 hr; 8 — 1900 to 0100 hr.

The significance of the migration figures and variations of migration velocity can be appraised from Figure 4, in which negative velocities signify descent and positive ones ascent.

The speeds of both descent and ascent increase with age. The maximum sinking velocity occurs in the hours preceding sunrise in all the groups. The descent is interrupted at about 0800 hr but is then resumed and continues approximately until noon, followed by the phase of ascent in all the groups. The maximum velocity of ascent occurs at around 1600 hr. The ascent of older individuals stops toward 1800 hr, i. e. at sunset, and the males even descend slightly. Young larvae complete their ascent toward 1900—2000 hr. At night, the migration velocity undergoes irregular fluctuations that are small in comparison to the measurement error.

* Data on the daytime depth of occurrence of the senior age groups can be regarded only as minimal, since at this time some of the individuals apparently sink below 300 m (see p. 245).

However, there is a slight descent at about 2200 hr in all the groups except the youngest larvae, followed by a similarly slight rise at around midnight, these movements being possibly analogous to the midnight sinking and pre-dawn rise of some pelagic organisms (see Cushing, 1951; Bainbridge, 1961).

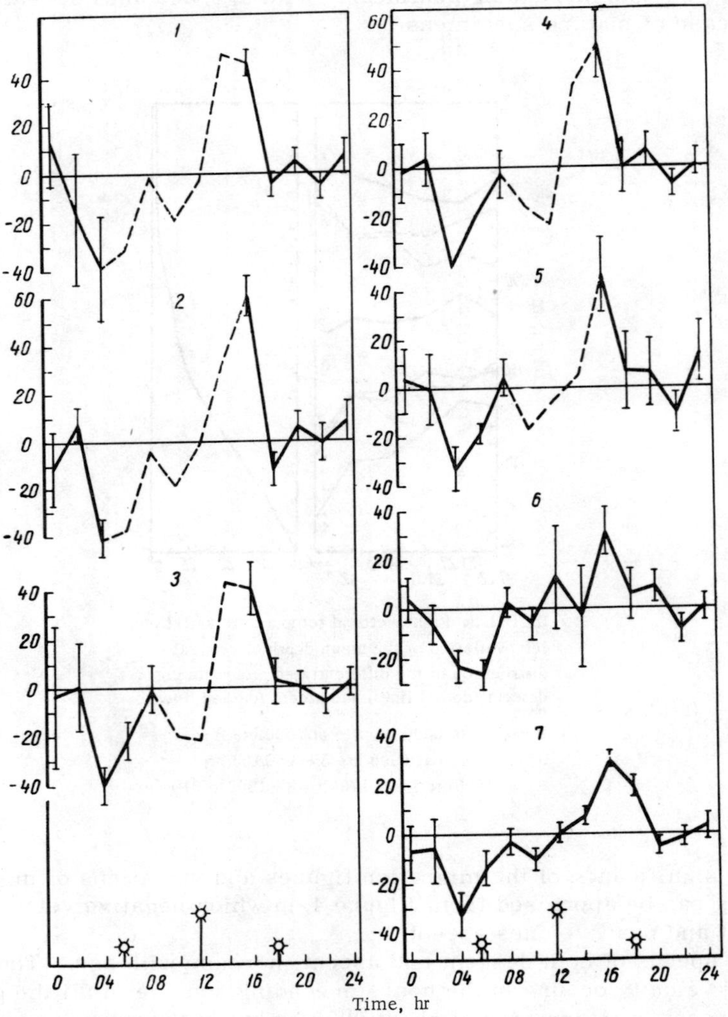

FIGURE 4. Variations of the mean velocity of migration of different groups of animals (m/hr):

Vertical lines indicate the confidence interval of velocity at significance level 5%. Dashed lines indicate variations of velocity the estimates of which may be shifted because some of the individuals sink below 300 m. The times of noon, sunrise and sunset are marked in the lower part of the figure; 1 – females; 2 – males; 3 – 7 immature individuals of the following size: 3 ,> 42; 4, 34 – 42; 5, 28 – 33; 6, 23 – 27; 7, 17 – 22.

As in certain other cases (see, for instance, Worthington, 1931; Vinogradov, 1968), the above migration procedure is difficult to explain from the standpoint of "light optimum" theory (Russel, 1927; Rose, 1925). The data fit much better the assumption (Rudjakov, 1970) that the diurnal vertical migrations derive from an alternation of phases of high and low locomotor activity. Organisms with negative buoyancy undergo passive sinking depth the phase of low activity. In contrast, during the phase of high activity they actively rise to the surface on account of their mechanisms of spatial orientation. In accordance with this hypothesis, the velocity variations depicted in Figure 4 can be roughly interpreted as variations of locomotor activity measured in certain relative units.*

An examination of the variation of migration velocity from this point of view reveals a phase of low activity from 0200 to 1200 hr. However, the sinking velocity in all the groups decreases abruptly around 0800 hr, again increases toward 1000 hr, and then drops to zero at about 1200 hr. This is followed, as already mentioned, by the phase of rising to the surface. The above-described halt can also be noted in Figure 3, which shows distinctly that from 0700 to 0900 hr the mean depth of occurrence of all the groups remains practically unchanged and then again begins to increase at a very considerable rate, at any rate in the older individuals. Although this delay occurs in the layer of maximum hydrological gradients (see Figure 3), it cannot be explained by a decrease in the passive sinking rate due to increased density and viscosity of the water, because if this were so, the sinking would continue at a progressively diminishing velocity. Apparently, the only explanation for delayed sinking in the thermocline, reported also by Vinogradov (1954) for certain zooplanktonic species in the Soviet Far Eastern seas, can be a change in the behavior of the animals (Banse, 1964; Beklemishev, 1969), manifesting itself in a temporarily increased locomotor activity resulting in an active braking of sinking. The direct causes of this increase are not known. Possibly, this is the manifestation of an endogenic rhythm of activity with two peaks which is observed in many animals (Aschoff, 1966), or else the influence of increased hydrostatic pressure (Knight-Jones and Morgan, 1966). However, the fact that the depth at which braking occurs and the depth of occurrence of the layer of maximum hydrological gradients coincide suggests an abrupt drop of temperature as the direct cause of the altered behavior. In fact, numerous investigations (Naumov, 1963) have shown that the activity of animals is at its minimum in the zone of the temperature optimum, increasing on both sides of it. We may assume that the sharp drop of temperature in the thermocline causes an increase in the activity of sinking animals which subsequently, following a period of acclimatization, again decreases until the onset of the phase of high activity, in accordance with the endogenic rhythm. The greater depth at which the delay occurs for the adults in comparison with the larvae (Figure 3) can be tentatively ascribed to the enhanced eurybionticity of mature organisms.

Besides the diurnal variation of the mean depth of occurrence of the different groups of individuals, there is also a diurnal variance and asymmetry of the vertical distribution.

* It will be possible to determine the position of zero on the activity scale after measurements of the passive sinking rate of *Cypridina* of different age. Apparently, it amounts to about 40 m/hr, because the sinking rate of male *Conchoecia spinirostris*, a pelagic ostracod with a fragile shell about 1 mm long, varied from 30.2 to 38.2 m/hr (Angel, 1970).

Because some individuals of the older age groups may descend below 300 m and the group classes are unequal with respect to depth at different times, the conclusions based on comparison of the variations of standard deviation (Table 4) cannot be reliable. The vertical dispersion increases with age, and it displays an appreciable upward trend in the daytime in the older individuals. However, in the case of the youngest larvae the dispersion increases only during vertical migrations and decreases appreciably at the depths of maximum ascent and descent.

TABLE 4. Diurnal variations of the standard deviation of vertical distrubution (m) of differentiated groups of *C. sinuosa*

Time, hr	Group							
	♀	♂	♀+♂	>42	34–42	28–33	23–27	17–22
23	37.20	28.03	30.96	25.00	25.35	26.70	37.39	23.91
01	11.58	12.50	21.65	31.46	25.89	28.35	18.16	18.90
03	56.70	8.33	39.66	18.82	25.81	31.46	15.39	29.91
05	48.42	16.67	45.14	15.39	21.44	14.86	27.07	26.83
07	?	?	?	39.52	40.90	27,25	28.53	28.53
09	?	?	?	42.12	44.79	33,33	18.90	31.73
11	?	?	?	?	?	?	27.86	18.20
13	?	?	?	?	?	?	56.76	21.20
15	44.10	40.10	44.72	49.00	44.90	48.79	30.46	19.61
17	33.99	41.36	36.29	41.15	33.97	34.30	42.12	25.76
19	27.35	21.24	25.74	30.61	28.81	47.19	29.94	24.41
21	26.56	33.28	29.30	29.17	28.59	30.62	28.72	24.52
23	31.66	31.24	31.34	23.23	25.82	29.99	29.31	25.48
01	31.90	23.20	29.34	30.23	25.99	45.88	24.41	35.36

The coefficient of skewness describes the skewness of the distribution. In the case of the vertical distribution of plankton, a negative coefficient indicates concentration of animals near the lower boundary of the layer in which they are located at a given moment, while a positive coefficient indicates concentration of the animals near the upper boundary. For a small number of group classes of individuals with respect to depth (3–4), calculation of the coefficient of skewness is of doubtful value, but variations of this parameter of the vertical distribution for the younger age groups, which do not descend deeper than 300 m in the daytime, appear to be quite regular (Figure 5). The vertical distribution is almost symmetric during the maximum migration velocity (in both descent and ascent). Maximum skewness occurs at migration velocities close to zero. The skewness is positive at nighttime and negative during the maximum daytime sinking.

Skewness of the vertical distribution may be due to the migrating group of animals reaching water layers in which the effect of certain factors counteracting further migration begins to manifest itself. When a population is sinking, such a factor may be the temperature gradient producing an increase in locomotor activity, as already pointed out. The positive skewness at night may be due to the ascending population's reaching the water surface. However, the crustaceans do not rise to the very surface (see Table 3), and therefore other explanations for the phenomenon must be sought.

According to an oral communication from E. G. Perueva, who performed experimental studies of the feeding of Calanus glacialis, the crustaceans cease active migration while they are feeding on captured algae. If such behavior is followed by zooplankters also under natural conditions, then the ascent of a migrating population must be limited by a layer with a high concentration of food items. Having reached this layer and found the food, an ascending animal discontinues or slows down its swimming movements and consequently begins to sink. After the food has been dealt with, active swimming is resumed until a new particle of food is encountered, and the process is repeated. Obviously, the higher the concentration of food items in a layer, the lower is the probability of an animal passing through the layer without encountering food. Therefore, an increased concentration of food in a layer should cause an increased positive skewness in the vertical distribution of ascending animals and a greater mean depth of occurrence during the maximum rise to the surface. Therefore, the nighttime depth of occurrence of migratory animals must depend on the location and distinctness of the layer with a high concentration of food particles and on the abundance of food in it. Variation of these factors may be one of the causes of discrepancies between the nighttime depth of occurrence of a population on different dates reported by Waterman et al. (1939) and revealed by a comparison of the vertical distribution of C. sinuosa on the first and second nights of our investigations.

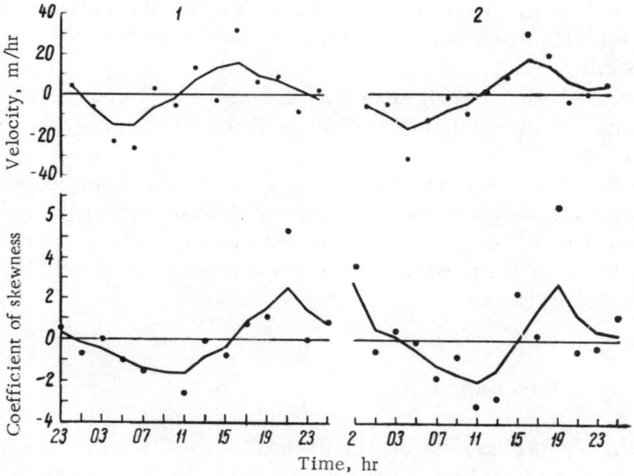

FIGURE 5. Variations of the migration velocity and skewness of vertical distribution for animals of sizes 23 – 27 (1) and 17 – 22 (2).

Approximate fitting was done by the method of sliding averages.

Bibliography

Publications in Russian

Arsen'ev, V. S., L. I. Galerkin, E. A. Plakhin, and V. V. Sapozhnikov. Osnovnye cherty gidrologii i gidrokhimii raiona issledovanii 44-go reisa "Vityazya" (Major Hydrological and Hydrochemical Features of the Region Investigated during the 44th Cruise of the "Vityaz"). (In this collection.)

Beklemishev, K. V. Ekologiya i biogeografiya pelagiali (The Ecology and Biogeography of the Pelagic). Moskva, "Nauka." 1969.

Bogorov, V. G., M. E. Vinogradov, N. M. Voronina, I. P. Kanaeva, and I. A. Suetova. Raspredelenie biomassy zooplanktona v poverkhnostnom sloe Mirovogo okeana (Distribution of Zooplankton Biomass in the Surface Layer of the World Ocean). — Doklady Akademii Nauk SSSR, Vol. 182, No. 6. 1968.

Naumov, N. P. Ekologiya zhivotnykh (Ecology of Animals). — Moskva, "Vysshaya shkolya." 1963.

Petipa, T. S. Pitanie i energeticheskii balans nekotorykh massovykh planktonnykh kopepod Chernogo morya otnosyashchikhsya k razlichnym ekologicheskim gruppirovkam (Feeding and Energy Balance of Some Mass Planktonic Copepoda in the Black Sea Belonging to Different Ecological Groups). Author's Summary of Doctoral Thesis. Moskva. 1965.

Vinogradov, M. E. Sutochnye vertikal'nye migratsii zooplanktona dal'nevostochnykh morei (Diurnal Vertical Migrations of Zooplankton in the Far Eastern Seas). — Trudy Instituta Okeanologii Akademii Nauk SSSR, Vol. 8. 1954.

Vinogradov, M. E. Vertikal'noe raspredelenie okeanicheskogo zooplanktona (Vertical Distribution of Oceanic Zooplankton). Moskva, "Nauka." 1968.

Vinogradov, M. E. and Yu. A. Rudyakov. Sutochnye izmeneniya vertikal'nogo raspredeleniya obshchei biomassy planktona v zapadnoi ekvatorial'noi Patsifike (Diurnal Variations of the Vertical Distribution of the Plankton Biomass in the Equatorial West Pacific). (In this collection.)

Publications in Other Languages

Angel, M. V. Observations on the Behavior of Conchoecia spinirostris. — J. mar. biol. Ass. U. K., Vol. 50, No. 3. 1970.

Aschoff, J. Circadian Activity Pattern with Two Peaks. — Ecology, Vol. 47, No. 4. 1966.

Bainbridge, R. Migrations. — In: The Physiology of Crustacea, 2. N. Y., London, Academic Press. 1961.

Banse, K. 1964. On the Vertical Distribution of Zooplankton in the Sea. — Progress in Oceanography, 2. Pergamon Press. 1964.

Cushing, D. H. The Vertical Migration of Planktonic Crustacea. — Biol. Rev., Vol. 26, No. 2. 1951.

Fager, E. W. and J. A. McGowan. Zooplankton Species Groups in the North Pacific. — Science. Vol. 140, No. 3566. 1963.
Gardiner, A. C. Vertical Distribution in Calanus finmarchicus. — J. Mar. Biol. Assoc. U. K. Vol. 18, No. 2. 1933.
Hardy, A. C. The Open Sea, Its Natural History, Vol. 1. The World of Plankton. Collins, London. 1956.
Knight-Jones, E. W. and E. Morgan. Responses of Marine Animals to Changes in Hydrostatic Pressure. — Oceanogr. Mar. Biol., Ann. rev., Vol. 4. 1966.
Moore, H. B., H. Owre, and T. Dow. Plankton of the Florida Current. III. The Control of the Vertical Distribution of Zooplankton in the Daytime by Light and Temperature. — Bull. Mar. Sci. Gulf and Caribbean, Vol. 3, No. 2. 1953.
Rose, M. Contribution à l'étude de la biologie du plankton. Le probleme des migrations verticales journalieres. — Arch. Zool. Exp. et Gén., Vol. 64. 1925.
Rudjakov, J. A. The Possible Causes of Diel Vertical Migrations of Planktonic Animals. — Mar. Biol., Vol. 6, No. 2. 1970.
Russel, F. S. The Vertical Distribution of Plankton in the Sea. — Biol. Rev., Vol. 2, No. 3. 1927.
Snedecor, G. W. Statistical Methods Applied to Experiments in Agriculture and Biology. Ames, Iowa State College Press. 1956.
Waerden, B. L., van der. Mathematische Statistik. Berlin, Springer. 1957.
Waterman, T. H., R. F. Nunnemacher, F. A. Chase and G. L. Clarke. Diurnal Vertical Migration of Deep-water Plankton. — Biol. Bull. Vol. 76, No. 2. 1939.
Worthington, E. B. Vertical Movements of Fresh-water Macroplankton. — Internat. Rev. Ges. Hydrobiol. Hydrogr., Vol. 25, No. 5/6. 1931.

VERTICAL DISTRIBUTION OF EGGS AND LARVAE OF FISH IN THE WESTERN TROPICAL PACIFIC

N. N. Gorbunova

Shirshov Institute of Oceanology of the Academy of Sciences of the USSR

The literature provides scarcely any information on the vertical distribution of fish eggs and larvae in tropical regions of the ocean; yet without studies of the depths at which spawning takes place and the larvae develop, the life cycles and biocenotic relations of fish cannot be understood.

Data on the depths inhabited by the larvae of certain species are found in several papers (Yabe and Ueyanagi, 1962; Strasburg, 1960; Mito, 1967; and others), but only Ahlstrom's work (1959) deals specifically with the vertical distribution of eggs and larvae in subtropical waters (California). This author found that larvae were most abundant in the upper 125-m layer, i. e. above the thermocline and partially in its upper layer. The abundance of larvae and the specific diversity are much lower in the thermocline itself and in deeper waters. In Ahlstrom's opinion, the temperature is the decisive factor controlling the vertical distribution of fish eggs and larvae.

Series of vertical catches in the upper 500 m were performed under stationary conditions in three test areas (polygons) during the 44th cruise of the "Vityaz." This paper describes the results of investigations at station 6033 (04°10' S, 168°31' E). Round-the-clock observations were conducted from December 17 to 25, 1968. The polygon was located at least 500 miles from the nearest islands. The depth was 3,430 m. Water temperatures averaged about 30° at the surface, 29.6 — 29.7° at depths of 50 m, about 28° at 100 m, and 25.6 — 26° at 150 m. Below this depth, the temperature dropped abruptly, to 18.5 — 21.5° at 200 m and to 10° at 300 m (Arsene'v et al., 1971).

The surface water mass, or the upper mixed layer, occupies the top 80 — 100 m and is followed by an intermediate layer of high salinity with a temperature gradient extending down to 270 — 300 m and possessing a stable maximum salinity core at 125 — 150 m.

MATERIAL

Eggs and larvae of fish were taken from catches with a Bogorov-Rass net (BR), the oceanic model of a Juday plankton net (JOM), and Savilov's pleuston net (PS). The ichthyoplankton net "BR," made of No. 14 plankton bolting cloth with a mouth area of 1 m^2, was used for the layers 0 — 25, 25 — 50, 50 — 100, and 100 — 200 m (13 samples taken from each layer) and

200 — 500 m (6 samples). JOM (No. 38 bolting cloth, mouth area 0.5 m^2) was used for the layers 0—25, 25—50, 50—75, 75—100, 100—150, 150—200, and 200—300 m, 24 samples being taken from each horizon. The PS pleustone net, used for the top 0.3—0.4 m, was used to obtain 7 samples.

The catches with the BR net (58 samples) were performed in the course of 3 diurnal periods. All the catches with JOM (168 samples) were performed within a single period of 24 hours, with synchronous operation of two winches.

The purpose of the investigations was to study the vertical distribution of the larvae and the rhythms of their diurnal migrations. Therefore the method of round-the-clock observations of the vertical distribution of plankton at a single station (Vinogradov and Rudyakov, 1971) proved very convenient. With vertical catches repeated 12 times during 24 hours it is possible to appraise the distribution of fish larvae with a high degree of reliability and to determine if they perform diurnal vertical migrations; simultaneous operation of two winches considerably enhanced the reliability of the results.

RESULTS

In all, 192 samples containing eggs and larvae of fish were examined. Larvae of epipelagic and mesopelagic species were encountered in the upper 500-m layer.

The epipelagic larvae in the catches with BR and JOM nets belonged to the following species: *Stolephorus zollingeri*, *Gempylus serpens*, *Thunnus alalunga*, *Katsuwonus pelamis*, *Euthynnus affinis*, *Antennarious* sp., and Stromateidae gen. sp. This list is supplemented with larvae of Exocoetidae, Oxyporhamphidae, Istiophoridae and Balistidae in the case of PS catches.

Larvae of mesopelagic fish were found only in the catches with BR and JOM nets. This group includes *Gonostoma elongatum*, *G. atlanticum*, *Cyclothone alba*, *Vinciguerria nimbaria*, *Yarella* sp., *Maurolicus* sp., Gonostomatidae gen. sp., *Symbolophorus evermanni*, *Lampanyctus* sp., *Diaphus* sp., *Diogenichthys atlanicus*, *Hygophum proximum*, *Myctophum* sp., *Lepidophanes pyrsobolus*, Myctophidae gen. sp., *Paralepis* sp., *Scopelarchus* sp., *Chauliodus sloani*, *Idiacanthus fasciola*, Melanostomiatidae gen. sp., *Argyropelecus* sp., and *Melamphaes* sp.

VERTICAL DISTRIBUTION OF LARVAE OF EPIPELAGIC FISHES

Larvae of epipelagic fishes accounted for only 3.5% of the total JOM and BR catches. This also includes the larvae of *Stolephorus zollingeri*, which are only a temporary element in the epipelagic ichthyoplankton. The true abundance of larvae of the epipelagic fishes is probably many times higher, since larvae living at the very surface were scarcely ever found in the catches obtained with the vertical nets.

The abundance of larvae of epipelagic fishes was low, not exceeding 10 specimens per catch, and often there was only the odd specimen. Only larvae of the anchovy *Stolephorus* were comparatively populous and of frequent occurrence. Eggs were caught along with the larvae. The morphological features of the eggs and larvae were in accordance with descriptions of the early stages of development of the Indian Ocean species *S. zollingeri* (Delsman, 1931).

The fact that the anchovy spawns in the open ocean is of great interest, because the range of this entire group of fishes is confined to the coastal zone. The eggs caught in the region investigated were at an early stage of cleavage of the embryonal plasma, and the larvae were 3 – 7 mm long, the frequency of occurrence of large larvae being considerably lower. We assume that adult spawning anchovies and also their larvae are carried out to the region from Melanesian coastal waters. This is borne out by a find, at the same station, of the neritic pelagic lancelet *Amphioxides pelagicus* 7.8 mm long.* In addition, the catches contained ostracods of the species *Cypridina sinuosa* which are usually confined to coastal regions.

Larvae of *S. zollingeri* were found in the upper 100-m layer (Table 1) but their highest abundance was limited to the depth of 75 m.

TABLE 1. Vertical distribution of larvae of *Stolephorus zollingeri* (according to JOM catches) and their size (data of all catches)

Horizon, m	Number of samples	Number of larvae per catch	Number of larvae per 100 m^3	Same, %	Size of larvae, mm
0 – 0.4	–	–	–	–	6.0 – 9.0
0 – 25	15	2.7	21.6	11.8	3.0 – 7.0
25 – 50	23	9.2	73.6	40.6	3.0 – 7.0
50 – 75	19	8.3	66.4	36.6	3.0 – 7.1
75 – 100	2	2.5	20.0	11.0	3.0 – 5.5

The density of concentration of the larvae exhibits some diurnal variation, 45 – 47 per m^2 at night and 36 – 38 specimens per m^2 during the day. The largest numbers of larvae were obtained from 0700 to 1500 hr in the daytime and from 1900 to 2200 hr at night (Figure 1). The vertical distribution of the larvae also undergoes diurnal variation (Figures 2 and 3). The rise to the surface occurs after 1500 hr. No larvae of this species were encountered in the upper 25-m layer between 0700 and 1500 hr. The larvae were also absent from daytime PS catches, but in contrast, 37 specimens were caught within a space of 20 minutes in an evening catch. The frequency occurrence of these larvae in the catches decreases with increasing depth at night but increases with increasing depth during the day (Figure 4).

Hence, larvae of *S. zollingeri* come up closer to the surface at night and sink somewhat below the upper 25-m layer in the daylight hours.

* Lancelets from the offshore zone of islands in the Melanesian Archipelago are believed to be carried out far into the ocean by streams of the South Equatorial Current (Lubny-Gertsyk, 1963).

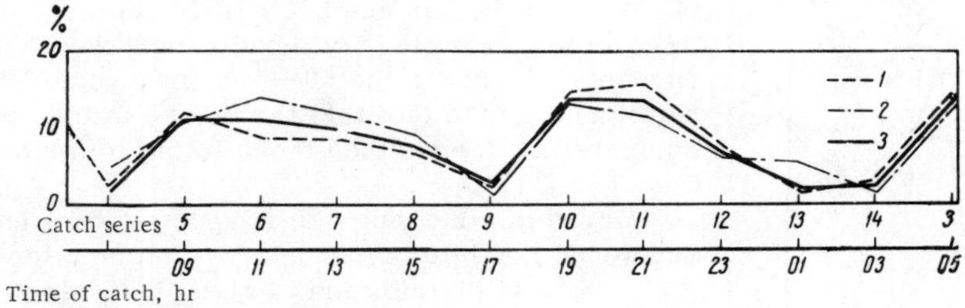

FIGURE 1. Diurnal fluctuations of the catches of Stolephorus zollingeri larvae: 1 — winch No. 4; 2 — winch No. 5; 3 — mean catch.

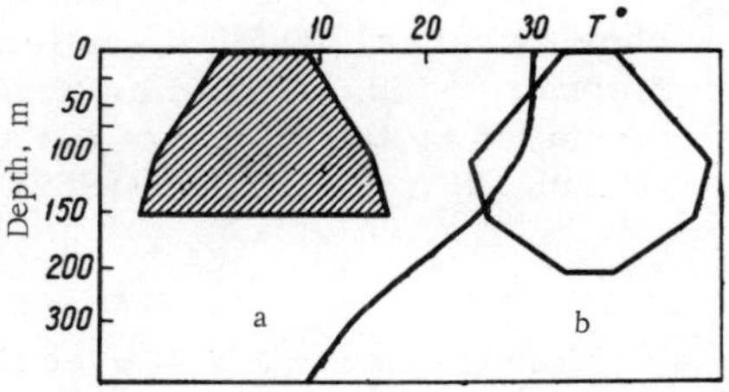

FIGURE 2. Vertical distribution of Stolephorus zollingeri larvae at night (a) and during the day (b):

T° — vertical distribution of temperature

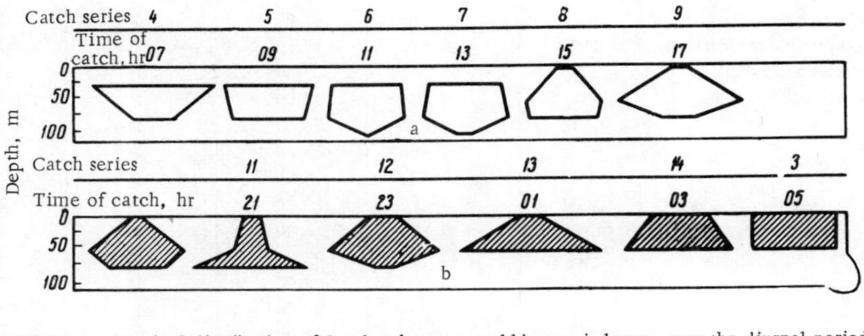

FIGURE 3. Vertical distribution of Stolephorus zollingeri larvae over the diurnal period: a — day; b — night.

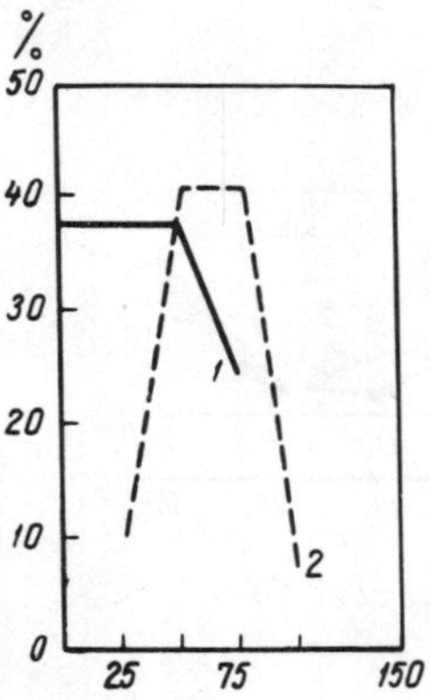

Depth, m

FIGURE 4. Frequency of occurence of Stolephorus zollingeri larvae with respect to depth at night (1) and during the day (2)

The embryonal and early postembryonal development periods of this species probably take place not at the surface but a little below 25 m. Eggs in the first stage of development were encountered only in the 50 – 100-m layer. This is confirmed by the fact that the larvae in the upper horizons are larger than those in the lower horizons (see Table 1).

Larvae of other epipelagic species were encountered much less frequently. Information on their vertical distribution is given in Table 2.

Tuna larvae occurred mainly above 75 m. The water temperature in this layer differs from the surface temperature only by 0.4–0.9°. Larvae of the Stromateidae were caught most frequently in the upper 25 m.

All larvae of epipelagic fishes remained for the most part above 75–100 m. Some of them (anchovies, tunas) perform vertical migrations over the diurnal period. Circadian behavioral rhythms are displayed by the larvae of most epipelagic species (Parin, 1967; Gorbunova, 1969).

TABLE 2. Finds of larvae of epipelagic fishes (apart from S. zollingeri)

Species	Net	Horizon, m	Time of catch, hr	Larvae	
				number per catch	length, mm
Gempylus serpens	BR	25–50	13.10	1	–
Thunnus sp.	BR	0–25	10.50	1	5.1
	JOM	0–25	13.00	1	5.7
	BR	25–50	02.00	1	3.6
	JOM	50–75	09.00	1	–
	BR	50–100	02.00	12	3.2–4.0
Th. albacares	BR	25–50	02.00	1	5.3
Katsuwonus pelamis	JOM	25–50	21.00	1	4.1
Euthynnus affinis	JOM	0–25	17.00	1	2.9
Antennarius sp.	BR	50–100	9.30	1	2.6
Stromateidae gen. sp.	BR	0–25	02.00	2	3.0–3.2
	JOM	0–25	05.00	1	3.0
	BR	0–25	08.00	2	3.0
	JOM	0–25	13.00	1	5.3
	JOM	0–25	15.00	1	4.0
	BR	0–25	15.00	7	2.5–3.0
	BR	25–50	02.00	2	2.8–3.3
	BR	25–50	08.00	1	3.0
	BR	25–100	22.00	1	29
	BR	50–100	02.00	2	3.0–3.5

VERTICAL DISTRIBUTION OF LARVAE OF MESOPELAGIC FISHES

Among the mesopelagic fishes, the most abundant larvae belonged to the Gonostomatidae and Myctophidae, 59 and 29% of the total catch, respectively. Larvae of other mesopelagic forms accounted for only 8.5%.

1. Family Gonostomatidae

Larvae of 9 species of Gonostomatidae were found in the catches. Larvae of Gonostoma elongatum and Cyclothone alba were most frequent in the samples.

Gonostoma elongatum. Larvae 2.0 – 6.8 mm long occurred mainly in the 75 – 150-m layer (Tables 3 and 4). Only sporadic specimens were caught at night above 75 m. Catches below 150 m were also very rare. Abundance was 9 – 14 specimens per m^2.

TABLE 3. Vertical distribution of larvae of Gonostoma elongatum (JOM catches)

Horizon, m	Night			Day		
	Number of positive catches	mean catch per 100 m^3	same, %	Number of positive catches	mean catch per 100 m^3	same, %
0 – 25	—	—	—	—	—	—
25 – 50	1	4	11.0	—	—	—
50 – 75	—	—	—	—	—	—
75 – 100	5	13.2	36.2	4	12	56.5
100 – 150	10	17.2	47.3	8	7.2	34.0
150 – 200	1	2	5.5	1	2	9.5
200 – 300	—	—	—	—	—	—

Early stages of G. elongatum larvae keep to the upper part of the thermocline and apparently do not perform diurnal vertical migrations (Figure 5). The frequency of occurrence and numerical abundance of the larvae in the 75 – 100 and 100 – 150-m horizons are nearly the same day and night. The larvae of this species are apparently typical inhabitants of the upper part of the thermocline and the waters above it at water temperatures of 25 – 29°.

TABLE 4. Vertical distribution of larvae of Gonostoma elongatum (BR catches)

Horizon, m	Night			Day		
	number of catches	number of positive catches	mean number of larvae per 100 m^3	number of catches	number of positive catches	mean number of larvae per 100 m^3
0 – 25	4	—	—	9	—	—
25 – 50	4	—	—	9	—	—
50 – 100	4	—	—	9	1	2
100 – 200	4	1	3	9	7	1.7
200 – 500	2	2	0.33	4	—	—

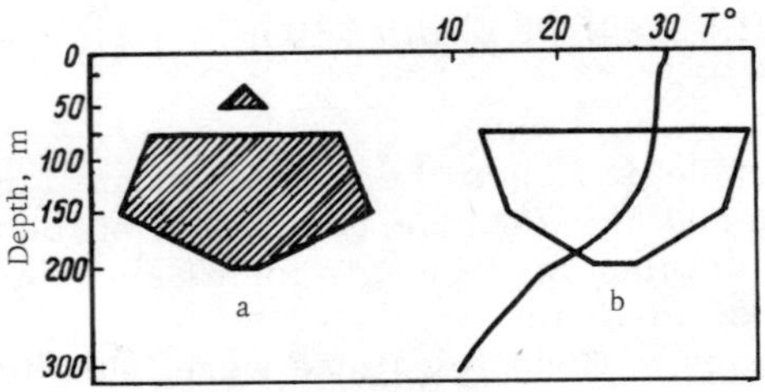

FIGURE 5. Vertical distribution of larvae of Gonostoma elongatum at night (a) and during the day (b):

T°, vertical temperature distribution.

Gonostoma atlanticum. Two larvae 3.0 mm long were caught in the 100–150-m horizon.

Cyclothone alba. Larvae of this species occurred no less frequently than those of G. elongatum. They were 3.5–6.0 mm long and kept mainly to depths of 25–150 m (Table 5). Their abundance is 8–10 specimens/m² according to JOM catches. Larger, older larvae occur below 150 m.

The larvae of C. alba were encountered more frequently and catches were larger at night than during the day.

TABLE 5. Vertical distribution of larvae of Cyclothone alba (JOM catches)

Horizon, m	Night			Day		
	number of positive catches	mean catch of larvae per 100 m³	same, %	number of positive catches	mean catch of larvae per 100 m³	same, %
0 – 25	–	–	–	1	8	28.6
25 – 50	2	8	24.7	–	–	–
50 – 75	7	10.4	32.2	1	8	28.6
75 – 100	2	8	24.7	3	8	28.6
100 – 150	2	6	18.4	1	4	14.2
150 – 200	–	–	–	–	–	–
200 – 300	–	–	–	–	–	–

The early stages of the larval development of C. alba take place in the upper water layers, mainly at depths of 25–150 m, i. e. above the thermocline and in its upper part at water temperature of 25–29° (Figure 6). The diurnal vertical migrations of these larvae are indistinct, although their highest frequency of occurrence is observed in the 50–75-m layer at night and, in the 75–100-m layer in daytime. Later stages descend deeper, to 200–500 m. Similar results were obtained by Mukhacheva (1967), who demonstrated that the size of C. alba increases with increasing depth.

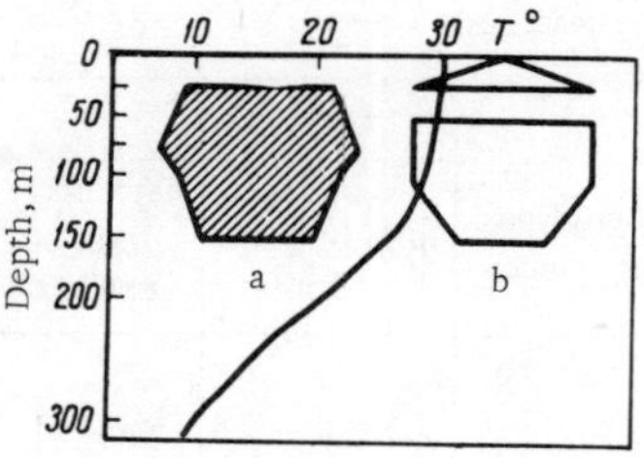

FIGURE 6. Vertical distribution of larvae of Cyclothone alba

Notation in Figure 5.

Other Gonostomatidae occurred much less frequently in the catches. Larvae of Vinciguerria nimbaria were found in only seven samples (catches of adult V. nimbaria with the Isaac-Kidd midwater trawl were also small in this region). Their abundance does not exceed 2—4 specimens per m². The larvae were caught in the 25—75-m layer at night (4 samples) and in the 200—150-m layer during the day (3 samples). Apparently these larvae are also confined to the upper 150-m layer. They may perform vertical migrations within this layer, but this cannot be confirmed because the number of positive catches is too small.

Twolarvae of Maurolicus sp. were caught in the 50—200-m layer and one larva of Danaphus oculatus in the 100—150-m layer.

2. Family Myctophidae (lanternfishes)

The catches contained larvae of Symbolophorus evermanni, Hygophum proximum, Diogenichthys atlanticus, Lepidophanes pyrsobolus, 3—4 species of the genus Diaphus, Myctophum sp., and Lampanyctus sp. The catches contained mainly larvae of the near-surface group of lanternfishes inhabiting the upper 500-m layer (Bekker, 1967). Larvae of S. evermanni and Diaphus. sp. were the most frequent and numerous in the samples, whereas larvae of the other species were less common.

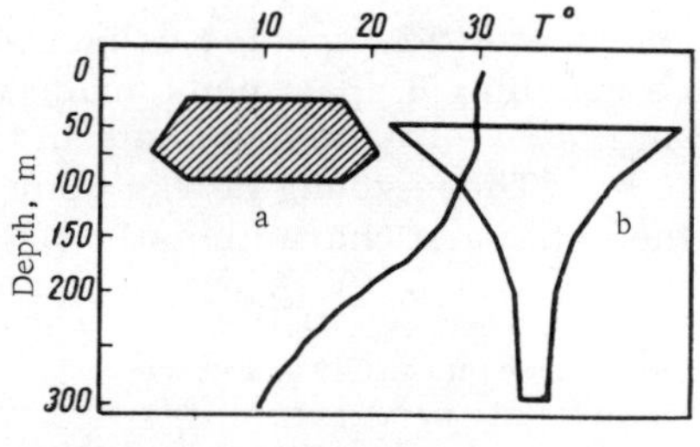

FIGURE 7. Vertical distribution of larvae of the genus Diaphus

Notation as in Figure 5.

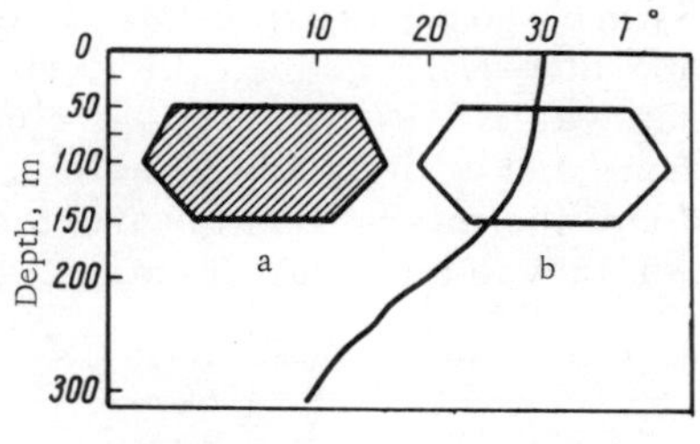

FIGURE 8. Vertical distribution of Symbolophorus evermanni larvae

Notation as in Figure 5.

Diaphus sp. Larvae of this genus 2.0—8.7 mm long were encountered at depths ranging from 25 to 150 m, reaching their highest abundance in the 50—75-m layer (Table 6, Figure 7). Abundance was 8—10 specimens/m². Larvae of the genus Diaphus perform diurnal migrations, like the adult fish. They rise to the surface at night, reaching the 25—0-m horizon, but they do not occur above 50 m in the daytime, and some descend even below 150 m.

TABLE 6. Vertical distribution of larvae of the genus Diaphus (JOM catches)

Horizon, m	Night			Day		
	number of positive catches	mean catch of larvae per 100 m^3	percentage ratio of catches	number of positive catches	mean catch of larvae per 100 m^3	percentage ratio of catches
0–25	–	–	–	–	–	–
25–50	3	10.4	29.5	–	–	–
50–75	5	14.4	41.0	2	16	53.4
75–100	3	10.4	29.5	3	8	26.6
100–150	–	–	–	2	4	13.3
150–200	–	–	–	–	–	–
200–300	–	–	–	1	2	6.7

The frequency of occurrence increases at night and decreases in the daytime, as is confirmed by BR catches (Table 7).

TABLE 7. Vertical distribution of larvae of the genus Diaphus (BR)

Horizon, m	Night			Day		
	total number of catches	number of positive catches	mean catch of larvae per 100 m^3	total number of catches	number of positive catches	mean catch of larvae per 100 m^3
0 – 25	4	–	–	9	1	4
25 – 50	4	–	–	9	–	–
50 – 100	4	2	8	9	–	–
100 – 200	4	1	1	9	1	1
200 – 500	2	–	–	2	–	

Symbolophorus evermanni. Larvae of this species occurred comparatively frequently, their abundance reaching 8 specimens under 1 m^2. The larvae 2.0–7.5 m long were caught in the 50–150-m layer both in the daytime and at nighttime (Table 8, Figure 8). These young larvae do not yet perform the migrations toward the surface that are characteristic of larger larvae and adult forms.

TABLE 8. Vertical distribution of larvae of Symbolophorus evermanni (JOM catches)

Horizon, m	Night			Day		
	number of positive catches	mean catch of larvae per 100 m^3	percentage ratio of catches	number of positive catches	mean catch of larvae per 100 m^3	percentage ratio of catches
0–25	–	–	–	–	–	–
25–50	–	–	–	–	–	–
50–75	2	8	33.9	1	8	30.7
100–150	3	10.4	44.0	2	12	46.2
150–200	3	5.2	22.1	2	6	25.1
200–300	–	–	–	–	–	–
	–	–	–	–	–	–

According to catches taken with the BR net of larvae S. evermanni occurred at 50 — 200 m, and one larva, 3.2 mm long, was caught in the 0 — 25-m layer.

Other lanternfishes. Larvae of Hygophum proximum about 4.5 mm long occurred in the upper 100-m layer at night, and a 13.2-mm-long fingerling of the species was caught in the 200 — 500-m layer also at night. A few larvae of the genus Myctophum 2.5 — 4.5 mm long were caught in the 50 — 150-m layer. Larvae of the genus Lampanyctus 2.5 — 4.5 mm long were encountered in the 25 — 100-m layer. A single larva of Lepidophanes pyrsobolus was caught at night in the upper 25 m. Larvae of Diogenichthys atlanticus 3.5 — 4.2 mm long occurred in the 100 — 200-m layer both during the day and at night (Table 9).

TABLE 9. Distribution of larvae of Myctophidae exclusive of Diaphus and Symbolophorus (JOM and BR catches)

Horizon, m	Night				
	Lepidophanes pyrsobolus	Myctophum sp.	Lampanyctus sp.	Hygophum proximum	Diogenichthys atlanticus
0—25	≡			≡	
25—50		≡	≡	≡	
50—75		≡	≡	≡	
100—150					
150—200					≡
200—300					
	Day				
	Lepidophanes pyrsobolus	Myctophum sp.	Lampanyctus sp.	Hygophum proximum	Diogenichthys atlanticus
0—25					
25—50					
50—75					
75—100			≡		
100—150		≡		≡	≡
150—200					
200—300					

Apparently, this group of Myctophid larvae, except for D. atlanticus, perform diurnal vertical migrations, rising to the surface at night and descending to the deep layers in the daytime.

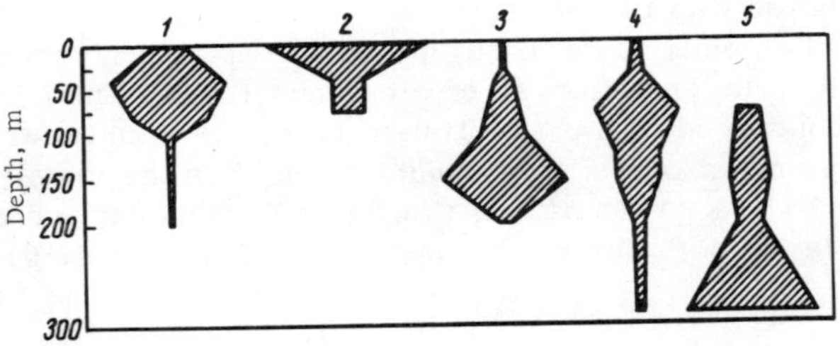

FIGURE 9. Vertical distribution of larvae:

1 – Stolephorus zollingeri; 2 – epipelagic fishes; 3 – 5 – mesopelagic fishes (3 – Gonostomatidae; 4 – Myctophidae; 5 – other species).

Larvae of near-surface lanternfishes in early stages of postembryonal development occur mainly in the upper 150 m (Figure 9); diurnal vertical migrations occur, but not in all species (e. g. S. evermanni).

3. Other mesopelagic fishes

The catches contained small numbers of larvae of mesopelagic fishes belonging to the families Paralepididae, Melanostomiatidae, Chauliodontidae, Idiacanthidae, Scopelarchidae, Sternoptychidae, and Melamphaeidae.

Larvae of this group occur below 75–100 m at relatively low water temperatures (Table 10).

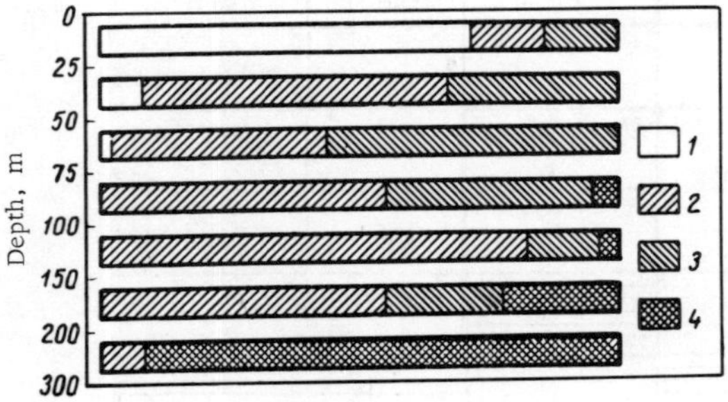

FIGURE 10. Frequency of occurrence of larvae of epipelagic (1) and mesopelagic (2–4) fishes:

2 – Gonostomatidae; 3 – Myctophidae; 4 – Other species.

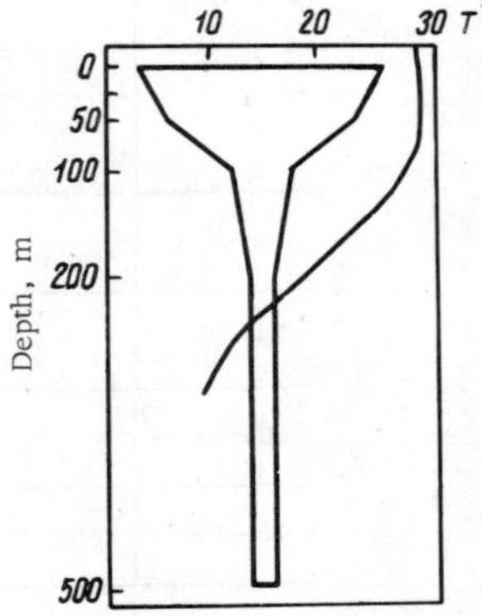

FIGURE 11. Vertical distribution of eggs:

T° – vertical temperature distribution.

TABLE 10. Distribution of larvae of mesopelagic fishes apart from the Gonostomatidae and Myctophidae (caught with the BR and JOM nets)

Horizon, m	Paralepididae	Scopelarchidae	Chauliodontidae	Indiacanthidae	Melanostomatidae	Melamphaeidae	Sternoptychidae
0—25							
25—50							
50—75							
75—100	≡		≡			≡	≡
100—150			≡	≡	≡		≡
150—200		≡	≡		≡		≡
200—300		≡	≡		≡		≡
Length of larvae, mm	3.0—6.0	3.2—7.7	5.5—7.2	5.5—7.0	4.0—11.0	4.0	5.6—17.5

The larvae developing in the upper 300-m layer may be classed in different ecological groups. Larvae of epipelagic fishes are predominant in the upper (50-m) layer; large numbers of larvae of mesopelagic forms are encountered below 50 m: the larvae of numerous Gonostomatidae and Myctophidae pass through the early stages of postembryonal development in the 50—150-m layer, while the larvae of the Sternoptychidae, Scopelarchidae, and Paralepididae become predominant below 75—100 m (Figures 9 and 10). It may be that this distribution pattern is confined to regions with predominant sinking of waters.

Eggs occurred from the surface down to 500 m, but the largest amount was obtained within the top 50 m (Figure 11). An interesting point is that some species spawn at depths of 200—500 m at comparatively low water temperatures.

CONCLUSIONS

1. Larvae of epipelagic and mesopelagic species were encountered in the upper 300-m layer of the open waters in the western tropical Pacific Ocean under conditions of predominant sinking of surface waters.

2. Besides representatives of the typical oceanic ichthyofauna, there were large numbers of larvae of the anchovy *Stolephorus zollingeri* and pelagic larvae of the lancelet *Amphioxides pelagicus*, which are more commonly found in coastal waters.

3. *Stolephorus zollingeri* occurs in the upper 150-m layer but mainly in the 25—75-m layer at a practically constant water temperature. The larvae rise close to the surface at night and as a rule sink somewhat below 25 m during the daylight hours.

4. Larvae of other species of epipelagic fishes occurred above the thermocline, from the surface down to 75 m but mostly in the top 25 m. Some of them (anchovies, tunas) perform diurnal vertical migrations.

5. Among the larvae of mesopelagic fishes, those of the Gonostomatidae (Gonostoma elongatum, Cyclothone alba) and Myctophidae (Symbolophorus evermanni, Diaphus sp.) were the most populous. The highest abundance of Gonostomatidae larvae occurred in the 50 — 150-m layer (9 — 14 specimens/m^2). Larvae of G. elongatum were found mainly in the 75 — 150-m layer. They keep above the thermocline and in its upper part in early stages of development and apparently do not perform diurnal vertical migrations. Larvae of C. alba spend their early larval period in the 25 — 150-m layer; diurnal vertical migrations are weak.

6. The highest abundance of the larvae of the near-surface Myctophidae, 8 — 10 specimens/m^2, occurred in the 50 — 150-m layer. Larvae of the genus Diaphus rise to the surface at night, reaching the 25 — 0-m layer and sink down to 50 — 150 m in the daytime.

Larvae of Symbolophorus evermanni 2.7 — 7.5 mm long keep to the 50 — 150-m layer and do not perform the diurnal migrations carried out by larger larvae and adults.

7. Other larvae encountered in the top 500 m belonged to the Paralepididae, Melanostomiatidae, Chauliodontidae, Idiacanthidae, Scopelarchidae, Sternoptychidae, and Melamphaeidae; the development of these forms goes on below 75 — 100 m.

Bibliography

Publications in Russian

Arsen'ev, V. S., L. I. Galerkin, E. A. Plakhin, and V. V. Sapozhnikov. Osnovnye cherty gidrologii i gidrokhimii raiona issledovanii 44-go reisa "Vityazya" (Major Hydrological and Hydrochemical Features of the Region Investigated during the 44th Cruise of the "Vityaz"). (In this collection.)

Bekker, V. E. Svetyashchiesya anchousy (sem. Myctophidae) (Lanternfishes (Family Myctophidae)). — In: Tikhii okean, Vol. 7. Biologiya Tikhogo okeana, Book 3. Moskva. "Nauka." 1967.

Gorbunova, N. N. Raion razmnozheniya i pitanie lichinok mech-ryby (Xiphias gladius L.) (Spawning Region and Feeding of the Larvae of Swordfish (Xiphias gladius L.)). — Voprosy Ikhtiologii, Vol. 9, No. 3. 1969.

Lubny-Gertsyk, E. A. Pelagicheskii lantsetnik (The Pelagic Lancelet). — Priroda, No. 2. 1963.

Mukhacheva, V. A. Tsiklotony (rod Cyclothone, sem. Gonostomatidae) (Genus Cyclothone (Family Gonostomatidae)). — In: Tikhii okean, Vol. 7. Biologiya Tikhogo okeana, Book 3. Moskva, "Nauka." 1967.

Parin, N. V. O sutochnoi izmenchivosti vstrechaemosti lichinok nekotorykh okeanicheskikh ryb u poverkhnosti vody (Diurnal Variation in the Frequency of Occurrence of Larvae of Some Oceanic Fishes at the Surface). — Okeanologiya, Vol. 7, No. 1. 1967.

Vinogradov, M. E. and Yu. A. Rudyakov. Sutochnye izmeneniya vertikal'nogo raspredeleniya obshchei biomassy planktona v zapadnoi ekvatorial'noi Patsifike (Diurnal Variations of the Vertical Distribution of the Plankton Biomass in the Equatorial West Pacific). (In this collection.)

Publications in Other Languages

Ahlstrom, E. Vertical Distribution of Pelagic Fish Eggs and Larvae off California and Baja California. — Fish Bull. U. S. Fish a. Wildlife Serv., Vol. 60. No. 161. 1959.

Delsman, H. Fish Eggs and Larvae from the Java Sea. The Genus Stolephorus. — Treubia, Vol. 13, No. 2. 1931.

Mito, S. Some Ecological Notes on the Planktonic Fish Larvae. — Inform. Bull. Planktol. Japan, No. 14. 1967.

Strasburg, D. Estimates of Larval Tuna Abundance in the Central Pacific. — Fish. Bull. U. S. Fish a. Wildlife Serv., Vol. 60, No. 167. 1960.

Yabe, H. and S. Ueyanagi. Contributions to the Study of the Early Life History of the Tunas. — Occas. report Nankai Reg. Fish. Res. Labor, Vol. 1. 1962.

VERTICAL DISTRIBUTION OF SUSPENDED ORGANIC MATTER IN THE TROPICAL PACIFIC WATERS OF THE SOUTHERN HEMISPHERE

Z. Z. Finenko and A. P. Ostapenya

Institute of Biology of the Southern Seas of the Academy of Sciences of the Ukrainian SSR and the Belorussian State University im. V. I. Lenina

It has been established that living organisms account for only a very small percentage of the total mass of suspended organic matter in the open seas and oceans (Sushchenya and Finenko, 1965; Finenko, 1965a, b; Finenko and Zaika, 1969; Ryther and Menzel, 1965), which for the most part consists of dead organic matter in various stages of decomposition. Part undoubtedly serves as the source of energy for bacteria and certain zooplankters. Daphnia, Artemia, and probably other filter feeders can subsist exclusively on dead organic matter for a long time (Esipova, 1969; Cushing, 1968).

Since under natural conditions, the consumers probably feed on both living plants and dead organic matter (Odum, 1968), studies of only the living part of the seston do not yield correct theories concerning the utilization of primary production by the following trophic levels, since detritus is ultimately formed from organic matter produced by plants. From the point of view of trophics it is very important to know the amount of dead organic matter that can serve as the source of energy for planktonic animals.

Pavlyutin and Ostapenya (see Sushchenya, 1968) showed that the caloricity of detritus declines with its aging, with a parallel decline of the coefficient of its utilization by filter-feeding animals. The detritus in the deep layers of the ocean is presumed to be older than that in the surface waters, but this pattern may be significantly disturbed by structural differences among water masses or by migrations of mass zooplanktonic species (Vinogradov, 1968).

The biological program for the 44th cruise of the "Vityaz" included determinations of the vertical distribution of suspended organic matter. This paper presents the results of determinations of organic carbon, phosphorus, the protein nitrogen in the seston, the rate of its decomposition and the main ratios of these parameters.

MATERIAL AND METHODS

The samples were collected in the western part of the South Equatorial Current (polygon No. 2, station 6033), in the north of the Coral Sea (polygon No. 3, station 6052), and in the South Equatorial Current (polygon No. 4, station 6064), with a 100-liter plastic sampler from 10–13 horizons from surface to bottom. The water was poured into polyethylene canisters, from which it passed to the filtration unit. The water samples were filtered on membrane ultrafilters coated with previously deposited powdered glass, prepared from Jena glass by thorough crushing in a mortar. The finely dispersed glass suspension was prepared by repeated sedimentation and elutriation in distilled water. The resultant suspension consisted mainly of particles of 1–15μ. After the filtration, the filters were washed with an isotonic solution of ammonium acetate to complete removal of chlorides and were then brought to air-dry state and placed in a desiccator. The sediment was carefully removed from the filter together with the glass and analyzed.

Organic carbon

Organic carbon in the samples was determined by wet digestion, which is extensively used in hydrobiological investigations for determination of dissolved and suspended organic matter in fresh waters and recently also in seas (Ostapenya and Kovalevskaya, 1965; Sushchenya, 1961; Sushchenya and Finenko, 1966; Finenko, 1965 a). The method of dichromate oxidation has been recommended for determination of suspended organic matter in the Practical Handbook of Seawater Analysis by Strickland and Parsons (1968).

The modified technique of wet digestion oxidizes organic matter in planktonic organisms to the average extent of 90% (Ostapenya, 1965). Oxidation of suspended organic matter, largely consisting of detritus in different stages of decomposition, raises the question of the completeness of oxidation of stable humified organic substances by the wet digestion method. Special experiments on destruction of organic matter of fresh-water plankton demonstrated the degree of oxidation of organic matter to be the same in different stages of decomposition (the experiment was conducted for 100 days at 20°). Extensive experience in applying the dichromate oxidation method to determination of humus in soil similarly indicates that underoxidation of stable humified substances must be very insignificant if it occurs at all.

Protein nitrogen

Protein nitrogen was determined by Lowry's method based on combined application of the biuret test and Folin's phenolic reagent (Vinberg, 1968; Modern Methods in Biochemistry, 1964). The samples were centrifuged for sedimentation of glass and then subjected to colorimetric determination in a photoelectric colorimeter with a red light filter. The content of protein nitrogen was calculated by dividing the amount of protein by 6.25.

Organic phosphorus

Determination of organic phosphorus is based on conversion of its compounds to mineral phosphate by oxidation with various strong oxidants. Determination of phosphorus in organic suspended matter followed the conventional method (Strickland and Parsons, 1968), with perchloric acid as the oxidant.

Dry weight

Water samples were passed through previously weighed No. 5 membrane filters (pores 1 — 1.2μ), after which the filters were washed with an isotonic solution of ammonium acetate, brought to constant weight in the drying oven at 70—80°, and weighed on an analytical balance.

Rate of disintegration of suspended organic matter

A measured volume of water (10—20 liters) was passed through a No. 5 membrane filter that had been boiled three times in distilled water. The filters with accumulated suspended matter were placed in flasks of volume 310 ml (the control was provided by filters without suspended matter). A preliminary experiment had shown that the conventional membrane filters remained practically unaffected by aquatic microflora for 8 days and consequently did not consume oxygen when incubated in water-filled flasks. It was therefore possible to keep the suspended matter accumulated on filters in the flasks together with the filters for BOD determination. The flasks were filled with water from the 300-m horizon that had been kept for 5 days in the laboratory prior to the experiment and kept in the dark at 30 ± 2°. Consumption of oxygen by the suspended matter was measured after 4 and 8 days. The experiments were performed in three replications.

Vertical distribution of suspended organic matter in the water column

The vertical distribution of suspended organic matter is decisively influenced by the nonuniform structure of water masses. According to our data (see figure), the quantitative vertical distribution of suspended organic matter and of its individual components is fairly uniform in the region investigated. The maximum concentration of suspended matter in the western part of the South Equatorial Current (polygon No. 2) occurred at 75—90 m, which was also the depth of occurrence of the maximum chlorophyll "a" concentration. The highest concentrations of organic phosphorus and nitrogen were also measured at the same depths (see figure, I).

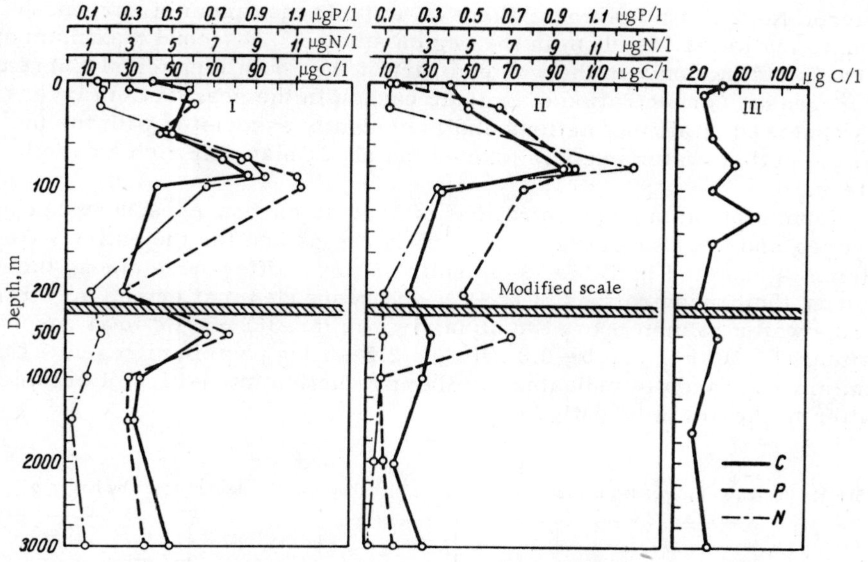

Vertical distribution of organic carbon (C), phosphorus (P), and protein nitrogen (N): I — polygon No. 2; II — polygon No. 3; III — polygon No.4.

The concentration of suspended organic matter decreased sharply in the subsurface layer of high salinity (100 — 200 m). In the intermediate layer of low salinity the maximum concentrations of C, P, and N were determined at 500 m, apparently owing to the high bacterial count in this horizon. The vertical distribution of the components of suspended organic matter was not proportional. The variation was less marked in the case of organic carbon than in the case of protein nitrogen, the concentration of which decreased gradually below 100 m to at most 1 μg/l at 1500—3000 m. The nitrogen concentration at 75 — 100 m was approximately one order of magnitude higher than its concentration in the near-bottom layers, whereas the concentrations of organic carbon and phosphorus at the same depths was 1/2 — 1/3 their concentrations in the 0 — 100-m layer. The overall vertical distribution pattern of suspended organic matter and its individual components in the north of the Coral Sea (polygon No. 3) was the same (see figure, II). The maximum concentrations of organic carbon, phosphorus, and nitrogen were again observed at depths with the maximum chlorophyll "a" concentration. In this region the concentration of organic matter below 100 m varied only slightly, while the concentrations of organic phosphorus and nitrogen gradually declined with increasing depth.

The lowest concentrations of organic carbon, not exceeding 50 μg/l, were determined in the upper 100-m layer in the South Equatorial Current, in polygon No. 4 (see figure, III). The maximum concentration, 100 μg/l, was observed in the subsurface high-salinity layer at 125 m. The vertical distribution of suspended organic matter in the lower horizons proved to be the same as in the western part of the South Equatorial Current (polygon No. 2). On the whole, there is a stable maximum of suspended organic matter at 75—90 m in the region studied. A second maximum at 500 m is distinct only in the western part of the South Equatorial Current. An increased concentration of organic carbon in the near-bottom layers was observed at all the stations and is evidently associated with the fact that the bottom water mass is formed mainly in plankton-rich antarctic waters.

Information on the concentrations of organic carbon, phosphorus, protein nitrogen, and total suspended matter by horizons and for the entire water column is supplied in Table 1. In spite of some differences among the regions, the concentrations of C, P, N, and suspended matter per m^2 throughout the water column proved to be fairly similar, 362—476 g total suspended matter, 83—101 g C_{org}, 0.5—0.8 g P, and 2.5—3.1 g protein nitrogen. This similarity apparently indicates similar production intensities of organic matter in the areas investigated.

TABLE 1. Contents of suspended matter, organic carbon, phosphorus, and protein nitrogen by horizons

Polygon No.	Layer, m	C	P	N	Matter	
					dry	organic
2	0 — 100	63.0	0.48	5.7	214	589.7
	100 — 200	35.5	0.68	4.0	125	332.2
	200 — 500	45.0	0.52	1.4	118	421.2
	500 — 1000	48.5	0.33	1.2	182	454.0
	1000 — 1500	33.5	0.30	0 6	158	313.6
	1500 — 3000	40.3	0.33	0.5	158	238.7
	Entire water column	101.0	0.77	3.1	476	950.1
3	0 — 100	65.0	0.69	4.7	205	608.4
	100 — 200	27.0	0.62	2.2	75	252.7
	200 — 500	27.0	0.27	1.1	107	252.7
	500 — 1000	30.0	0.07	1.0	125	280.8
	1000 — 2000	29.0	0.09	0.7	114	271.4
	2000 — 3000	29.0	0.12	0.3	126	271.4
	Entire water column	90.3	0.50	2.5	362	844.2
4	0 — 100	37.5	—	—	—	349.1
	100 — 200	56.7	—	—	—	527.8
	200 — 500	33.7	—	—	—	313.7
	500 — 1500	28.5	—	—	—	265.3
	1500 — 3000	23.6	—	—	—	219.7
	Entire water column	83.5	—	—	—	777.2

Note. The contents of C, N, P, and dry matter by horizons are stated in mg/m^3, those for the entire water column in g/m^2. The contents of organic matter by horizons are stated in cal/m^3, those for the entire water column in $kcal/m^2$.

RATIOS OF ORGANIC CARBON, PHOSPHORUS, AND PROTEIN NITROGEN

More and more attention is now being paid to determinations of the quantitative ratios of individual components in the suspended matter, because these data permit an indirect estimation of the degree of mineralization and the food value of suspended organic matter at different depths. For living plankton, C:N:P =10:17.7:2.45 (Fleming, 1940). Similar ratios were obtained for zooplankton, mainly copepods, 100:23:2.2 (Beers, 1963). The ratios determined by Antia et al. (1963) and McAllister et al. (1960) were 100:21:3 for phytoplankton in coastal waters with a slight admixture of detritus and 100:23:3 for phytoplankton in the open waters.

Information available on the chemical composition of algae (Lanskaya and Pshenina, 1961, 1963) makes it possible to calculate the C:N ratio in different species of diatomaceous algae grown in the laboratory. It is 100:15 — 20. The C:N ratio in organic matter was somewhat higher, 100:12, in the net collections during the "bloom" of *Chaetoceros curvisetus* and *Skeletonema costatum* in Sevastopol Bay, evidently because the amount of detritus in the samples was small. The relative nitrogen content of small flagellate algae and dinoflagellates is approximately half that in the diatoms. The above values refer to total nitrogen, protein nitrogen accounting for 60—70% of the total nitrogen (Barashkov, 1963; Bogdanov et al., 1968).

The C:N ratios determined by us for suspended organic matter differ considerably at different depths. The ratios of organic carbon, protein nitrogen, and organic phosphorus in suspended matter are as follows:

Depth, m	C:N:P (polygon No. 2)	C:N:P (polygon No.3)
10	100: 3.4:2.7	100: 2.9:0.3
10	100: 2.7:0.5	—
25	100: 2.7:1.1	100: 2.2:1.3
75	100:11.2:1.1	—
80	—	100:13.3:1.0
90	100:10.3:1.2	—
100	100:15.1:2.5	100: 9.4:2.0
200	100: 4.6:0.9	100: 4.0:2.1
500	100: 2.2:1.1	100: 3.0:2.2
1000	100: 2.8:0.9	100: 3.5:0.3
1500	100: 0.6:0.7	—
2000	—	100: 3.4:0.7
3000	100: 1.8:1.2	100: 0.7:0.5

The maximum C:N ratios, 100:9—13, were observed at 75—100 m. They suggest that the suspended organic matter at that depth consisted mainly of live plankton. The percentage of protein nitrogen gradually decreased below 100 m, and its concentration relative to that of organic carbon in the near-bottom layers averaged one order of magnitude lower than in the 75—100-m layer, meaning that a considerable part of the suspended matter below 100 m is dead organic matter. A similar pattern of variation of the relative concentration of protein nitrogen was observed in the Kurile-Kamchatka Trench (Bogdanov et al., 1968), where protein nitrogen accounted for 12.7 — 14.2% of total nitrogen on the surface, but did not exceed 1% at 4,000—6,000 m, also pointing to significant destruction of protein. The relative content of protein nitrogen in suspended matter evidently provides an adequate criterion for its degree of decomposition, since a decrease of the relative content of protein in organic matter indicates that most of the organic compounds in the detritus are in a state of destruction.

The content of organic phosphorus in suspended matter also decreased with increasing depth, although not so sharply as that of nitrogen. The slight decrease of phosphorus in suspended matter in comparison to live plankton indicates that detritus may contain considerable amounts of organic phosphorus. Its rate of decomposition is somewhat higher than that of carbon. If the rates of decomposition of organophosphorus compounds and carbon were the same, the C:P ratio would remain constant and equal to that in live plankton, but in fact the ratio varies down to 500 m. Below this depth the variation becomes insignificant, possibly because the rate of decomposition of both elements in organic matter which has been mineralized to a considerable extent is the same.

There are few data in the literature on the relative contents of organic phosphorus in suspended organic matter. Recent determinations of C, P, and N in dissolved and suspended organic matter in subtropical Pacific waters (Holm-Hansen et al., 1966) revealed an insignificant variation of the relative content of P in the suspended matter down to 1,300 m. The C:N:P ratio for the entire water column averaged 100:8:2 in suspended matter and 100:11:1 in dissolved matter. Hence, the relative content of phosphorus and total organic nitrogen in suspended and dissolved organic matter is approximately half that in live plankton. There is probably very little difference between the degrees of mineralization of dissolved and suspended organic matter.

RATE OF DECOMPOSITION OF SUSPENDED ORGANIC MATTER

The rate of incorporation of suspended organic matter into production cycles can to a certain degree be characterized by the consumption of oxygen by the microflora associated with the suspended particles. Our method for determining of the rate of destruction of suspended matter obviously suffers from shortcomings that are too numerous to yield data characterizing the absolute values of the rate of destruction but it is adequate for determination of relative values and for comparison of the rate of destruction of suspended organic matter at different horizons.

The experimental results of determinations of the rate of oxygen consumption by organic suspended matter are shown in Table 2. The constant of the rate of oxygen consumption calculated by the monomolecular reaction equation decreased with depth. At 2000 m it was 6.5 times smaller than at 75 m. According to Drachev (1964), the constant of the rate of oxidation of humified compounds is 0.001; this value is one order of magnitude below that calculated for samples taken from 3000 m, which shows that there are substantial amounts of oxidizable organic matter even at the very bottom.

From data on oxygen consumption over a certain period and the oxygen consumption rate constant it is possible to calculate the maximum amount of organic matter that can be oxidized; 1 mg organic suspended matter from the horizons 75, 1500 and 3000 m absorbed 0.563, 0.476, and 0.221 mg O_2, respectively, over an experimental period of 4 days. From these values and the amount of weighed organic matter in the same horizons we can readily calculate the amount of oxygen that must be consumed under natural conditions in 4 days, in mgO_2/l. In order to calculate the maximum amount of oxidizable organic matter, we assume that in nature the unstable fraction of suspended organic matter is completely oxidized over an infinitely long time, $t = \infty$. Then, basing ourselves on the monomolecular reaction, the maximum amount of oxidizable organic matter a (stated in mgO_2/l) can be calculated using the formula

$$a = \frac{x}{1 - 10^{-kt}}.$$

TABLE 2. Consumption of oxygen by suspended organic matter

Horizon, m	Concentration of suspended matter, µg C	Exposure, days	O_2 consumption, mgO_2/m	K at t = 30°	K at t = 20°
75	0.36	4	1.89	0.30	0.13
		8	2.00		
1500	0.37	4	1.82	0.17	0.07
		8	1.92		
3000	0.83	4	1.71	0.04	0.02
		8	2.88		

* The oxygen consumption rate constant was scaled to 20° by Krogh's curve.

By the calculated data (Table 3), the amount of oxidizable matter constitutes 60 – 70% of the total amount of suspended organic matter, at all depths. These values are close to those based on a large number of experimental data for decomposing plankton (Skopintsev, 1949). About 2/3 of the total suspended matter can evidently potentially participate in the production cycle.

TABLE 3. Ratio of suspended to oxidizable organic matter

Horizon, m	Concentration of suspended matter, mgO_2/l	Concentration of oxidizable matter	
		mgO_2/l	%
75	0.108	0.065	60.1
1500	0.084	0.050	59.5
3000	0.096	0.068	70.8

Knowing the constant of the rate of oxygen consumption and the amount of oxidizable matter, it is possible to calculate the time required for the entire unstable organic suspended matter to be oxidated. The entire unstable organic matter in the upper layer at 30° decomposes in 1.5 days. At depths of 1500 and 3000 m at 2 – 3° it decomposes in 27 and 150 days. These calculations yield only the order of magnitude, but nevertheless they show that the oxidation processes in the sea go on more rapidly than hitherto assumed.

Bibliography

Publications in Russian

Barashkov, G. K. Khimiya vodoroslei (The Chemistry of Algae). Moskva, Izdatel'stvo Akademii Nauk SSSR. 1963.

Bogdanov, Yu. A., Yu. A. Grigorovich, and M. P. Shaposhnikova. Opredelenie belka v vodnoi vzvesi (Determination of Protein in an Aqueous Suspension). — Okeanologiya, Vol. 8, No. 6. 1968.

Cushing, D. H. Modelirovanie produktsionnogo tsikla v more (Simulation of the Production Cycle in the Sea). — In Sbornik: "Osnovnye problemy okeanologii". 1968. Moskva, Izdatel'stvo "Nauka." 1968. [Russian translation.]

Drachev, S. M. Bor'ba s zagryazneniem rek, ozer i vodokhranilishch promyshlennymi i bytovymi stokami (Control of Pollution in Rivers, Lakes, and Reservoirs Caused by Industrial and Domestic Effluents). Moskva, "Nauka." 1964.

Esipova, M. A. Rost i razmnozhenie Daphnia magna i D. longispina pri pitanii detritom (Growth and Reproduction of Daphnia magna and D. longispina Feeding on Detritus). — Gidrobiologicheskii Zhurnal, Vol. 5, No. 5. 1969.

Finenko, Z. Z. Soderzhanie organicheskogo veshchestva v sestone Chernogo i Azovskogo morei (Content of Organic Matter in the Seston of the Black and Azov Seas). Kiev, "Naukova dumka." 1965a.

Finenko, Z. Z. Otnoshenie mezhdu pervichnoi produktsiei, khlorofillom i organicheskim uglerodom v morskom planktone (Ratio of Primary Production, Chlorophyll, and Organic Carbon in Marine Plankton). — In Sbornik: Voprosy gidrobiologii. I s"ezd Vsesoyuznogo gidrobiologicheskogo obshchestva. Tezisy dokladov. Moskva, "Nauka." 1965b.

Finenko, Z. Z. and V. E. Zaika. Soderzhanie vzveshennogo organicheskogo veshchestva v vodakh Araviiskogo morya (The Content of Suspended Organic Matter in the Arabian Sea). — Okeanologiya, Vol. 11, No. 4. 1969.

Lanskaya, L. A. and T. I. Pshenina. Soderzhanie belka, zhira, uglerodov i zoly v nekotorykh massovykh planktonnykh vodoroslyakh Chernogo morya, vyrashchennykh v kul'turakh (The Content of Protein, Fat, Carbohydrates, and Ash in Some Cultured Mass Planktonic Algae of the Black Sea). — Trudy Sevastopol'skoi biologicheskoi stantsii, Vol. 14. 1961.

Lanskaya, L. A. and T. I. Pshenina. Sravnenie khimicheskogo sostava nekotorykh vidov diatomovykh v kul'turakh i more (Comparison of the Chemical Compositions of Some Species of Diatoms, in Cultures and in the Sea). — Trudy Sevastopol'skoi biologicheskoi stantsii, Vol. 16. 1963.

Odum, E. Ekologiya (Ecology). Moskva, "Prosveshchenie." 1968.

Ostapenya, A. P. Polnota okisleniya organicheskogo veshchestva vodnykh bespozvonochnykh metodom bikhromatnogo okisleniya (Completeness of Dichromate Oxidation of the Organic Substance of Aquatic Invertebrates). — Doklady Akademii Nauk BSSR, Vol. 9, No. 4. 1965.

Ostapenya, A. P. and R. Z. Kovalevskaya. O soderzhanii vzveshennogo organicheskogo veshchestva v poverkhnostnom sloe morskikh vod (On the Content of Suspended Organic Matter in the Surface Layer of the Sea). — Okeanologiya, Vol. 5, No. 4. 1965.

Sovremennye metody v biokhimii (Modern Methods in Biochemistry). Moskva, "Meditsina." 1964.

Skopintsev, B. A. O skorosti razlozheniya organicheskogo veshchestva otmershego planktona (On the Rate of Decomposition of Organic Matter of Dead Plankton). — Trudy Vsesoyuznogo gidrobiologicheskogo obshchestva, Vol. 1. 1949.

Sushchenya, L. M. Nekotorye dannye o kolichestve sestona v vodakh Egeiskogo, Ionicheskogo i Adriaticheskogo morei (Some Data on the Amount of Seston in Waters of the Aegean, Ionian, and Adriatic Seas). — Okeanologiya, Vol. 1, No. 4. 1961.

Sushchenya, L. M. Detrit i ego rol' v produktsionnom protsesse v vodoemakh (The Role of Detritus in the Production Process in Water Bodies). — Gidrobiologicheskii Zhurnal, Vol. 4, No. 2. 1968.

Sushchenya, L. M. and Z. Z. Finenko. Soderzhanie vzveshennogo organicheskogo veshchestva v vodakh tropicheskoi Atlantiki i nekotorye kolichestvennye sootnosheniya mezhdu ego komponentami (Content of Suspended Organic Matter in Tropical Atlantic Waters and Some Quantitative Ratios of its Components). — Okeanologiya, Vol. 6, No. 5. 1966.

Vinberg, G. G. (ed.). Metody opredeleniya produktsii vodnykh zhivotnykh (Methods of Determining the Production of Aquatic Animals). Minsk, Izdatel'stvo "Vysheishaya shkola." 1968.

Vinogradov, M. E. Vertikal'noe raspredelenie okeanicheskogo zooplanktona (Vertical Distribution of Oceanic Zooplankton). Moskva, "Nauka." 1968.

Publications in Other Languages

Antia, N. J., C. D. McAllister, T. R. Parsons, K. Stephens, and J. D. H. Strickland. Further Measurements of Primary Production using a Large-Volume Plastic Sphere. — Limnol. and Oceanogr. Vol. 8, No. 2. 1963.

Beers, J. R. Studies on the Chemical Composition of the Major Zooplankton Groups in the Sargasso Sea off Bermuda. — Woods Hole Oceanographic Institution. Ref. 64 — 84, appendix 1. 1963.

Fleming, R. H. The Composition of Plankton and Units for Reporting Population and Production. — Proc. 6th Pacific. Sci. Congr. Vancouver, Vol. 3. 1940.

Holm-Hansen, O. J., J. D. Strickland, and P. M. Williams. A Detailed Analysis of Biologically Important Substances in a Profile off Southern California. — Limnol. and Oceanogr., Vol. 11, No. 4. 1966.

McAllister, C. D., T. R. Parsons, and J. D. H. Strickland. Primary Productivity at Station 2P in the North-East Pacific Ocean. — J. Cons. Internat. Explor. Mer., Vol. 25, No. 3. 1960.

Ryther, J. H. and D. W. Menzel. On the Production, Composition and Distribution of Organic Matter in the Western Arabian Sea. — Deep-Sea Res., Vol. 12, No. 2. 1965.

Strickland, J. D. H. and T. R. Parsons. A Practical Handbook of Seawater Analysis. — Bull. Fish. Res. Board Canada, No. 167. 1968.

BIOCHEMICAL OXYGEN DEMAND

A. P. Ostapenya
Belorussian State University im. V.I.Lenin

Destruction of organic matter in water caused by the life activity of microorganisms is accompanied by consumption of dissolved oxygen, and therefore the rate of destruction and the amount of mineralized organic matter can be determined from changes of the oxygen concentration. The method of determining the biochemical oxygen demand (BOD) is extensively used in practical hydrobiological investigations. The principle of the method consists in exposing water samples isolated in flasks in the dark at a constant temperature. The rate of oxygen consumption, which is proportional to the rate of destruction of organic matter, is calculated from the loss of oxygen in the exposed flask.

Several authors have shown by experiment that the biochemical oxygen demand has, as a rule, a diminishing absolute rate. The kinetics of the process can be described by the following first-order reaction equation:

$$X_t = L_0(1 - 10^{-kt}), \qquad (1)$$

where X_t is the biochemical oxygen demand over time t; L_0, total oxygen demand of the water, or the amount of oxygen required for oxidation of readily oxidizable organic substances at the beginning of destruction process; k, constant of the rate of the process of oxygen consumption; t, time.

The oxygen consumption rate constant k can be readily calculated from data on oxygen consumption at two observation times. The calculation formula is

$$K = \frac{1}{t} \lg \frac{X_{t_1}}{X_{t_2} - X_{t_1}},$$

where X_{t_1} and X_{t_2} are the respective BOD values over periods t_1 and $2t_1$.

The method of BOD determination has often been used to characterize the decay of organic matter in seawater. Experience has shown that determinations of BOD in seawater involve several difficulties. Because of the comparatively low concentration of oxidizable organic substances, the BOD values in seawater, especially in oligotrophic oceanic zones, are so small that they often lie at the sensitivity limit of the method used for determination of oxygen. The consequences of the so-called "flasks effect," which to a certain extent distorts the course of the processes involved in the destruction of organic matter in water, are greater in oligotrophic waters (Skopintsev, 1948), and therefore the use of BOD values to characterize the dynamics of decay of organic substances in marine waters should be used with caution.

The results of calculations of the amount of readily oxidizable organic matter (L_0) from BOD are much more representative. The suspended and dissolved organic matter is represented by two fractions. One of them is readily assimilated by aquatic organisms and thereby the energy accumulated in this organic matter is incorporated in the production processes. The second fraction consists of stable organic matter which is not readily oxidized and is virtually unavailable to aquatic organisms. The rate of incorporation of this fraction of organic substances into the production process is vanishingly low. Obviously, the two fractions play very dissimilar roles in the turnover of matter and energy, hence the importance of elucidating the amount and distribution of readily assimilable organic matter.

The water samples for BOD determinations in the 44th cruise of the "Vityaz" were taken with a sterile 7-liter sampler. In order to equalize the water and air temperatures, the samples were placed in carboys which had been processed with hydrochloric acid. BOD determination was carried out in flasks with volume 310 ml. The flasks containing the water samples were exposed in the laboratory in a dark polyethylene tank with water at 30 ± 2°. As a rule, BOD was determined twice, usually after 5 and 10 or after 3 and 6 days after the beginning of the experiment. All determinations were carried out in 2 or 3 replications.

The oxygen concentration in the flask was determined after Winkler. Gas bubbles appeared in some flasks (those with samples obtained from deep horizons). To avoid losses of oxygen, the flasks were fixed in the stand in an oblique position, so that the gas bubble remained in the flask when the stopper was opened, and the oxygen in it was fixed with suitable reagents.

A 210-ml portion from each flask was titrated with 0.01 N hyposulfite. The burette scale value was 0.01 ml. With a BOD of the order of magnitude 0.1 mgO_2/l, the error of determination was approximately 10%.

In most cases, oxygen consumption in the experiment amounted to tenths of a milligram per liter. The determinations of BOD were therefore sufficiently accurate for an approximate appraisal of the processes of oxygen consumption.

The results of the determinations are shown in Table 1.

The absolute oxygen consumption rate declined with the passage of time at all the stations except at station 6064 (polygon No. 3) and therefore equation (1) could be used in the processing of data.

The data were used to calculate the oxygen consumption rate constant (Table 1), which to a certain extent characterizes the availability of organic matter for the microflora oxidizing it. With a relatively large quantity of readily oxidized unstable organic matter in the water, the BOD rate and correspondingly the value of k will be larger than if there is mostly organic matter that is difficult to oxidize.

The oxygen consumption rate constant was found to be 0.158 as an average from all the data.

TABLE 1. BOD and oxygen consumption rate constant by horizons

Station No.	Horizon, m	Exposure days	BOD, mgO_2/l, over the exposure time	Rate constant at 30°	Rate constant at 20°
6024	0	5	0.25	0.083	0.039
		10	0.34		
	750	5	0.78	0.102	0.045
		10	1.02		
6025	0	5	2.80	—	—
6027	0	5	0.43	—	—
6028	0	5	0.16	—	—
6033	0	3	0.29	0.186	0.082
		6	0.37		
	220	3	0.39	0.248	0.110
		6	0.46		
	400	3	0.34	0.223	0.101
		6	0.41		
	750	3	0.40	0.141	0.062
		6	0.55		
6052	15	3	0.14	0.281	0.124
		6	0.16		
	70	3	0.13	0.111	0.049
		6	0.19		
	100	3	0.11	0.181	0.083
		6	0.14		
6057	85	3	0.13	0.111	0.049
		6	0.19		
	130	3	0.13	0.011	0.005
		6	0.25		
6059	25	3	0.16	0.378	0.168
		6	0.19		
6064	70	3	0.05	—	—
		5	0.10		
	400	3	0.09	—	—
		5	0.09		
	1500	3	0.22		
		5	0.45		
6065	25	3	0.29	0.127	0.056
		6	0.41		

Most of the published data on the BOD rate were yielded by experiments carried out at 20°. The ***k*** constant values were thus scaled to 20° by means of Krogh's normal curve for comparison with the literature (Vinberg, 1956). The mean oxygen consumption rate constant at 20° is 0.07, with standard deviation 0.03.

Data currently available suggest that the constant may vary over a considerable range depending upon the nature of the organic matter undergoing oxidation. The constant is close to 0.001 for stable humified substances (Drachev, 1964), while values of 0.1 — 0.3 are usual for effluents (Cherkinskii, 1951). The oxygen consumption rate constant for Caspian waters is 0.03 — 0.05 (Skopintsev, 1938). The rate constant for the investigated Pacific region, 0.07, is close to values characterizing the decay of planktonic organic matter (Skopintsev, 1950). Much higher values have been reported in the literature for marine waters. For example, the rate of relative oxygen consumption in the Atlantic is close to 0.20 (Keys et al., 1935, cited by Skopintsev, 1948).

It is currently presumed that stable, difficult to oxidize organic matter must accumulate in deep waters. Consequently, the oxygen consumption rate constant should diminish with increasing depth. However, the results obtained during the cruise showed that far from diminishing with increasing depth, it even somewhat increases. This rate constant averages 0.066 for the 0—200-m horizon at 20° and 0.078 for the 200—750-m horizon. The scatter of values is too large to be able to speak of a significant increase of the constant with increasing depth, but the fact that it does not diminish with depth deserves serious attention.

TABLE 2. Total oxygen requirement of the water and BOD in the first diurnal period

Station No.	Horizon, m	Total oxygen requirement of the water, mgO_2/l	BOD in the first diurnal period, mgO_2/l	Station No.	Horizon, m	Total oxygen requirement of the water, mgO_2/l	BOD in the first diurnal period, mgO_2/l
6024	0	0.39	0.07		750	0.64	0.18
	750	1.12	0.23	6052	15	0.16	0.08
6025	0	3.34	1.02		70	0.24	0.05
6027	0	0.51	0.15		100	0.15	0.05
6028	0	0.19	0.05	6057	85	0.24	0.05
6033	0	0.40	0.14		130	1.78	0.04
	220	0.47	0.20	6059	25	0.20	0.08
	400	0.43	0.17	6065	25	0.49	0.12

BOD values for the first diurnal period were calculated for comparison of data elicited in different areas and with different exposure times (Table 2). For stations with a single BOD determination (stations 6025, 6027, 6028) the diurnal BOD values were calculated from the mean oxygen consumption rate constant 0.158.

The BOD values ranged between 1.02 and 0.04 mgO_2/l at different stations. Comparatively high diurnal values were observed in the vicinity of the Gilbert Islands (stations 6024 and 6027) and in polygon No. 2 (station 6033).

Unexpected results were obtained in a section near Butaritari Atoll. The maximum BOD value 1.02 mg O_2/l was determined at station 6025, which was the farthest from the atoll. The BOD values diminished regularly on approaching the atoll, down to only 0.05 mg O_2/l at the entrance to the lagoon (station 6028).

The BOD values were appreciably smaller in polygon No. 3 (station 6025) than in polygon No. 2. It was impossible to calculate the diurnal BOD values for polygon No. 4 (station 6064) because the results could not be described by first-order reaction equations. Nevertheless, the three-day BOD values (Table 1) suggest that the BOD of water in the polygon was comparatively small.

The diurnal BOD values were low, 0.04—0.08 mg O_2/l, in Australian coastal waters (stations 6057 and 6059).

The BOD value was comparatively high, 0.12 mg O_2/l, in Kieta Bay (Bougainville Island, station 6065).

BOD values increase somewhat with increasing depth. Unless this is an artefact, the increase of BOD values with increasing depth, along with the increasing oxygen consumption rate constant, may indicate that there is relatively less decomposed organic matter in the deep layers of the areas investigated.

The experimental data were used to calculate the complete BOD values in order to estimate the amount of organic matter available to the aquatic microflora (see Table 2). The distribution of the amounts of readily oxidizable organic matter by stations approximately corresponds to the distribution of BOD values.

The concentration of readily assimilable organic matter stated in terms of oxygen was found to be 0.49 mg O_2/l as an average from the entire body of data exclusive of station 6025. With the oxygen equivalent of organic matter 1.50 mg O_2/mg (Skopintsev, 1950) and carbon content of 50% in the organic matter, it follows that the concentration of readily oxidizable organic matter stated in terms of carbon is 0.16 mg C/l. It is of interest to compare this quantity with the total amount of organic matter dissolved and suspended in the water. The concentration of organic matter is approximately the same in different regions and at different depths in the ocean, usually 0.5 — 3.0 mg C/l (Menzel, 1964; Plunkett and Rakestraw, 1955). The value of 1.5 mg C/l can be adopted as the average for the World Ocean (Skopintsev, 1961). Since our investigations were conducted in comparatively unproductive tropical waters, the concentration of organic matter may be taken to be 1.0 mg C/l. Therefore, the fraction of organic matter that is available to aquatic heterotrophs is 16%.

Bibliography

Publications in Russian

Bruevich, S. V. Gidrokhimicheskie raboty Gosudarstvennogo okeanograficheskogo instituta v Barentsevom more v 1927—1930 gg. (Hydrochemical Studies by the State Oceanographic Institute in the Barents Sea in 1927—1930). — Doklady 1 sessii Gosudarstvennogo okeanograficheskogo instituta (GOIN), No. 1. 1931.

Cherkinskii, S. N. Sanitarnye usloviya spuska stochnykh vod v vodoemy (Sanitary Regulations for Discharging Effluents in Water Bodies). Moskva. 1951.

Drachev, S. M. Bor'ba s zagryazneniem rek, ozer i vodokhranilishch promyshlennymi i bytovymi stokami (Control of Pollution in Rivers, Lakes and Reservoirs Caused by Industrial and Domestic Effluents). Moskva, "Nauka." 1964.

Skopintsev, B. A. Organicheskoe veshchestvo v Kaspiiskom more (Organic Matter in the Caspian Sea). — Doklady Akademii Nauk SSSR, Vol. 18, No. 7. 1938.

Skopintsev, B. A. O potreblenii kisloroda v prirodnykh vodakh stoikimi organicheskimi veshchestvami (On the Consumption of Oxygen in Natural Waters by Stable Organic Substances). — Gidrokhimicheskie materialy, No. 16. 1948.

Skopintsev, B. A. Organicheskoe veshchestvo v prirodnykh vodakh (Vodnyi gumus) (Organic Matter in Natural Waters (Aquatic Humus)). — Trudy Gosudarstvennogo okeanograficheskogo instituta, Vol. 17(29). 1950.

S k o p i n t s e v, B. A. Nekotorye rezul'taty izucheniya organicheskogo veshchestva v morskikh vodakh primenitel'no k poznaniyu protsessov osadkoobrazovaniya (Some Results of Studies of Organic Matter in Marine Waters with Reference to Sedimentation Processes). — In Sbornik: "Sovremennye osadki morei i okeanov." Moskva, Izdatel'stvc Akademii Nauk SSSR. 1961.

V i n b e r g, G. G. Intensivnost' obmena i pishchevye potrebnosti ryb (Metabolism and Food Requirements of Fish). Minsk, Izdatel'stvo Belorusskogo gosudarstvennogo universiteta. 1956.

Publications in Other Languages

K e y s, A., E. C h r i s t e n s e n, and A. K r o g h. The Organic Metabolism of Sea Water with Special Reference to the Ultimate Food Cycle in the Sea. — J. Mar. Biol. Assoc. U. K. Vol. 20, No. 2. 1935.

M e n z e l, D. W. The Distribution of Dissolved Organic Carbon in the Western Indian Ocean. — Deep-Sea Res., Vol. 11, No. 5. 1964.

P l u n k e t t, M. A. and N. W. R a k e s t r a w. Dissolved Organic Matter in the Sea. — Deep-Sea Res., Suppl. to Vol. 3. 1955.

S e i w e l l, H. R. Consumption of Oxygen in Sea Water under Controlled Laboratory Conditions. — Nature, Vol. 140, No. 3546. 1937.

W a k s m a n, S. A. and C. E. R e n n. Decomposition of Organic Matter in Sea Water by Bacteria. — Biol. Bull., Vol. 70, No. 3. 1936.

Z o B e l l, C. E. The Effect of Oxygen Tension on the Rate of Oxidation of Organic Matter in Sea by Bacteria. — J. Mar. Res., Vol. 8, No. 3. 1940.

THE SPATIAL STRUCTURE OF COMMUNITIES IN THE EUPHOTIC ZONE OF THE OCEAN TROPICS

M. E. Vinogradov, I. I. Gitel'zon, and Yu. I. Sorokin
Shirshov Institute of Oceanology of the Academy of Sciences of the USSR, and
Institute of the Biology of Inland Waters of the Academy of Sciences of the USSR

Although information on the biology of marine organisms and their relationships with the living and dead environment is now plentiful, our knowledge of the structure and life activity of oceanic communities is still incomplete, chiefly because of the extreme complexity of interactions in actual systems. In order to understand the makeup and functions of communities and the features of their production cycle, it is useful to consider the interaction of only the major factors, i. e. to deal with a model of the community instead of the community itself. Obviously, all the major relations and processes in the community must be taken into account in order to construct a sufficiently complete model. Furthermore it is important that the model include an equilibrium stage in the existence of the system, in which the gain and loss of energy should be balanced.

The best studied oceanic communities are the pelagic communities of the temperate-coldwater zone, for which the largest number of attempts at simulating the production cycle have been made (Riley, Stommel, and Bumpus, 1949; Cushing, 1959; Steele, 1962; Vollenweider, 1965; Dugdale, 1967; and others). A complete review of these was recently published by Patten (1968). However, working with communities of boreal waters involves considerable difficulties in view of the significant seasonal variation of production. The production cycle encompasses the entire year (at any rate, most of it) and therefore a model should take into account data for the different seasons; however, it is practically impossible thereby to obtain the characteristics of a single specific community.

The communities of open regions of the ocean tropics have the advantage of being subject to much smaller seasonal variations, since the system may be considered to be at an equilibrium at each moment of time without significant error. This applies primarily to regions that are not affected by monsoons. The influx of biogens from the deep layers is here related to quasi-stationary upwelling and not to seasonal mixing, while the intensity of solar radiation and the wind regime remain more or less stable over the year. The system's production cycle may therefore be appraised by means of comparatively brief observations which can be performed on a single community. Still, the actual structural and functional features of pelagic communities in the tropical zone have been much less thoroughly studied than those in the temperate-coldwater regions, and this makes modeling very difficult.

The energic and structural approaches to productivity studies of aquatic communities call for the fullest possible reckoning with their spatial distribution, numerical abundance, biomass and productivity within the investigated biotope. The difficulty with pelagic communities is that organisms do not have a uniform vertical distribution in the water column due to abiotic and probably also biotic factors. There are layers with a high concentration of organisms and others with very low concentrations (Barnard, 1958; Semina, 1967; Vinogradov, 1968; and others).

Standard methods for estimating the abundance and productivity of populations that are nonuniformly distributed in the water column yield either mean data for extensive layers (hauls with plankton nets) or data for discrete depths (bathometer samples). Neither of these techniques is capable of producing the real pattern of distribution of the organisms, and thin layers with especially high concentrations are missed or "smeared" over the extensive layer of the catch. The danger of overlooking the maxima of concentration and production for forage organisms of the first link in the food chain (algae and bacteria) is especially serious in this case, because it is these organisms which largely govern the production capacities of the community as a whole. Consumer animals are capable of active movement. Migrating in the water column, they find concentrations of forage organisms and form concentrations at such spots.

Attempts have been made to discard the standard collection horizons in distribution studies of phytoplankton and bacteria. The depths of collecting were chosen in accordance with the pattern of stratification of waters at the places of observation and the distribution of temperature and density gradients. The temperature distribution was determined by sounding (continuous recordings). The concentrations of phytoplankton and bacteria were assumed to be located at the depth of the largest temperature or density gradient and in this way it was possible to disclose layers in which the concentration of active phytoplankton or the bacterial production were tens of times greater than in the adjacent waters (Sorokin, 1959, 1964). However, even this technique is rather arbitrary and it does not guarantee that the samples will really be obtained from the layer of highest concentration. Although there is a relationship between highest density gradients and layers with the highest concentration of living suspended matter, it may not be in evidence in every case. The problem cannot be solved without continuous sounding of a biological parameter with simultaneous aimed sampling, and until recently no sounding methods for biological parameters were available. Now, however, apparatus for continuous recording of bioluminescence has been developed at the Institute of Physics of the Siberian Branch of the Academy of Sciences of the USSR.

This paper presents the first results of studies of the vertical distribution of plankton in the euphotic zone of the tropical Pacific waters by the method of continuous sounding of the bioluminescence field with simultaneous aimed sampling. Comparison of these data with the distribution of environmental factors elucidates some features of the spatial structure of pelagic communities in the tropical waters and makes it possible to construct a functional model.

METHODS

Bioluminescence intensity was sounded with a bathyphotometer with a special attachment for stimulating bioluminescence and for limiting the volume in which the measurement is performed. The sounding device connected with the recording unit on board by multicore cable was lowered from the sampling winch. The depth sensor produced a continuous signal of the depth of occurrence of the bathyphotometer. The recordings were carried out with an ink XY-recorder, which produced a graph of bioluminescence intensity as a function of depth directly during the movement of the sounding device. A single sounding from 200 m to the surface took 3 minutes. Samples of typical records of the depth variation of bioluminescence are shown in Figures 1 and 2. For a detailed description of the apparatus and sounding techniques the reader is referred to the paper by Gitel'zon et al. in this collection.

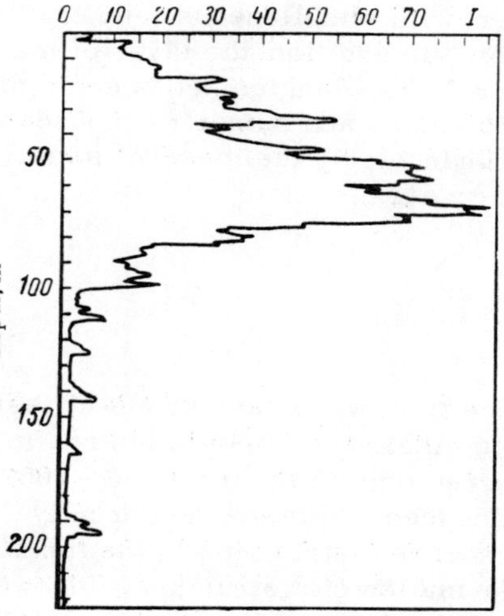

FIGURE 1. A bathyphotogram produced by the XY-recorder

A bimodal bioluminescence maximum.

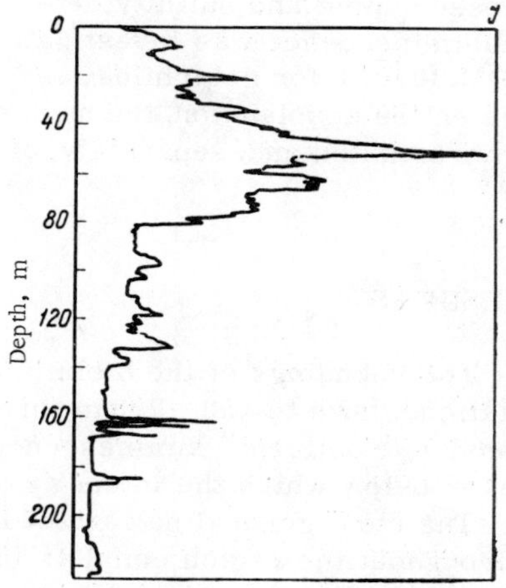

FIGURE 2. Bathyphotogram.

A unimodal bioluminescence maximum.

The analytical samples were taken with 5-liter plastic samplers sterilized with HCl. The sampler was suspended from the cable of the bioluminescence probe one meter from the instrument's sensor. The samples were taken at night after a series of recordings of the vertical profile of bioluminescence intensity had been obtained down to 150 m. Tentative sampling depths were chosen by means of the bathyphotograms. Then the instrument was lowered to a predetermined depth with simultaneous recording of bioluminescence, the predetermined bioluminescence density maximum or minimum was found and, at the command of the operator monitoring the recording, the sampler was shut at the required depth. This was followed by another control bioluminescence recording while the instrument with the shut sampler was raised and lowered in order to ascertain the correct location of the sampler in the predetermined horizon.

The profiles of bioluminescence intensity so obtained characterized the relative distribution of luminescent plankton correlated with the abundance distribution of the entire plankton. The water samples were used for analysis of the vertical distribution of forage organisms and environmental factors. The relative distribution of vitally active phytoplankton (K_d) was determined by means of C^{14} (Sorokin, 1959). The numerical abundance of phytoplankton and its biomass were determined in a Nojotte chamber. The total bacterial count was determined by direct microscopic examination on VUFS membrane filters with pores of 0.1 — 0.3 μ. The activity of heterotrophic microflora was determined from the rate of assimilation of C^{14}-tagged dissolved organic matter by bacterioplankton (Sorokin, 1971).

The bacterial production was measured from the intensity of heterotrophic assimilation of CO_2 by the radiocarbon method (Romanenko, 1964). The same samples were used for determinations of the content of total and mineral nitrogen, phosphorus, silicon, and oxygen, with simultaneous temperature measurement and salinity determination. In addition, the layer of maximum bioluminescence was investigated with a Juday plankton net, oceanic model, JOM 80/113, for determinations of the biomass and numerical abundance of the entire zooplankton, the numbers of undoubtedly luminescent forms, of phytoplankton and, separately, of Pyrocystis.

RESULTS

The soundings of the bioluminescence field were carried out at test stations down to 150 — 200 m through the night at 5 — 10-min intervals. Intensive "confluent" luminescence occurred only in the upper 90 — 100-m layer below which the intensity of luminescence diminished abruptly.

The most general pattern of luminescence distribution in the top 100 m throughout the region consists in the formation of a structure with two maxima. The lower, usually more distinct, maximum occurred at depths of 60 — 100 m, while the upper maximum, distinctly separated from the lower, was situated at 20 — 50 m. It was more variable with respect to its magnitude and position, and was even absent in some soundings (compare Figures 1 and 2). The vertical extent of the maximum does not usually exceed 10 m (Gitel'zon et al., 1971).

Aimed sampling was conducted as station 6052 in the Coral Sea and station 6064 in the open waters of the South Equatorial Current. Comparison of samples taken at station 6052 at the surface and in and below the lower layer of maximum luminescence at once revealed a strikingly nonuniform distribution of life in the water column (see table). Phytoplankton activity in the maximum luminescent layer proved to be ten times as high as that in deeper water and at the surface. The total bacterial count and bacterial biomass in the maximum zone were 5 — 8 times those in the overlying and underlying layers, approaching those in mesotrophic waters. The bacterioplankton biomass in the maximum layer was 130 mg/m^3, 2 — 4 times the average phytoplankton biomass for waters in this region. Bacterial

Concentrations of biogenic elements, activity of phytoplankton (K_d) and microflora, and luminescence intensity in aimed samples from and outside the layer of maximum bioluminescence (92 m) at station 6052

Depth, m	Temperature, °C	Temperature gradient, degrees/m	O_2, mg/l	N – NO_3, µg-at/l	P, µg-at/l	Relative bacterial activity (consumption of tagged organic matter), %	K_d, %	total count, thous/ml	Bacteria: biomass, mg/m³	Bacteria: production, mg/m³	P/B coefficient	Relative intensity of luminescence (% of maximum)
6	29.3	0.01	4.5	0	0	100	100	88	16	24	1.5	7.7
92	25.1	0.035	3.6	0.4	0.20	1000	1260	430	130	74	0.6	100
106	24.2	0.032	3.22	1.0	0.34	28	140	56	10	18	1.8	4.6
126	23.7	0.03	3.17	2.4	0.42	36	210	91	17	17	1.0	1.5

production in the layer of maximum luminescence was 74 mg/m³, or about 8 mg C/m³, 3 – 4 times the diurnal production of phytoplankton in the surface layer.

Detailed investigations carried out at station 6064 (series 6) made it possible to discover still higher phytoplankton and bacterioplankton concentrations in the zone of the principal (lower) luminescence maximum (see Figure 3, a). A trend toward a similar stratification is traced in the distribution of zooplankton. The results of aimed sampling of zooplankton in the layer of maximum luminescence at station 6064 (the luminescence maximum during the catches was located at a depth of about 60 m) were as follows (percentages of the maximum):

Layer investigated, m	9 – 35	50 – 71	78 – 101
Relative distribution of emitted energy (sounding data)	31	100	49
Zooplankton biomass	37	100	48
numerical abundance	51	100	37
Numerical abundance of luminescent copepods Lucicutia and Pleuromamma	22	100	17

The activity of planktonic algae and the concentration and production of microorganisms in the layer of maximum luminescence are many times higher than in the overlying and underlying layers. Thus, at 115 m, i. e. 55 m below the layer of maximum luminescence intensity, the biomass and especially the activity of bacteria were already vanishingly small, similar to the values that can only be encountered at depths of 700 – 800 m. The concentration of bacterioplankton in the maximum layer (total count one million cells/ml; biomass, 300 mg/m³) approaches the feeding optimum of filter-feeding calanids. The experiment suggests that bacterioplankton is fairly efficiently utilized by calanids at a concentration of 250 – 300 mg/m³ (Pavlova, Petipa, and Sorokin, 1971; Petipa, Pavlova, and Sorokin, 1971). Their intensity of feeding (C_{assim}/C) on bacterioplankton at its optimum concentration approaches that on algae.

Hence, even the oligotrophic tropical waters of the open ocean contain layers with sufficient quantities of forage organisms for filter-feeding invertebrates. The distribution curves for luminescence intensity suggest that a large part of the zooplankton is concentrated in just these layers. This explains the situation, always observed in oligotrophic tropical waters, that the zooplankton goes along with mean biomasses of 20 — 40 mg/m^3 of forage organisms, which is far below the threshold concentration below which the filter feeders do not make up their metabolic losses (Jørgensen, 1962). Hence, the very existence of the communities of oligotrophic tropical regions of the ocean depends on the presence of layers with considerable concentrations of producers and consumers in comparison to the background.

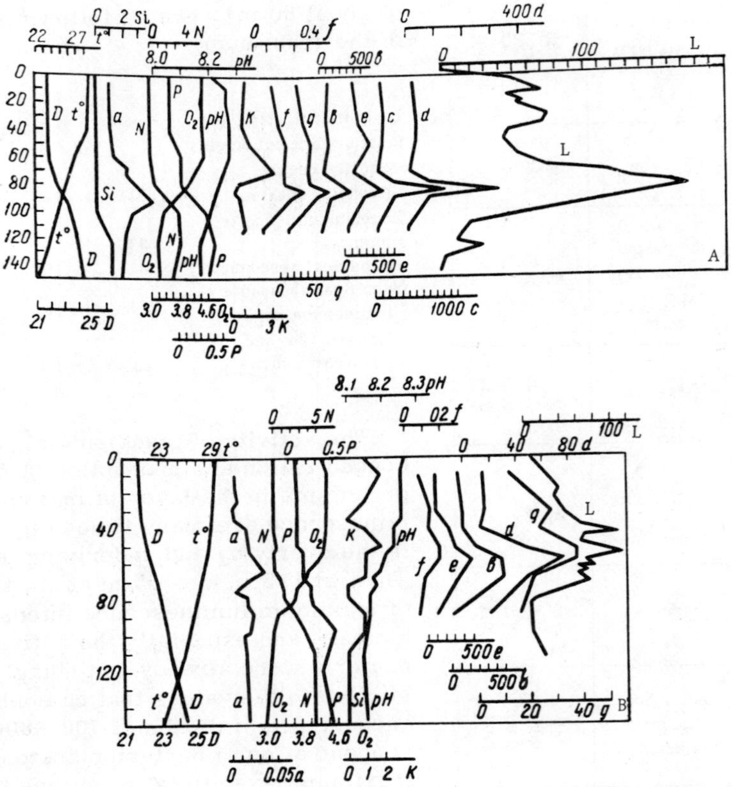

FIGURE 3. Stratification of physical, chemical, and biological parameters by results of aimed sampling based on simultaneous bathyphotometric sounding:

A — station 6064-6 (09°45' S, 164°09' E); B — station 6064-1 (09°30' S, 164°09' E). t° — temperature; D — density; α — density gradient; Si — silicon, μg-at/l; N — mineral nitrogen, μg-at/l; P — phosphorus, μg-at/l; O_2 — oxygen, ml/l; pH — concentration of hydrogen ions; b — relative abundance of phytoplankton, %; c — total bacterial count, thous · ml; d — bacterial biomass, mg/m^3; e — relative bacterial activity, %; f — CO_2 assimilation by bacteria, mg C/m^3 · day; g — bacterial production, mg/m^3 · day; K — P/B coefficient for bacteria; L — intensity of bioluminescence.

It is to be noted that the layer of maximum bioluminescence is at the same time a zone of abrupt decrease in the concentration of biogenic elements (Figure 3). The biogens are intensively absorbed by this layer of concentrated life. Besides the phytoplankton, the direct consumers of biogens may include bacterial cells, if the organic substance on which they subsist is deficient in certain biogens (probably nitrogen and phosphorus).

The layers enriched in algae and microflora may become diffuse, gaining in vertical spread and correspondingly becoming less distinct with respect to the concentrations when the stability of stratification of water masses is disturbed. At station 6064 (series 1) the luminescence intensity maximum is vertically diffuse, occupying a layer about 40 m thick. The other biological indexes also display a broad maximum in a layer about 30 m thick. The concentration of bacterioplankton in this layer is only 1/3 — 1/5 that in a layer with a sharply expressed maximum (see Figure 3, b).

DISCUSSION

The discovery of a distinct and stable stratification in the distribution of plankton in the trophogenic layer allows us to form several concepts concerning its ecological role and certain features of the spatial structure and functioning of the communities inhabiting that layer. The part played by the lower stable maximum, which brings together all three links of the community's food chain — producers, consumers and reducers — may be especially important. As already mentioned, this maximum constitutes a sharp boundary between waters that are rich and poor in life and is probably responsible for the disappearance of mineral forms of biogens from the layers above.

As to the causes of such a markedly nonuniform distribution of living suspended matter in the eutrophic zone, we are not inclined to reduce them to the simple physical phenomenon of the plankton's being suspended in the zone of maximum density gradient. In the first place, cells should become concentrated in the zone of their neutral buoyancy, i. e. they should follow the absolute density values. The density gradient affects the width of the concentration zone: the higher the gradient, the narrower is the concentration zone of cells. However, the width of this zone also depends on the range of variation of the density of the cells themselves. Consequently, we should not always expect concentrations of cells in the maximum density gradient, although the probability of encountering concentrations here is at a peak. The bioluminescence maxima we recorded generally did not coincide with the upper boundary of the pycnocline.

The formation of the lower maximum life concentration may have a biological explanation. The phytoplankton, and consequently also the bacterioplankton and zooplankton, finds optimum conditions for development in a comparatively thin layer which is limited by a shortage of light below and by a shortage of biogens above (Sorokin, 1959). The phytoplankton achieves a high intensity of photosynthesis in this layer, its magnitude being controlled by the dynamic equilibrium of the two critical factors, penetration of light from above and rate of influx of the biogens from the underlying layers. This equilibrium also governs the depth of occurrence of the layer, which need not necessarily coincide with the maximum density gradient.

The advantages offered in this layer with respect to the growth rate of cells result in high concentrations of cells and prevent dispersion of the aggregations by turbulent mixing.

Development of bacteria in the layer of the lower maximum is possibly due not only to the high seston concentration but also to the influx of dissolved organic matter into this layer in the course of vertical water exchange. Experiments have shown that the cold intermediate and deep oceanic waters contain a large stock of dissolved organic matter which is resistant to bacterial decay at low temperatures but can be oxidized by bacteria with a rise of temperature to 25—30° (Sorokin, 1971).

It may also be that the microflora subsisting on this organic matter develops more rapidly in the maximum layer under the influence of the influx of biogens that are essential to the bacteria no less than to the phytoplankton. At any rate, their combined activity in the main maximum zone has the result that the biogens arriving from below are consumed in this layer and scarcely penetrate above it (see Figure 3); in other words, this layer cuts the overlying part of the community off from the source of biogens. Actually, the isolation is not complete but we are using this expression to emphasize the characteristic feature of the situation, although as yet we are unable to state it in quantitative terms. Biogenic elements may be transported, for instance, by migrating zooplankton and by other biological means.

It is important to determine the quantitative extent of this isolation or, which comes to the same thing, to compare the intensity of the vertical flux of biogenic elements against the requirement of the population of the overlying layers for these elements. This is the key aspect to be considered when constructing a functional model of this community.

In the model of the ecosystem of surface oceanic tropical waters elaborated prior to the 44th cruise of the "Vityaz" (Lyapunov, 1971) it was considered that the functioning of a pelagic biogeocenosis in the low latitudes of the open ocean, far from powerful upwellings of deep waters, can be simulated in only two dimensions, depth and time. It was assumed that the processes directed along the vertical axis ensure the equilibrium and stability of the system in question. In its most elementary representation, the system was regarded as a steady-state cycle of productive and destructive processes in which the loss of biogens with sinking living organisms, their residues, and the products of their vital activity is compensated by their influx from below due to turbulent diffusion.

We now see that the influx of biogens from below is blocked by the lower layer of concentration of the consumer organisms, so that the system turns out to be unbalanced. There is no need to consider the isolation as absolute, but it is enough to admit the inequality of the downward and upward fluxes of biogens.

It becomes obvious that the population of the layers above the lower maximum, including the population of the upper maximum, subsists also on account of some other source of biogens. A probable source is the transport of biogens by a horizontal flow of water from upwelling zones (chiefly from divergence zones of currents) toward zones of sinking on convergences. The biogens rising into the surface layers are then repeatedly utilized in the production-destruction cycle in the system phytoplankton—zooplankton—bacterioplankton.

The biogenic background of the surface layer becomes depleted by the constant outflow with downward-migrating animals and sinking phytoplankton and detritus. This is the pattern of succession of the biocenosis along the path of the flow of water from a divergence toward a convergence (Margalef, 1967; Timonin, 1969). Correspondingly, the magnitude of the upper maximum, existing chiefly on account of horizontally transported biogens, must gradually decline until its virtual disappearance with increasing distance from the zone of upwelling. At the same time, the lower maximum, owing its existence to the utilization of biogens brought upward by turbulent mixing, is preserved. The same pattern was observed in the Atlantic Ocean (Sorokin, 1959).

The possibility of a "recycling" of biogens in the surface layer is quite obvious even for coldwater regions (Ketchum, 1962; Cushing, 1968; and others). At the Second Oceanographic Congress Cushing mentioned that he suspected that the winter maximum of nitrates and phosphates performed several cycles before their concentration in the water reaches zero. Each cycle involves a partial loss in one of the dead ends, like animal tissues or suspended organic detritus. The rate of this turnover is stepped up by the high temperature of the heated surface waters in the tropical zone.

Besides the biogens, the same turnover involves dissolved organic matter brought up by the deep waters; this is intensively utilized by bacterial flora in the heated surface layers, and this flora is then consumed by the zooplankton. This is an extremely important source of energy supply to the community of the surface layers which plays no less an important part than the photosynthesis of primary organic matter by algae.

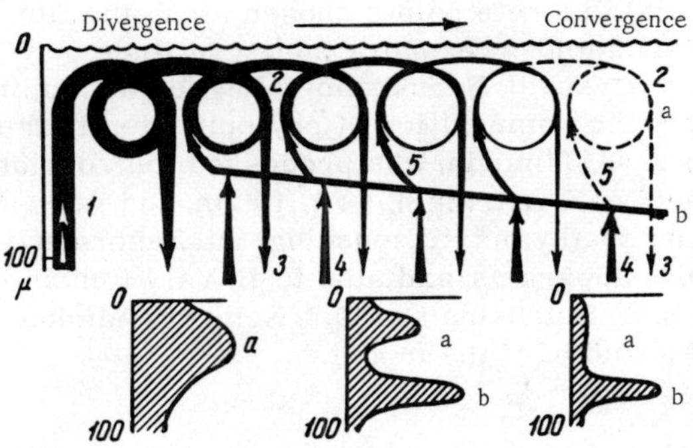

FIGURE 4. Diagram of the cycle of biogenic elements and organic matter in the succession of the pelagic community in the oceanic tropical waters:

Top: 1 — upwelling of biogenic elements and dissolved organic matter from deep layers in the upwelling zone; 2 — their repeated utilization in the production-destruction cycles of the community of the surface layers; 3 — loss with sinking organic residues and migrating organisms; 4 — turbulent rise of biogenic elements and dissolved organic matter from deep layers and their retention in the zone of the lower maximum; 5 — incorporation of the biogenic elements rising from the lower maximum layer with migrating organisms into the production cycles of the surface community, which does not compensate for the losses of biogens with the downward flow. Bottom: vertical variation in the distribution pattern of relative phytoplankton abundance (K_d) at different distances from the upwelling zone (a — upper, b — lower concentration maximum).

Figure 4 is a diagram illustrating the turnover of biogens and dissolved organic matter involved in the trophic chains of the pelagic community in the oceanic tropical waters, i. e. the mechanism maintaining the above-described vertical structure of the community. The pattern shown in the figure is probably apt to be modified under the influence of several factors. For instance, it depends on the time elapsing in different cases between the upwelling of water to the surface and its sinking. However, if the diagram is basically correct, the succession of the community through the entire turnover cycle of water should be regarded as the elementary simulated ecosystem, or at least in that period when the water is in the surface layer. In other words, the model requires the addition of a third dimension, namely along the horizontal axis along the movement of "aging" water. Observations of variations in the community should likewise be conducted along the water flow.

Obviously, the observation techniques become extremely complicated in this case, and the mathematical model of the system will also become very complex. However, we may simplify the problem by confining ourselves to studies of the different states of the system succeeding one another. Such sections through the system may be provided by characteristics of its different bidimensional (depth, time) components. A series of such sections performed for different states of the system, for instance, for its state in the upwelling zone, in the sinking zone, and at 2 or 3 points between them, may provide a fairly good idea of the state of the entire system, i. e. the structural variations of the community over its succession period. This approach will also substantially facilitate investigation of the system, which may be conducted at several discrete points chosen along the flow of "aging" water.

In conclusion the authors express their gratitude to V. V. Sapozhnikov for the chemical analyses, I. N. Sukhanova for her determinations of the abundance and specific composition of phytoplankton in the samples taken with the sampler, A. G. Timonin, who processed the zooplankton from aimed catches with the net, A. S. Artemkin, L. A. Levin, and A. P. Shevyrnogov, who carried out the photobathymetric sounding, Yu. Zagorodnyi for his help in building the special apparatus, and also to B. Ya. Vilenkin, B. M. Mednikov, Yu. A. Rudyakov, E. A. Shushkina and O. I. Koblents-Mishke for their participation in the discussion of the model.

Bibliography

Publications in Russian

Cushing, D. H. Modelirovanie produktsionnogo tsikla v more (Simulation of the Production Cycle in the Sea). — In Sbornik: "Osnovnye problemy okeanologii." Moskva, "Nauka." 1968. [Russian translation.]

Gitel'zon, I. I., L. A. Levin, A. P. Shevyrnogov, R. N. Utyushev, and A. S. Artemkin. Batifotometricheskoe zondirovanie pelagiali i vozmozhnosti ego primeneniya dlya issledovaniya prostranstvennoi struktury biotsenoza (Pelagic Bathyphotometric Sounding and its Possible Application to Studies of the Spatial Structure of Biocenoses). (In this collection.)

Lyapunov, A. A. O postroenii matematicheskoi modeli balansovykh sootnoshenii v ekosisteme tropicheskikh vod okeana (Construction of a Mathematical Model of Balance Relationships in the Ecosystem of Tropical Oceanic Waters). (In this collection.)

Pavlova, E. V., T. S. Petipa, and Yu. I. Sorokin. Rol' bakterioplanktona v pitanii morskikh pelagicheskikh organizmov (Bacterioplankton in the Diet of Pelagic Marine Organisms). (In this collection.)

Petipa, T. S., E. V. Pavlova, and Yu. I. Sorokin. Izuchenie pitaniya massovykh form planktona tropicheskoi oblasti Tikhogo okeana radiouglerodnym metodom (Radiocarbon Studies of Mass Plankton Forms in the Tropical Zone of the Pacific). (In this collection.)

Romanenko, V. I. Geterotrofnaya assimilyatsiya CO_2 bakterial'noi floroi vody (Heterotrophic Assimilation of CO_2 by Bacterial Aquatic Flora). — Mikrobiologiya, Vol. 33, No. 4. 1964.

Semina, G. I. Fitoplankton (Phytoplankton). — In: Tikhii okean, Vol. 7. Biologiya Tikhogo okeana, Book 1. Plankton. Moskva, "Nauka." 1967.

Sorokin, Yu. I. O vliyanii stratifikatsii vodnykh mass na pervichnuyu produktsiyu fotosinteza v more (On the Effect of Stratification of Water Masses on Primary Production of Photosynthesis in the Sea). — Zhurnal Obshchei Biologii, Vol. 20, No. 6. 1969.

Sorokin, Yu. I. Kolichestvennaya otsenka roli bakterioplanktona v biologicheskoi produktivnosti tropicheskikh vod Tikhogo okeana (Quantitative Evaluation of the Role of Bacterioplankton in the Biological Productivity of Tropical Pacific Waters). (In this collection.)

Timonin, A. G. Struktura pelagicheskikh soobshchestv. Kolichestvennoe sootnoshenie razlichnykh troficheskikh gruppirovok planktona v frontal'nykh zonakh tropicheskoi oblasti okeana (Structure of Pelagic Communities. Quantitative Ratios of Different Trophic Groups of Plankton in Frontal Zones of the Tropical Region of the Ocean). — Okeanologiya, Vol. 9, No. 5. 1969.

Vinogradov, M. E. Vertikal'noe raspredelenie okeanicheskogo zooplanktona (Vertical Distribution of Oceanic Zooplankton). Moskva, "Nauka." 1968.

Publications in Other Languages

Barnard, F. Plankton et benthos observés durant trois plongées en bathyscaphe au large de Toulon. — Ann. Inst. Océanogr. Monaco, n. s., Vol. 35, fasc. 4. 1968.

Cushing, D. H. On the Nature of Production in the Sea. — Fish. Invest, ser. II, Vol. 22, No. 6. 1959.

Dugdale, R. C. Nutrient Limitation in the Sea: Dynamics, Identification and Significance. — Limnol. and Oceanogr., Vol. 12, No. 4. 1967.

Jørgensen, C. B. The Food of Filter Feeding Organisms. — Rapp. Cons. Internat. Explor. Mer., Vol. 153. 1962.

Margalef, R. D. The Food Web in the Pelagic Environment. — Helgoländer Wiss. Meeresunters., Vol. 15, No. 1 — 4. 1967.

Patten, B. C. Mathematical Models of Plankton Production. — Internat. Rev. ges. Hydrobiol., Vol. 53, No. 3. 1968.

Riley, G. A., H. Stommel and D. F. Bumpus. Quantitative Ecology of the Plankton of the Western North Atlantic. — Bull. Bingham. Oceanogr. Coll., Vol. 12. 1949.

Sorokin, J. I. On the Primary Production and Bacterial Activities in the Black Sea. — J. Cons. internat. Explor. Mer., Vol. 29, No. 1. 1964.

Steele, J. H. Environmental Control of Photosynthesis in the Sea. — Limnol. and Oceanogr., Vol. 7, No. 1. 1962.

Vollenweider, R. A. Calculation Models of Photosynthesis-depth Curves and Some Implications Regarding DayRate Estimates in Primary Production Estimates. — Mem. Ist. Ital. Idrobiol., Vol. 18, Suppl. 1965.